Handbook of Commercial and Industrial Facilities Management

Other Books of Interest from McGraw-Hill

AVALLONE & BAUMEISTER • *Marks' Standard Handbook for Mechanical Engineers*

BHUSHAN & GUPTA • *Handbook of Tribology*

BRADY & CLAUSER • *Materials Handbook*

BRALLA • *Handbook of Product Design for Manufacturing*

BRINK ET AL. • *Handbook of Fluid Sealing*

BRUNNER • *Handbook of Incineration Systems*

CORBITT • *Standard Handbook of Environmental Engineering*

EHRICH • *Handbook of Rotordynamics*

ELLIOT • *Standard Handbook of Powerplant Engineering*

FREEMAN • *Standard Handbook of Hazardous Waste Treatment and Disposal*

GANIC & HICKS • *The McGraw-Hill Handbook of Essential Engineering Information and Data*

GIECK • *Engineering Formulas*

GRIMM & ROSALER • *Handbook of HVAC Design*

HARRIS • *Handbook of Acoustical Measurements and Noise Control*

HARRIS & CREDE • *Shock and Vibration Handbook*

HICKS • *Standard Handbook of Engineering Calculations*

HODSON • *Maynard's Industrial Engineering Handbook*

JONES • *Diesel Plant Operations Handbook*

JURAN & GRYNA • *Juran's Quality Control Handbook*

KARASSIK ET AL. • *Pump Handbook*

KURTZ • *Handbook of Applied Mathematics for Engineers and Scientists*

MASON • *Switch Engineering Handbook*

NAYYAR • *Piping Handbook*

PARMLEY • *Standard Handbook of Fastening and Joining*

ROHSENOW ET AL. • *Handbook of Heat Transfer Applications*

ROHSENOW ET AL. • *Handbook of Heat Transfer Fundamentals*

ROSALER & RICE • *Standard Handbook of Plant Engineering*

ROTHBART • *Mechanical Design and Systems Handbook*

SCHWARTZ • *Composite Materials Handbook*

SCHWARTZ • *Handbook of Structural Ceramics*

SHIGLEY & MISCHKE • *Standard Handbook of Machine Design*

TOWNSEND • *Dudley's Gear Handbook*

TUMA • *Engineering Mathematics Handbook*

TUMA • *Handbook of Numerical Calculations in Engineering*

WADSWORTH • *Handbook of Statistical Methods for Engineers and Scientists*

WALSH • *McGraw-Hill Handbook of Machining and Metal Working*

WOODRUFF, LAMMERS, & LAMMERS • *Steam-Plant Operation*

YOUNG • *Roark's Formulas for Stress and Strain*

Handbook of Commercial and Industrial Facilities Management

William Wrennall

Quarterman Lee
Editors in Chief

The Leawood Group, Ltd.
Leawood, Kansas

McGraw-Hill, Inc.

New York   San Francisco   Washington, D.C.   Auckland   Bogotá
Caracas   Lisbon   London   Madrid   Mexico City   Milan
Montreal   New Delhi   San Juan   Singapore
Sydney   Tokyo   Toronto

Library of Congress Cataloging-in-Publication Data

Handbook of commercial and industrial facilities management / William
Wrennall, Quarterman Lee, editors in chief.
    p.   cm.
   ISBN 0-07-071935-7
   1. Factory management. 2. Facility management. 3. Plant
engineering. I. Wrennall, William. II. Lee, Quarterman.
  TS155.H28135 1994
   658.2—dc20   93-17498
                                              CIP

1 2 3 4 5 6 7 8 9 0 DOC/DOC 9 9 8 7 6 5 4 3

ISBN 0-07-071935-7

*The sponsoring editor for this book was Robert W. Hauserman, the
editing supervisor was Caroline Levine, and the production supervisor
was Donald F. Schmidt. It was set in New Century Schoolbook by
McGraw-Hill's Professional Book Group composition unit.*

Printed and bound by R. R. Donnelley & Sons Company.

Contents

Foreword

McGraw-Hill, Inc., is to be commended for its initiative in publishing the *Handbook of Commercial and Industrial Facilities Management.*

Productivity as a profession and operational principles in all industrial and service-oriented operations have made marks over the last 30 years; this text gives us a foundation of the successes.

The Japanese organized their productivity effort beginning in the 1950's and the result over the years has been quite astonishing. It is needless to emphasize the strength of Japan's industry and global commercialization. However, less emphasis has been given to the global consequence of their efforts. Today the industrialized Far East (Taiwan, South Korea, Singapore, and quite recently Malaysia and Indonesia) show astonishing ability in implementing rationality and productivity management throughout their industrial operations.

Another consequence of the productivity skill of the East is that it has caused the United States, and to a somewhat lesser degree, Europe, to recognize clearly their shortcomings in industrial facilities management.

In Japan, implementation of productivity management was logically communicated through quality groups and constant modernization of their organizations within the corporations. The learning capacity and adaptation of skills from near and foreign sources have been paramount in productivity evolution and commercial elegance.

As the entire world now requires this knowledge of commercial and industrial facilities management, the Handbook, edited by William Wrennall and Quaterman Lee, is a much-needed and most timely gift to the world at large. The 18 chapters cover most of the principles and techniques for world class manufacturing and may, indeed, be a breakthrough in the educational and operational sphere of productivity management. The World Academy of Productivity Science finds it gratifying that this valuable contribution to our field of profession has been provided.

Martin T. Tveit, Ph.D., M.Sc., DIC
Chairman, World Confederation of Productivity Science

Preface

All of us have, at one time or another, struggled with the frustration of trying to perform world-class work in a building or with a layout that was unsuitable. This may have meant repeatedly walking long distances and trying to communicate with people and departments far removed. Or it may have meant managing an organization that was splintered and isolated by the physical boundaries of its buildings and land. Perhaps this Handbook can make a small contribution towards reducing such frustrations by providing organizations with ideas for facilities that enhance their productivity, profitability, and competitive advantage.

Facilities design has evolved from *edict* and *template shuffling* through *systematic layout planning* into a *strategic process*. It is no longer enough to be a "facilities planner." Today's business engineer knows that facilities are intertwined with processes, organization, and even the social systems of a firm. Facilities design cannot be divorced from socio-technical systems design, environmental engineering process design, or business strategy.

As you can see from scanning the Contents, the body of knowledge for this field is substantial. We are therefore pleased to have had the opportunity to edit this *Handbook of Commercial and Industrial Facilities*, the first comprehensive work on the subject.

The Handbook contains chapters on subjects from site selection to facilities maintenance, and from designing clean rooms to managing a construction project. It provides tools, techniques, ideas, and procedures that have been accepted and proven by world-class organizations. We want you to draw on it in designing or reviewing the contribution of your facilities to your total productivity.

We thank the 18 authors who have shared their knowledge and experience with us. They range from chief executives of large corporations to consulting firm partners; from simulation specialists to directors of quality in health care companies; from college professors to consulting engineers. The authors' wide range of commercial and industrial experience has been invaluable.

Acknowledgments

All of us have drawn on the experience of our friends in government, industry, and commerce, and we thank them.

The editors wish to thank in particular the following individuals who have supported our efforts in teaching, managing, consulting, and researching this Handbook.

Dr. Martin T. Tveit, Chairman of the World Confederation of Productivity Science.

Rob Dowdy and his team at Variform, Inc.

Dick Fent, Plant Manager, FMC.

Bob Jergens, Director, Real Estate Facilities Management, Marion-Merrell Dow.

Rudy Bouie, Director of Facilities, Argonne National Laboratory.

Bill Budnovich, President, Devine Design.

We appreciate the support for work on this project given to us by the directors of The Leawood Group Ltd.

We thank Diane Elston and Kathleen Bell of The Leawood Group staff, who have tolerated our impatience and supported us in preparing the manuscripts with their hundreds of figures. Their efforts in coordinating with our authors and publishers have earned our lasting respect.

William Wrennall
Quarterman Lee

1

Facilities Design for World Class Competition

William Wrennall
President, The Leawood Group, Ltd.
Leawood, Kansas

Quarterman Lee
Vice President, The Leawood Group, Ltd.
Leawood, Kansas

The Importance of Facilities

Facilities provide the capacity for operations. They are an important strategic element for most businesses. They are among the most costly investments and compete with marketing, product development, and other business activities for scarce resources.

Facilities often require long lead times for construction. In a competitive environment, where new products and customers constantly emerge, timing becomes critical. A facility that is built too soon can drain a company's financial resources; a facility built too late can lose sales and market share and may relegate an otherwise fine product to permanent second place.

Facilities last for many years, usually much longer than either products or processes, yet their fundamental configuration is often permanent. A purpose-built facility designed around a specific process technology may become a white elephant when the technology changes. On the other hand, a plant designed for flexibility in process and product may not compete well on cost with optimized, purpose-built facilities.

The American steel industry, for example, had the world's largest capacity in open-hearth furnaces at the end of World War II. But the introduction overseas of basic oxygen furnaces rendered much of this capacity obsolete. In response, United States Steel built a huge integrated mill in Burns Harbor, Indiana, in the 1960s, with the entire site being designed around the basic oxygen process. Shortly thereafter, integrated steelmaking with basic oxygen was rendered obsolete by specialized minimills using electric arc technologies. This illustrates the hazards of making facilities decisions without considering marketing, changes in technology, and the larger environment.

The best facilities are an integral part of their company's financial, marketing, and product strategies. When properly designed within the strategic framework, they can support and enhance the firm's competitiveness. When designed around purely technical considerations, they can drain a firm's financial and managerial strength and erode the competitive position.

The Historical Development of Industrial Facilities

Industrial facilities are the product of the Industrial Revolution. The Industrial Revolution gained momentum during the middle years of the eighteenth century and dominated the nineteenth century. It was a unique phenomenon in the history of humankind, and its repercussions have spread throughout the world.

The Industrial Revolution

It was much more than a revolution in technology—the techniques of making things; it was a revolution whose cultural, social, economic, and environmental implications altered the way in which people thought and lived. Our society's prosperity is based on over two centuries of progressive industrialization, during which cottage industries were replaced with factories. The evolution of industrial facilities is tabulated in Fig. 1.1.

The term Industrial Revolution was coined in the 1830s or earlier. Britain was the first industrial country; the reasons why are not agreed upon by historians. What appears to have taken place is that about the middle of the eighteenth century Britain reached a point of "take off into self sustained growth"[1] based on radical changes in methods of production. These changes, occurring over a relatively short period, had widespread effects on society, the national economy, and the landscape.

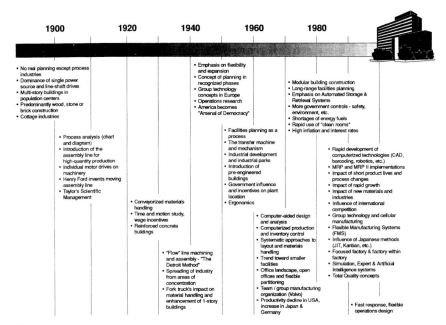

Figure 1.1 Facility planning chronology.

The Industrial Revolution has been described as a technological revolution in which new manufacturing machines were driven by new sources of power in new large-scale units called *factories*. The year 1830 is considered by Cossons to mark the end of the Industrial Revolution and the beginning of the period of true industrialization. Railways were beginning to gain acceptance, and Michael Faraday produced a continuous electric current in 1831.

The first Industrial Revolution resulted in the British factory system. This was followed by the new methods of production in the United States called *mass production* or the Ford, or Detroit, method. The economic benefits in the United States followed a similar pattern to those that occurred in Britain, resulting in the United States becoming the most powerful country in the world.

The facilities required for the new manufacturing were very different from the early European designs. With more sophisticated technology in equipment and supporting services, and more demanding environments, factory layouts were adapted to suit. Since World War II, other countries have industrialized and re-industrialized to what Doll[2] calls the "Post-Industrial Stage." The introduction of group technology (GT) and team building in Europe was followed by just-in-time (JIT), or the Toyota system. With the design of fast-response biotechnical zero-defect systems, further changes in facilities design have occurred. Those companies who have made the shift to becoming competitive are known as "world class."

The development of industrial facilities

Some of the early factories were in the textile industry and were known as *mills*. An experimental mill was built in Birmingham, England, about 1741. Water-powered mills were developed by Hargreaves in Nottingham and by Arkwright in Derbyshire in the 1790s. The first American cotton mill, Slater's Mill, opened in 1793 at Pawtucket, Rhode Island. It was built by Samuel Slater, who had been trained by Arkwright.

These early water-powered mills were multistory and located over streams or rivers. Water power supplied approximately one-third of the power used in cotton mills in Britain as late as 1834. The steam engine had been patented by James Watt and introduced into a cotton mill in Nottingham in 1785, but it was not until the 1840s that steam engines were being built to exceed the horsepower of the largest waterwheels.

Early facilities designs were influenced by the sources of power to drive the machines of the Industrial Revolution. Later, water and wind power were replaced by steam and then electric power. The generation of power became detached from the facility as coal, oil, gas, or nuclear energy was generated and the power distributed to factories and offices. The move from labor-intensive to capital-intensive manufacturing methods has reduced direct labor costs to about 6 percent or less in world class companies.

Labor and scientific management developments

Another movement was occurring at the end of the eighteenth century. Specialization of labor organization was reported by Adam Smith in *The Wealth of Nations* in 1776. Dividing pin manufacture into four operations was the beginning of functional department layouts. Others used division of labor to increase productivity at that time. Around 1800 Boulton and Watt made organizational changes in their foundry in Soho, England.

Frederick Taylor began experimenting in metal cutting in the Midvale Steel Company in 1881. He is also responsible for separating planning from doing, resulting in "functional foremanship." Fayol in France extended management theory to the general management level, while Taylor concentrated on manual operations.

Taylor found that the workers at the Bethlehem Steel Works wasted a great deal of time moving from one job to another because the storage yard was about two miles long and almost half a mile wide. To control the waste time, Taylor made a map (layout) of the steel-yard for planning jobs in advance. The improvements Taylor made resulted in the first functional layouts with support departments such as:

- Tool and storage rooms
- Personnel departments
- Training departments
- Planning departments
- Payroll departments

The layout of manufacturing facilities became important for ensuring that the movement of materials and workers was minimized and the new departments were properly located.

Operations Strategy

Why strategy wins business wars

"Strategy" derives from the Greek word *strategos. Strategos,* translated literally, is "the generals' art." The word has been applied and misapplied in many contexts. As used in business, strategy has the following characteristics:

- *Extended time horizon.* Strategies extend over a long time horizon, years or decades for most industries. Because a strategy pervades the business and organizational culture, it does not change easily or quickly.

- *Theater of operations.* Business strategy, like military strategy, involves large segments of the competitive environment rather than a single customer or item.

- *Pervasiveness.* Strategies pervade a wide spectrum of activity, from daily interactions with customers to repairs on equipment to capital expenditure decisions.

- *Impact.* Strategies have major, long-term consequences. These may not become apparent, however, for many years. In war, strategy determines the fate of armies and nations; in business, it determines the fate of corporations and industries.

- *Pattern of behavior.* Strategy is a guide for decision and action throughout the organization for an extended time period. It is the element that gives an organization its characteristic pattern of behavior.

- *High level.* The nature of strategy requires that top managers (the generals) develop it. Top management may accept or abdicate the task, but they cannot abdicate the responsibility.

Tactics are the actions and decisions which derive from strategy. A tactic usually has a:

- Short time horizon
- Limited operational scope
- Limited organizational effect
- Transient impact
- Low-level development

While each tactical decision or action is of limited nature, the cumulative effect, over time and throughout the organization, mirrors or even defines actual strategy.

Every manufacturing organization has a strategy. That strategy may be:

- Explicit or implied
- Consistent or incongruous
- Logical or irrational
- Sound or defective
- Planned or ad hoc

A sound operations strategy derives from and supports corporate and marketing strategies. Conversely, the development process often contributes to the marketing strategy. For example, manufacturing might find a method for customizing a product. This ability then opens new markets which were previously inaccessible.

Figure 1.2 illustrates these various levels of strategy and tactics. Corporate strategy is, or should be, conceived at the highest executive levels. Corporate strategy pertains to issues of finance and long-term growth. It also pertains to the way various divisions of the firm interact for balance and synergy. It considers the firm's financial strength and sources of capital.

Marketing strategy addresses narrower concerns than corporate strategy. In larger firms, there will be multiple marketing strategies corresponding to the various divisions and markets. Traditionally, marketing strategy concerns itself with the "four P's": Product, Price, Promotion, and Place. Some typical issues are:

- What products do we offer?

Figure 1.2   Strategies, missions, and tactics.

- What price structure do we use?
- How do we advertise, sell, distribute?
- When, where, and how do we deliver?

Operations strategy first translates the marketing strategy into manufacturing terms. It then describes, in a broad way, how manufacturing intends to supply the product(s). Not all pieces of the marketing strategy are relevant to manufacturing. Promotions, for example, are not primarily a manufacturing concern unless they place unusual stress on the system.

Strategic development

Figure 1.3 shows the normal strategic developmental sequence. Corporate strategy evolves at the highest corporate levels. These strategies derive from traditions developed and decisions made over many years. They are often difficult to change radically. They consider issues of capitalization and finance.

Marketing strategies, ideally, incorporate and support the corporate strategy. A large firm usually has marketing strategies for each division and perhaps for each market segment.

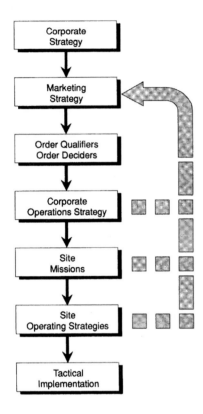

Figure 1.3 Strategic development.

With marketing strategies defined, the next step is to identify the "deciders," "qualifiers," and "losers" for each. Divisions or segments with similar market criteria are candidates for a common facility and common operations strategy. The corporate-level operations strategy should address this. It should also address issues of site focus and overall organization. It should identify the size, location, products, processes, and missions of each facility.

When a site has an identified mission, plan the site-level operations strategy. This usually is a two- to five-page statement of mission and policy. Tactical implementation includes decisions on equipment purchases, layouts, organization structure, and individual task assignments. Small organizations usually combine several steps.

The link to marketing

Operations strategy links to marketing through deciders and qualifiers. Each customer makes a buying decision on one or more factors. These factors are *market criteria*. Among the possible market criteria are:

- Quality
- Price
- Product features
- Customization
- Delivery speed
- Delivery reliability
- Service
- Advertising
- Reputation
- Financing

Operations may have no influence on some of these factors. Market criteria which involve only financing or advertising, for example, would not enter into operations strategy development.

Factors are either deciders or qualifiers. Customers use *deciders* when choosing between purchasing options. Improved performance on a decider brings additional orders.

For some factors, customers may require a minimum performance level before considering a product, but performance above this minimum brings no additional benefit. Such factors are *qualifiers*. They qualify a product to be in the market. All potential competitors must meet the same qualification.

With the market criteria identified, the operations strategy team can design a manufacturing system to deliver the required performance. California Microdat will require a flexible facility that produces the highest quality with dependable output. Cost will be secondary and pressures to maintain a full product line can be resisted. We will soon return to this example.

How a defined strategy improves facility layout

The layout reflects and affects each element of a business enterprise, particularly the operations strategy. An effective layout supports a firm's strategy, while an ineffective layout hinders it. The layout analyst who is ignorant of his or her firm's strategy may design a layout that is controversial and fails to meet the needs of both the business and the customer.

A framework helps plan your strategy

A business operations system is complex. It has many disparate elements of equipment, people, buildings, organization, procedures, soft-

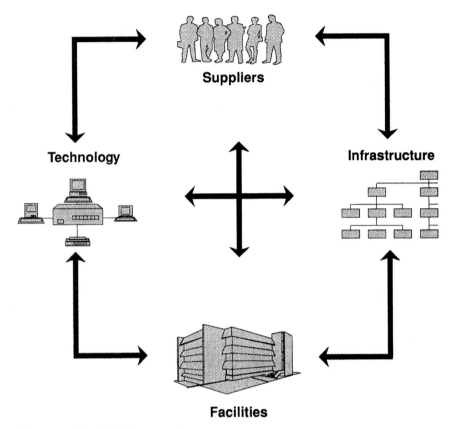

Figure 1.4   The holistic perspective.

ware, and knowledge. These elements must coordinate and effectively work together to serve the customer and meet corporate goals.
Figure 1.4 illustrates the holistic effect.

The operations strategy statement should address all of the elements.
When complete and approved, it guides design, decisions, and action.

We identify two broad areas for strategic development: *process* and
infrastructure. The process includes equipment, operations, and people who add value to the product. It consists of those elements which
directly transform raw materials and purchased items into something of increased value to the customer. Infrastructure supports the
process but does not directly participate in it. Infrastructure includes
people, information systems, equipment, buildings, and many other
elements. For convenience, we divide infrastructure into physical and
nonphysical components. Physical infrastructure is, essentially, the
facilities. Nonphysical infrastructure includes people, organization,

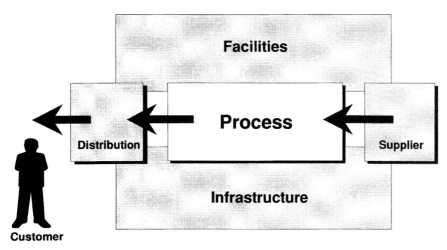

Figure 1.5   Manufacturing strategy—integration.

knowledge, and the intangible and semitangible elements. Software is a semitangible example.

Figure 1.5 illustrates how infrastructure, physical and nonphysical, supports the process. Figure 1.6 is an outline for an operations strategy summary. Use this as the skeleton when preparing a summary for your layout project. Each topic requires one to five sentences to identify the issues and preferred solutions.

Your site mission.   The site mission states the purpose of each plant site in a concise statement. In an organization with only one site, the site mission directly reflects the corporate and marketing strategies. With several sites, their missions are an important piece of operations strategy.

Process.   Based on marketing strategy and the deciding criteria, the summary identifies preferences and absolutes for equipment and arrangement. It also identifies the legitimate uses of inventory and inventory levels.

Nonphysical infrastructure.   In this section, the summary identifies policies and preferences for the supporting organization. Included are personnel, organization structure, information systems, production control, accounting, engineering, maintenance, and all other supporting services.

Infrastructure departments are often at odds with other aspects of an operations strategy. Specialists often staff such departments,

1. SITE MISSION
Site Focus
  Products
  Markets
  Volumes
  Geography
Multi-Site Integration
  How does this site fit with others
Key Manufacturing Task
External Strategic Issues
  Political
  Environmental
  Community Involvement
  Other...

2. PROCESS
Production Mode(s)
  Project
  Functional
  Cellular
  Toyota
  Line
  Continuous
Process Scale
  Large - High Capital
  High Volume
  Long Changeover
  Large Lots
  Inflexible
  Low Direct Cost
Setup/Lot Size
Capacity
  Timing
  Lead/Track/Lag
  Reserve
Quality Capability
Technology Level

3. NON-PHYSICAL INFRASTRUCTURE
Quality Approach
  Quality Police
  Quality at Source
Personnel Policies
  Technical Skill Depth
  Technical Skill Breadth
  Interpersonal Skills
  Employment Security
  Compensation
  Training
  Performance Measurement
  Safety
  Ethics

Organization Structure
  Organization Focus
    (Functional / Product / Other)
  Depth
Organization Style
  Exploitive
  Bureaucratic
  Consultative
  Participative
Accounting Policies
  Process / Job Costing
  Time-Based Accounting
  Overhead Allocation
  Decision Criteria
  Knowledgebase Investments
  Inventory Accounting
Production Control
  Trigger
    Make-to-Orders
    Make-to-Stock
    Kanban
  Type
    Physical Link
    Broadcast
    Kanban
    MRP
    Re-Order Point
Supplier Policies
  Selection Criteria
  Single / Multiple Sources
  Contract Time Horizons
  Scheduling Approach
  Shipping Policies

**4. FACILITIES
(Physical Infrastructure)**
Site Focus
  Product
  Process
  Market
  Geographic
  Other
Site Location & Size
Transportation Access
Utility Systems
Expansion Policies
New Product / Process Flexibility
Resale / Disposal Policy
Hazardous Waste Policy
Environmental Issues

Figure 1.6   Operations strategy summary.

bringing viewpoints which may be at odds with others. These viewpoints then appear in the design of their systems and the decisions they make. Lack of goal congruence for the business, as seen by functional heads, illustrates this point.

Facilities. Facilities should integrate with and support the other elements of a manufacturing system. If new products are important market criteria, the facility and utility systems should be highly flexible. If low cost is a primary decider and new products are rare, then less expensive special-purpose systems are appropriate.

The total number of sites, their locations, size, and predominant focus, are issues. These are among the most important strategic decisions a firm will make.

Facilities are a major determinant of capacity. Facilities policies and decisions therefore have significant strategic consequences. These decisions are among the most important that a firm makes. They are not always treated as such.

Operations Focus

In 1974, Wickham Skinner's seminal article "The Focused Factory" put forth the concept that large factories with many products and processes rarely perform well. The range of problems they were forced to address was too large and the problems of coordination too great. But focused factories were contrary to the conventional wisdom of the time; the conventional wisdom held that larger factories spread overhead cost and provided economies of scale.

The economy-of-scale theory holds that larger processes, equipment, organizations, and buildings are always more efficient. Proponents of this idea cite many examples, often from early or middle periods of the Industrial Revolution. Henry Ford's River Rouge complex is a favorite.

Many business processes do benefit from larger volume. But extending this idea to all processes, organizations, and situations ignores the diseconomies of scale, or economies of scope, as well as other considerations. For example:

- *Diminishing returns.* The beneficial effects of larger volume typically diminish beyond a certain point.

- *Bureaucracy.* Larger facilities and organizations beget increasing hierarchies, staff specialists, policies, procedures, and rules. This increasing bureaucracy consumes costly resources. It slows the organization's response to market and other external changes. It focuses attention inward and away from the customers.

- *Confusion.* The increased complexity of larger enterprises is a source of confusion. Communication efforts are proportional to the square of the site population. Doubling the number of people quadruples coordination and control costs. This is above and beyond the issue of bureaucracy.

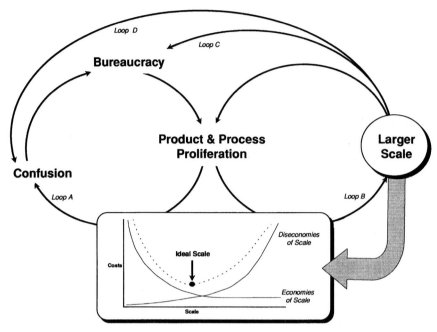

Figure 1.7 Dynamics and economics of scale.

- *Product/process proliferation.* Most firms do not grow larger with the same products and processes. Typically, increased size comes from additional products and the processes to make them. There is also pressure from the bureaucracy to integrate vertically.

The diminishing benefits of increased scale are quickly overcome by changeover costs and reduced experience curve effects. These effects are interrelated and dynamic. Figure 1.7 shows how product/process proliferation leads to confusion, which in turn leads to bureaucratic attempts to organize the confusion. Bureaucracies tend to grow in accordance with Parkinson's law. This leads to pressure for additional product/process proliferation in a recursive loop. This is labeled "Loop A" in Fig. 1.7.

The pattern has other self-reinforcing loops. Product/process proliferation also produces financial and other pressures for larger scale. Larger scale often demands additional products to keep new processes and bureaucracies busy. These are noted as "Loop B" and "Loop C." Larger scale also increases confusion, "Loop D."

With the resulting growth, initial economies of scale are overtaken by the diseconomies as shown in the graph in Fig. 1.7. The dynamic effects are slow, often occurring over years or decades. Because of the lengthy time delay, management rarely recognizes the self-reinforcing patterns. The result of slowly decreasing margins is usually attributed to other immediate causes with bureaucratic solutions.

The focused factory endeavors to reduce confusion, bureaucracy, and product proliferation. It limits the number of key tasks the organization must perform. It reduces bureaucracy by simplifying work and reducing the number of people, products, and processes.

Arbitrarily reducing product range or size is unlikely to help. Competitive advantage comes from focusing facilities and organizations on only one or two manageable and consistent tasks. These *key operations tasks* directly support the firm's marketing strategy.

Scope and scale

Scale refers to the size of a given plant or manufacturing unit. *Scope* refers to the number of significantly different products, processes, markets, and regions which the facility serves. In an unfocused facility, increased scale comes largely from increased scope.

Focus criteria

Reducing scale is easy. Reducing scope is more difficult: It requires a basis for segregating operations. This basis, or set of *focus criteria,* may take several forms. The more common are:

- *Process.* A particular process may require high investment and large volume for economic operation. Alternatively, special skills may dictate a concentration based on a particular process. Process-focused plants concentrate technical expertise but are difficult to coordinate and control. Process-focused facilities are organized around particular functions. Their *space-planning units* (SPUs) have names such as Grinding, Paint, or Wave Solder Department. Many and different products move through each department. A single product must visit several and perhaps many departments before completion. Such layouts inherently provide manufacturing flexibility, but they present scheduling difficulties and require higher work-in-process inventories and associated space.

- *Product.* A dedicated facility manufactures a product or group of similar products. Product focus promotes quality, reduces inventory, and improves response to changing demand. It often requires broader skills and may reduce flexibility when introducing new products. Product-focused layouts are organized around a product or small group of similar products. Lines and workcells are usually product focused. The SPUs carry names such as Light Cable Department, Circuit Board Department A, or Series A2 Department. Product-focused layouts are usually customer-responsive and produce high quality. They are simple to schedule and control compared to a corresponding process-focused layout.

- *Market.* Market segments are a basis for focus when customer demands differ substantially. Quality, delivery, option, variety, and order size may differ between market segments. If so, these market criteria provide a basis for focus. Specific customers or classes of customers are the basis for SPUs in such an operation. Names such as Wholesale Products Department and Midwest Motors Department predominate. In practice, market focus and product focus usually result in similar layouts and groupings.

- *Volume.* Similar or identical products may require separate processes and infrastructure for high and low delivery volumes. Production volume can therefore dictate focus. Layout cells and SPUs carry names like Short-Run Department or High-Speed Operations.

- *Geography.* Physical proximity to customers, to supplies, or to special skills may be necessary. Here, geography may be a basis for focus. This often arises where delivery speed is mandatory. It also occurs internationally when import restrictions apply. While geographic focus is often important at the site level, it is rarely appropriate within a site.

- *Infrastructure.* Support systems are, by definition, peripheral to the manufacturing process. However, they and other dominant site factors can impact organization structures and the associated facilities requirements. An SPU built around a material-handling system or electrical distribution system might have an infrastructure focus. Equipment, systems, or people may support production but do not directly process the product. Such elements are part of the infrastructure. Infrastructure may be physical (facilities) or nonphysical (people-information-organization). While somewhat unusual, infrastructure can be a focus criterion.

Specialized and expensive facility requirements may necessitate a facility focus. Aircraft manufacturers, for example, bring together operations requiring high bay facilities. Such disparate operations as assembly, welding, machining, and test may require a high-headroom facility. These factors can dictate an infrastructure focus.

Nonphysical infrastructure may also be a basis for focus. A worldwide manufacturer of toiletries concentrated perfume blending at a single site. The specialists there possessed olfactory skills essential to both new-product development and production quality control.

An argument can be made that only market focus is acceptable since customer satisfaction is the primary goal. However, a pure market focus may not account for the real limitations of manufacturing processes and infrastructure. While it provides little basis for the planning and layout of a production facility, it can have considerable impact on what is done where.

Focus levels

Although the concept of operations focus originally evolved at the site or factory level, it also applies at other business unit levels. Focus dictates, or should dictate, which site addresses what product, process, market, geography, or volume. Each site would specialize along the selected focus dimension.

At the regional or global level, particular countries or regions might specialize. For many years the large automobile firms built their large cars in North America and small cars in Europe. Now the focus has changed dramatically.

Just below the site level, a facility might have two or more *plants-within-a-plant* (PWPs). Each PWP is the equivalent of an independent factory, with its own infrastructure.

Focus also applies below the PWP level. A layout cell is a space which contains a set of complementary machines, fixtures, activities, or people. Layout cells often, but not always, correspond to organizational departments. Departments or cells may specialize by process, product, or other suitable focus criterion.

Focus can develop at the workstation level. Individuals may specialize by product, process, customer, or any suitable focus criterion.

In planning our operations strategy, we should examine focus issues at each level since they often differ. A company may decide, for example, on a geographic focus at the site level. At the cell level it may choose a functional focus.

Site-level focus decisions may involve the full range of focus criteria. The plant layout designer must eventually arrange machines and equipment. Such arrangements require decisions between product and process focus.

Degrees of focus

Discussion of focus leads us into a discussion of the following modes:

- Continuous
- Line
- Toyota (linked cellular)
- Cellular
- Functional

■ Project

Only line and continuous production can achieve pure product focus. Only the functional mode can achieve a pure process focus. Cellular and Toyota modes are intermediate. Line and continuous modes can be intermediate between product and process focus when they produce more than one product. The project mode is a special case which fits neither the product nor process category.

There are many ways to mix focus. Parts and materials may move through a process-focused machine shop, for example. Product-focused assembly lines might then perform final assembly.

In Fig. 1.8, we illustrate the advantages and disadvantages of product and process focus.

A focus algorithm

In many factories, focus is nonexistent, inappropriate, or inconsistent. In the usual case, a functional layout with process focus is producing high- or medium-volume products. Most manufacturing, and other organizations as well, display a strong bias for the old-style, long-established functional mode.

Manufacturers misuse the functional mode most often. However, any mode has the potential for misapplication. In one situation, a Detroit-style assembly line builds massive off-road vehicles at 1.5 per day. The results are poor.

How can the layout designer achieve an optimum degree and mix of manufacturing focus? The algorithm of Fig. 1.9 can help. The product-focused modes, especially line and continuous, offer many advantages in quality, low inventory, and efficiency. The process-focused functional mode is most frequently misapplied. For these reasons, the algorithm starts with a pure product focus and line production. It then backtracks through the Toyota, cellular, and functional modes toward a feasible alternative. The focus algorithm has the following steps:

Step 1: Chart operations.

Step 2: Identify dedicated focused factories.

Step 3: Identify trial group of focused factories.

Step 4: Test group of focused factories for adequate volume.

Step 5: Identify trial group technology (GT) cells.

Step 6: Test trial GT cells for adequate volume.

Step 7: Verify product assignments.

Start with an operation process chart for each distinct product. Any industrial engineering handbook or text shows the conventions for

Disadvantages

Product Focus

- Some Duplication of Key Management Positions
- Potential for Lower Machine Utilization
- Traditional Managers are Often Uncomfortable with this Focus
- Sometimes has Trouble Handling Very Different New Products
- Shop Floor "Monuments" Must Still be Shared

Process Focus

- Increased Friction Between Functional Managers
- Higher Control Costs & Increased Probability of Error
- Product Costing Difficult
- Effective Quality Improvement Efforts are more Difficult
- Difficult to Downsize
- Less Flexible than a Product Focused Plant

Advantages

Product Focus

- Simplified Cost & Plant Controls
- Greater Cooperation Across Functional Disciplines
- More Responsive Purchasing & Support Services
- Flatter Organization Structure
- Greater Accountability & Responsibility Lower in Organization
- Faster Product Velocity through Greater Inventory Turns
- Simplified and More Focused Quality Improvement Efforts
- Used in Situations with Lower "Economies of Scale"
- More Responsive to Customer Needs
- Overall Overhead Typically Reduced
- More Amenable to being Managed as a Profit Center

Process Focus

- Traditional Managers Comfortable with this Focus
- Ability to Share Scarce Resources & Increase Utilization
- Typically Used in Industries Like Chemical Processing
- Suited to Situations with a Heterogeneous Labor Force
- Usually, the Units are Cost Centers

Figure 1.8   Focus comparison.

constructing these charts. Line the charts up, side by side as shown in Fig. 1.10. Each operation on a product is a *product operation* (PO). In the typical situation, a single PO requires too little time and equipment for a dedicated workstation or department. The layout designer must somehow group various POs.

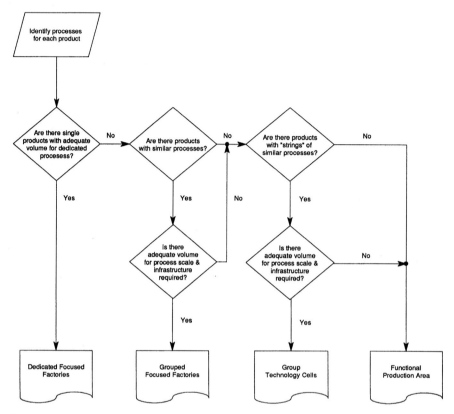

Figure 1.9   Manufacturing focus.

One method is to group all POs which require the same equipment or type of process. Envelopes 1 and 2 in Fig. 1.10 illustrate this. Such grouping provides a pure process focus. At the other extreme, a designer might group all operations required by a single product. He or she would then locate these in a single product department or workstation. Envelope 3 illustrates this point. This is a pure product focus.

In Step 2, the designer examines each product for a trial product-focused grouping. There are, we believe, only two valid reasons for rejecting a pure product focus:

1. The available feasible processes have a large scale with respect to product volume. They cannot economically process a single product. For example a small turned pin requires only 0.25 machine hours per week for the anticipated product volume.

2. Some element of the infrastructure has a large scale with respect to product volume. It cannot reasonably serve a single product. An example is when a highly skilled electronics technician calibrates

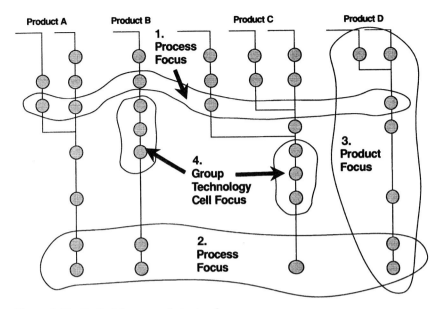

Figure 1.10   Optimizing manufacturing focus.

a circuit board. Production requires the technician's skills for only about two hours per week.

If a trial product passes both tests, it should have its own manufacturing area and possibly its own PWP. Such products can be removed from further consideration.

In Step 3, the designer examines each remaining product. The designer identifies subsets of the manufacturing operations (strings) which are common between two or more products. When such a string occurs, the designer tests it for adequate volume vis-à-vis process and infrastructure scale. If passed, such operation strings become GT cells as in Envelope 4 in Fig. 1.10.

Step 4 aggregates all remaining operations into functional areas. The identified areas become layout cells. Such layout cells may be GT, functional, or product cells.

An analyst might perform this task on paper as illustrated. More often, the algorithm structures the thought process, thus arriving at a cell definition without formally charting all operations.

The designer may also use the *production flow analysis technique* developed by Burbidge or the techniques of a coding and classification approach. These techniques are particularly useful when dealing with hundreds or thousands of parts or products. Traditional grouping techniques should supplement rather than displace the above algorithm.

Operations focus and operations strategy

The focus criteria selected should be consistent with corporate goals, market strategy, operations processes, and infrastructure. Developing focus is one of the most important elements of operations strategy.

Capacity

Facilities are a primary element of capacity. They are usually the most expensive element and the most difficult to change. Facilities decisions and capacity decisions often go together.

At the strategic level, a capacity policy and the resulting decisions should address at least the three factors of timing, increment, and location.

Timing specifies when the capacity will be available and, by inference, when to make decisions and start projects. At a policy level, specify timing as a function of the long-term demand curve and cyclical changes.

Practical capacity may be available only in large increments. In a plastics extrusion facility, extrusion machines can produce 200,000 pounds of product per week. With slowly increasing demand, an additional extruder may not have full utilization for some time. Of course, smaller and larger machines are often available, but these may not be ideal from a technological or financial viewpoint.

Figure 1.11 shows a rising demand curve and an available capacity line with substantial capacity increments and, presumably, substantial investments. In Fig. 1.11a, the capacity is in place prior to demand. In Fig. 1.11b, capacity follows the demand curve as closely as the increments allow. In Fig. 1.11c, availability of capacity lags behind the demand curve. The combination of increment and timing decisions brings either a capacity shortfall, capacity excess, or both.

When an industry faces decline, Fig. 1.11 is inverted. The issue then involves shedding capacity in the most effective way.

The capacity increment decision involves technological, cost, and focus issues. Small increments of capacity can mean small factories or small machines. Small increments and closer tracking of demand lend themselves to product-focused strategies. Larger capacity units sometimes have advantages in quality and production cost.

The timing decision involves both internal and external issues. Typical internal issues are:

- Availability of capital
- Utilization policies
- Skills and competencies

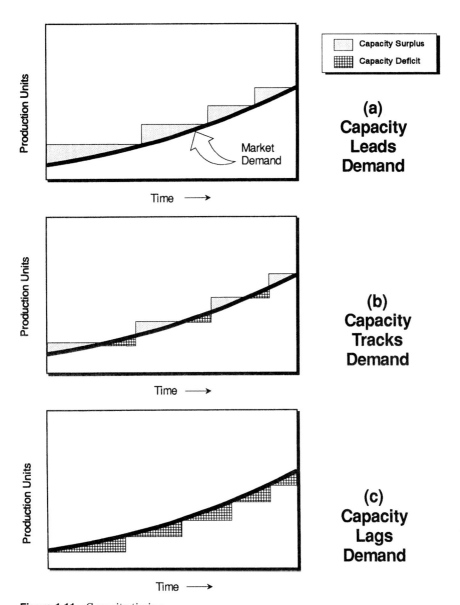

Figure 1.11   Capacity timing.

External issues involve the larger environment. They are either customer or competitor oriented. Government, social, and political factors can also influence capacity decisions.

The most important customer-oriented issue involves the effects of a capacity shortfall. In some situations, a capacity shortfall is disastrous. For example, a firm has developed an important new product, with or without patent protection, which will attract imitations. The capacity for production must be in place before the market demand. If it is not, the firm may lose customers and market share. If an important element of the firm's marketing strategy is customer service, failure to meet demand can negate the entire marketing strategy.

Other firms compete in commodity-like markets. In such markets there is little product differentiation and customers buy primarily on price. Failure to meet demand in such markets may lose sales and associated profits short-term, but the longer-term consequences are insignificant. These producers are often subject to periodic overcapacity and undercapacity situations which affect the entire industry. In these situations, overcapacity severely affects prices and thus profits for everyone. Undercapacity tends to raise prices and ensures a ready market for those who can respond.

Most firms follow one of six policies, either by intention or default. These policies are:

- *Demand lead.* When the benefits or risks of excess capacity outweigh the ill effects of insufficient capacity, demand lead is the strategy of choice. Demand lead policies attempt to ensure that sufficient capacity is always in place well in advance of customer demand. It applies where excess capacity is not financially crippling, margins are high, product features are important, and product life cycles are short. An example is the electronics industry. Figure 1.11a might reflect a demand-lead policy.

- *Demand lag.* Demand lag is appropriate for situations opposite to demand lead. Firms with small margins and large capital requirements may be unwilling or unable to build in advance of demand. Here the risk and consequences of excess capacity are high. The firm wants to ensure sales for all the capacity it might have available. Figure 1.11c illustrates this. A halfway compromise between "lead" and "lag" is shown in Fig. 1.11b.

- *Industry preemption.* In commodity-like markets with few competitors, preemption may be the policy of choice. This approach builds capacity well in advance of demand. The intent, however, is different from that in a market-lead policy. Here, the intent is to prevent other firms in the industry from building and depressing prices for everyone. If demand continues to grow, the preempting firm gains additional business and market share. Preemption strategies require strong financial resources and strong nerves.

- *Industry coordination.* In stable industries with few competitors and commodity-like products, coordination may be an appropriate policy. Here, firms take turns expanding so that excess capacity from several firms does not appear simultaneously. Formal coordination may have legal consequences. Informal methods of coordination such as press releases well in advance are acceptable. Such signals inform other firms of an intent to expand.

- *Investment timing.* Most of the above policies can lead to expansion at the top of a business cycle. Such expansions are expensive since construction costs are greater. Moreover, the capacity may come on stream just as demand turns down. Investment timing policies attempt to predict demand and initiate expansions near the bottom of cyclical demand.

- *Capital expenditure minimization.* The drivers for this scenario are internal, financial, and quantitative. The strategy tries to minimize and postpone capital expenditure and maximize utilization.

Types of capacity expansions

There are many ways to increase capacity. Different types of capacity expansions require different levels of investment, increments, and capacity lead time. Among the choices for capacity increase are:

- Hiring
- Out-sourcing
- Process improvements
- Operating additional shifts
- Equipment additions
- Facility additions
- New sites

Increasing capacity may be as simple as hiring more people, either temporarily or as regular employees. Out-sourcing products or processes may be an option. Process improvements include equipment updates and modifications. Working additional shifts can improve equipment and facilities utilization. New equipment added to the same facility might be feasible. New equipment coupled with a facility expansion is usually possible. A new site with facilities and equipment is the last option.

At a particular time, most firms can choose from several options. These may include some or all of the above types. Consider each option in turn. Each one usually requires increasing capacity lead time, cost, and increment size.

Hiring has a capacity lead time of days or weeks. The investment cost can be as little as a few hundred dollars and the capacity increment, for most firms, is very small.

Out-sourcing usually requires weeks or months. The investment may range from zero to a few thousand dollars and the capacity increment is frequently variable.

Process improvements can often upgrade existing equipment and operations for additional capacity. These changes have lower cost than new equipment and have shorter lead times.

If extra shifts will be working longer-term, this will take time to implement. Extra supervision, maintenance, and support personnel may be needed.

Equipment additions involve new equipment installed in current facilities. If room is available, the cost is much lower than an entirely new and outfitted facility. Lead time is usually less as well.

Facility additions will accommodate new equipment when no room is available in the current facility. Additions on an existing site are usually a much smaller undertaking than building on a new site. They generally make use of existing infrastructure such as maintenance and administrative services. The lead time for a facility expansion is also less than that for bringing a new site on stream. New sites are the most expensive type of capacity addition. They have the longest lead time and the largest capacity increment.

Firms which adopt a capital minimization policy focus on the lowest-cost, smallest-increment option. Since they often delay capacity decisions, they also need the shortest-lead-time option. But the number of low-investment options is limited. With hiring, people cease to be the limiting capacity factor. Out-sourcing is often practical with the development of strategic relationships, but this can become difficult or unwise when strategic technologies are involved.

Squeezing in new equipment may overfill or congest the current buildings. Figure 1.12 illustrates this scenario. This approach seems good at first because it allows the firms to track demand closely and avoid high-investment major projects. However, once the available short-term increases are in place, the only remaining options are high-investment and long-lead projects. At this point, the firm now faces a decision for a major facility expansion or a new site. Both of these are long-lead, high-investment, large-scale options. The company is probably in considerable difficulty, being unused to making long-range decisions, and often procrastinates.

A more rational approach would first survey all conceivable capacity expansion options, then rank them for scale, lead time, and investment cost. Shorter-term improvements would be held in reserve for

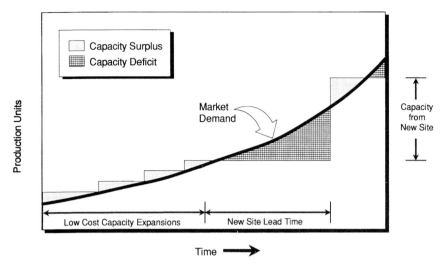

Figure 1.12 Typical capital-minimization strategy.

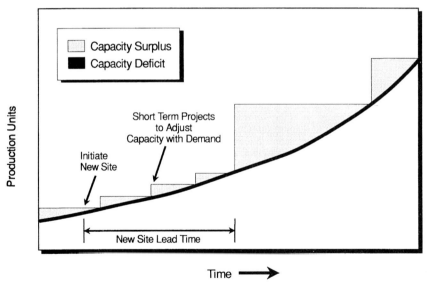

Figure 1.13 Balanced capacity—lead strategy.

the time when a sudden upturn in demand required fast action. Figure 1.13 shows how this scenario tracks demand better over the long term with less risk of a major capacity shortfall.

Facilities are among the most important strategic decisions for a company to make. The firm which approaches facilities decisions with full knowledge of their implications and consequences can position itself for long-term competitiveness. The firm which focuses entirely on internal pressures and narrow issues may find itself with resources which are ill suited to its marketing and business strategies.

Focus Examples

Product focus

California Microdat manufactures minicomputers. While designing a new facility, their consultants presented the directors with several layout options. One option featured focused factories and product-focused cells. Other options simply improved the existing functional layout.

The product-focused layout promised quantum improvements in quality and delivery reliability, which were deciding market criteria. This layout, however, would require major changes in the firm's organization and culture. Such changes are difficult and often fail, but the product-focused layout would not work without them. Product focus would thus require a radically different operations strategy.

The directors debated the issues. The product-focused layout carried a high risk of immediate failure if the organization could not or would not change. The traditional layouts would work without organizational changes, but they carry a long-term peril—*failure to meet customer requirements with consequent loss of market share and profitability.*

Microdat management chose product focus for their new facility. They immediately began to develop and implement the associated elements of a JIT-based operations strategy. The results, in this instance, were outstanding. But if other parts of the strategy implementation had failed, Microdat's new factory would have been a disaster.

Process focus

Northstar Aluminum Products makes castings for many customers. They specialize in medium-precision sand castings using green-sand and jolt-squeeze molding equipment. Other processes such as die casting, investment, automatic molding, and lost-foam are available, but are more suited to higher volume. Hand ramming produces molds for prototype or very low volume work.

Casting metal in sand is an ancient art and simple in principle. Execution, however, requires skill, experience, and close control of many variables. Technically, it is more difficult than many so-called "high-tech" processes. Accordingly, Northstar has chosen to focus its plant not just on the product but on a select subset of casting processes.

Commercial Fixtures Corp. manufactures a variety of lighting products which incorporate many precision, low-volume castings.

Their foundry, however, was too small to purchase supplies effectively. Neither could it support the pattern makers and process engineers required for cost-effective operation. Cost, quality, and productivity problems caused Commercial Fixtures to close its casting operation. The firm then contracted with Northstar Aluminum for its cast components.

The two situations described are examples of process focus based on the necessity to concentrate expertise on a highly technical function. Also, cooperation between two companies resulted in more cost-effective focused operations for both.

Product focus

Inter-Defense Industries (IDI), a manufacturer of armored vehicles, supplies many foreign armies from its factories in Western Europe. High variety, low volume, and the unpredictable fortunes of politics and war had previously led this firm toward a process focus.

Their new Mark VIII Medium Tank is a major success, with long-term contracts and stable delivery schedules. Despite the high volume and predictable schedules which they had always desired, manufacturing could not cope. High inventories, poor quality, delays, and coordination problems threatened a reputation that the company had been building for eighty-two years.

Because of the higher volume, production stability, and higher complexity, IDI segregated the new Mark VIII in a separate facility.

Within this overall *product focus,* some components had dedicated processes such as welding, machining, and assembly. They were close to the assembly areas. Like a river, the final assembly line was fed by branches where major components were assembled. These branches were, in turn, fed by the confluence of smaller production streams.

One major tributary of the final assembly stream was the turret. The turret had guns, missile launcher, fire control, command systems, and armor. It is the most complex single subassembly except for the hull itself.

A focused factory built turrets on the same site. All equipment and operations necessary to make the turret, its subassemblies, and many components were in this *product-focused PWP.*

At the workstation level, day-to-day schedules stabilized as each operation produced only enough for immediate downstream requirements. Inventories shrank and quality problems surfaced quickly. Significant improvements in quality, productivity, and delivery resulted.

While process and product focus are most common, other criteria may be important. Examples of market, geographic, and infrastructure focus follow.

Market focus

Northwest Specialty Steel, a steel founder, focused two facilities on markets. The alloys and size of castings in both foundries were similar. However, their military armor market requires high quality and extensive documentation. The contracts are long-term and allow stable schedules. Experience in casting armor is an important selling feature.

The firm's General Industry (GI) Division makes castings for construction equipment, oil-field equipment and general industry needs. Quality standards are lower and procedures different. These markets have many new products and unpredictable orders.

The Armor Division hires experienced foundry workers, pays well, and has a stable employment policy. The GI Division pays lower wages and frequently changes employment levels as products vary. Only a cadre of senior workers can depend on steady employment.

As a result of these differences in apparently similar castings, the Armor and GI Divisions evolved different production control systems, quality assurance organizations, and suppliers. They have focused two divisions with similar products and volumes around different markets.

Geographic focus

The following considerations, among others, can indicate the advantages or opportunities for a focus with a geographic emphasis:

- Short shelf life
- Hazardous materials or processes
- Special support
- Climatic advantages
- Geological conditions
- Regional consumer factors

Grandma's Pies manufactures fresh pies for restaurants. Its drivers deliver many varieties of high-quality pies to medium-sized, upscale restaurants who special-order 24 hours in advance. Requirements for freshness and personal contact led Grandma's to a geographic focus having all customers within a 150-mile radius.

Growth will replicate this focus in other geographic areas, possibly with clones of the initial design.

Summary

In this chapter, we have outlined the impact of focus and strategy on facilities design and thus on a firm's competitiveness. We have reviewed how wide-spectrum focus leads into operations strategy and helps determine facilities design. We show how these designs must support the business strategy.

Factories have only existed for two hundred and fifty years, yet manufacturing systems have changed several times. The evolutionary pace is accelerating and world class companies are transforming their operations to outpace the competition. This includes facilities design and implies flexibility. The interaction of the elements of operations requires a holistic approach. The following chapters provide information on many of the aspects of facilities design for world class operations.

References

1. Cossons, Neil, *Industrial Archaeology,* David & Charles, North Pomfret, VT, 1975.
2. Doll, William J., and Mark A. Vonderembse, "The Evolution of Manufacturing Systems: Towards the Post-Industrial Enterprise," *Proceedings of the 5th International Conference of the Operations Management Association,* University of Warwick, 1990.

Further Reading

Adair, John: *Effective Teambuilding,* Pan Books, London, 1987.
Buffa, Elwood S.: *Meeting the Competitive Challenge,* Dow Jones-Irwin, Homewood, IL, 1984.
Emerson, Howard P., and Douglas C. E. Naerhing: *Origins of Industrial Engineering,* Industrial Engineering & Management Press, Norcross, GA, 1988.
Ford, Henry: *Today and Tomorrow,* Doubleday, Page & Company, Garden City, NY, 1926.
Hays, Robert H., and Steven C. Wheelwright: *Restoring Our Competitive Edge,* Wiley, New York, 1984.
Hicks, Philip E.: *Introduction to Industrial Engineering and Management Science,* McGraw-Hill, New York, 1977.
Hill, Terry: *Manufacturing Strategy,* Macmillan Education, London, 1985.
Kantrow, Alan M.: *Survival Strategies for American Industry,* Wiley, New York, 1983.
Liberman, Marvin B.: "Strategies for Capacity Expansion," *Sloan Management Review,* vol. 28, no. 4, Summer 1987.
Miller, Stanley S.: *Competitive Manufacturing,* Van Reinhold, New York, 1988.
Porter, Michael E.: *Competitive Strategy,* The Free Press, New York, 1980.
Senge, Peter M.: *The Fifth Discipline,* Doubleday, New York, 1990.
Shinohara, Isao: *NPS New Production System,* Productivity Press, Cambridge, MA, 1988.
Skinner, Wickham: *Manufacturing: The Formidable Competitive Weapon,* Wiley, New York, 1985.

2

Location Decisions: Situation and Site Selection

Howard A. Stafford
Professor of Geography
University of Cincinnati
Cincinnati

Business activity can be viewed as a "box" within which is created form, time, place, and/or ownership utility, through the application of labor and energy. Materials (inputs) flow into the box from suppliers, and products and/or services flow out to consumers, who are often other manufacturers and producers. Most of the inputs will come from somewhere else. Likewise, the outputs will go elsewhere. The basic question addressed in this chapter is: "Where should the business—the 'box'—be best located relative to the many locations of the many inputs and customers?" More succinctly, the chapter is concerned with the "Where?" question. It focuses on locational decision making for manufacturing plants and retail outlets.

Location Principles

Frictions of distance and economies of scale

Tensions. All business locations are consequences of the tensions between the friction of distance and economies of scale. If, for example, there were no friction of distance and only economies of scale, then there would be only one grocery store in the world, and its location would not matter to either customers or to management. Conversely, if there were no economies of scale, and only friction of distance, then there would be no grocery stores—every family would be self-sufficient. Obviously, the sizes and spread (locations) of activities are linked issues.

Internal and external economies of scale.   Economies of scale may be both internal to the firm and external to the firm. The classic illustration of internal economies of scale is the decline in per unit costs with increasing plant size. External economies of scale are the advantages of spatial agglomeration accruing from nearness to related (linked) industries (localization economies) or from access to shared and jointly paid for infrastructure (urbanization economies). Just-in-time (JIT) inventory principles emphasize the importance of proximity to suppliers, distributors, and customers, and influence local labor-management relations.[1] Of course, there also may be diseconomies of scale, when operating units become too large to manage effectively, or areas become too crowded.

Measuring the friction of distance.   The friction of distance can be measured in many ways, including physical distance (as miles or kilometers), time, or money. It is the costs of shipping supplies to the plant or outlet, or of shipping products to customers, or the costs to customers of getting to stores. It also can be the distances workers must or will travel. These are important to the feasible or best spatial arrangement of the firm; relevant measures might include perceived distances, safety, and convenience.

Economies of scale and number of units.   The smaller the efficient economic size of unit (lower economies of scale), the greater the number of discrete units to be located and the greater the firm's locational flexibility. Conversely, for individual units, a good location is usually more important for small units than large. Because of technological advances, managerial efficiencies, and out-sourcing of some activities formerly conducted in-house, the average number of employees per fabricating plant is decreasing. A rule of thumb is that 500 to 1000 employees is desirable.[2] The opposite trend, toward larger stores— especially in terms of floor area—appears to be the case for outlets of some major retail chains, such as supermarkets and discounters of several types.

Costs, revenues, and profits over space

Profits are the difference between revenues and costs. If revenues do not vary by location, then the least-cost location also will be the maximum-profit location. Conversely, if costs do not vary over space, then the maximum-revenue location is the maximum-profit location. Of course, in reality both costs and revenues vary simultaneously, and location solutions are only approximations of the best combinations of least costs and maximum revenues.

Retailers maximize revenues.   Retailers emphasize revenues; they will carry any locational costs (rents, wages, advertising, etc.) so long as the volume of business and margin justify it. Locationally, they are market-oriented.

Manufacturers not least-costers.   There is a common (mis)conception that manufacturers seek lowest costs. Manufacturers may even think so themselves, but it is not entirely true. They rarely locate in the absolute lowest-cost place for any one factor, or even in the lowest-cost place for a combination of factors. What is true is that manufacturers often attempt to minimize costs within specific geographic contexts, but the locations of markets (therefore, revenues) are usually critical in establishing the more general geographic region within which costs are minimized. It is common for manufacturers to be market-oriented at the national and international scales, and labor-oriented at the regional and local scales. Market-oriented locations may decrease transport costs, but they can be even more important if sales are increased and service times are decreased. Also, the critical labor issues often are availability and productivity rather than simple wage rates, and better-trained labor often is easier to obtain in the same areas that contain the major markets.

Profits, growth, and stability.   The motives of businesspersons are many. Two motives of locational importance are profits and the growth of the firm. Of course, the continued existence and possible growth of the firm are dependent on satisfactory profits, but professional managers are likely to stress the long-run stability and security of the firm over maximum short-run profits (if the shareholders will allow, and if sufficient revenues are generated for current operations). Growth of the firm often means spatial expansion. Thus, a firm may tolerate low returns on a strategically placed outlet because it is "important to have a presence there." Conversely, relatively successful operations may be discarded because they do not fit longer-term strategies; locations may be irrelevant in such situations.

The geography of the firm

Organizations have geographies. Imagine a map on which were shown the locations of the firm's

Headquarters

Production facilities

Warehouses

District sales offices

Sales territories

Major suppliers

Major customers

And also imagine on the map arrows, perhaps of varying widths, indicating the directions and volumes of flows between the locations, flows of

Inputs

    Materials

    People

    Ideas/information

Outputs

    Materials (products)

    Services

all as suggested in Fig. 2.1.

Spatial compactness. Other things being equal, the more compact the corporate space, the more efficient and secure the firm. Links

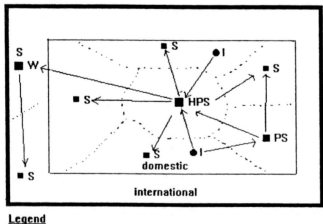

Legend
H=headquarters   I=major supplier   S=sales office
P=plant   W=warehouse   ···· =sales territory

Figure 2.1   The geography of a firm: Locations and flows.

between company units are facilitated, as are links with key markets and suppliers. Effective management is easier. Distances between plants within an organization and between operations and headquarters may be crucial. Townsend cautions organizations *to have respect for geography:*

> If your business is in Cleveland, start or acquire an operation in Santa Barbara at your peril. Absentee management is fatal. And the disaster potential is equal to the square of the distance—measured in hours—between your home base and the new plant. No matter how determined you are to visit it frequently, you'll discover that your capacity to find last-minute reasons not to go is unlimited. If the new operation is in Europe or the Far East, the problems increase by cube functions. It is twenty-seven times harder to cope with an operation in Hong Kong than one in Duluth.[3]

Spatial increment strategy. An effective spatial expansion strategy for a firm often is a spatial increment policy, where new units are located on the periphery of the company's existing space, rather than being jumped into new territories.[4] For retailers, this strategy has the added benefit of borrowing consumer awareness from nearby areas; in new and separated territories, store and brand awareness often must be built from the ground up. Distance from the nearest warehouse may be a limiting factor. Of course, there are other situations when leaps into nonadjacent areas are warranted, especially when markets have matured to a size which will support a local operation.

Geographic inertia. Geographic inertia exists; it is a powerful influence on the general economic landscape, and on the locational patterns of individual firms. Things at rest tend to stay at rest. Managements are reluctant to abandon fixed capital or to admit to locational mistakes. The changing geography of the marketplace may go undetected until late in the game. The majority of plant closures occur because of economic failure (rather than the redeployment of assets within the firm in search of even greater profits). Most new branch plants are built, and thus new locations are selected, because of intrafirm demands for new productive capacity. Locational inertia is much less pronounced in retailing than in manufacturing; the fixed investments are smaller, and the marketplace sends signals to misplaced outlets relatively quickly and forcefully. However, even in retailing there is more inertia than commonly realized. A losing store may be kept open because of long-term obligations (e.g., a 20-year lease), or to help carry market overheads, such as advertising. In any event, a decision to stay in place also is a location decision, even if often not recognized as such. It is a decision not to go elsewhere. It implies that the current place is at least as good as any other place, or that the costs of search and transfer are higher than the perceived benefits of another location. In truth, of course, many good spatial reorganization opportunities are unrecognized.

Location decisions

Perfect locations.   There are *no* perfect locations. More precisely, it is impossible to determine the optimum locations because of lack of data, models which become too complex when burdened with too many variables, and because of uncertainty—uncertainty about the future, uncertainty about the actions of competitors, suppliers, and customers. It is possible, however, by careful analysis and reasoning to increase the odds of picking winning locations and not going to losing places.

Where this? What here?   There are two basic types of location decision. The first is "I have an activity, where shall I put it?" A good example is the firm which has decided to build a branch plant or open a new retail outlet, which already has decided on its size and product mix, and now needs to select a good location. The second type of location decision is "I have a place, how shall I use it?" Examples include landowners such as farmers deciding what to plant, or developers deciding between building a shopping mall or an industrial park on an already owned property. This chapter focuses on the first of these two questions, but it should be kept in mind that after initial location it is possible (though often expensive) to make adjustments to the reality of the area via product mix changes or contraction or expansion of the operation.

Spatial variability and complexity.   Location decisions must take into account many variables. As important as the magnitude of the variable is its spatial variability. For example, even if the firm's largest outlay is for labor, if these costs do not vary significantly over space then labor is not a locational variable.

Search costs.   There are search costs. Good locations enhance the operation and profitability of the firm, but they cannot be selected randomly, by the tossing of a dart at a map. Their selection requires talent and time. There are the costs of time for the people involved, and there can be time lost vis-à-vis competitors. The trick is to decide when the marginal search costs begin to exceed the marginal improvements in the location decision and the search process should be brought to a conclusion.

Rare events. Most businesses have little experience in making location decisions. It is a relatively rare event for most. The importance of location often is underestimated, and too little is spent on the search. Least sophisticated in locational decision making are the single-site retail operations (which is one of the major reasons for the high rates of business failures within this group). Multisite manufacturing firms are somewhat better at assessing locations and locational strategies. The most sophisticated situation and site selectors, and the most successful, are the multioutlet major retail chains; for them, choosing and assessing locations is a continuous activity. However, even chain-store retail firms often are reacting to the proposals of others (e.g., developers) rather than being proactive; a good strategic plan is useful in assessing proposals which come to the firm.[4,5]

Expansion vs. new site. The most common location decision is, in fact, the decision to expand, contract, or change the product mix of production in existing facilities at existing locations. If there is an active consideration of location and it is decided that the best course of action is to make major changes within the existing spatial framework (e.g., build an addition onto a plant), then an in-site decision has been made. Conversely, if the decision is to locate something elsewhere, then a new-site decision has been made. There are subtypes of new sites. A "brownfield" location is one where an existing industrial site is reused by a new owner/operator (e.g., buying an existing plant and refitting it). A "greenfield" location involves using a land parcel for a very different type of use than its previous use (e.g., locating a manufacturing plant on a former farm); greenfield sites often require zoning changes. Industrial parks, which are now the most popular venues for small- and medium-scale manufacturers, have some of the characteristics of both. Like greenfield sites, the buildings are often new and purpose-built. Like brownfield sites, the land is already serviced (utilities, roads, etc.) and zoning is set.

Situation and site. Locations have two major aspects, site and situation. Most of the above refers to situation. A land area's *situation* is its location *relative to* other important spatial distributions, such as nearness to the market or centrality within an adequate trade area, or accessibility by a labor force, or nearness to corporate headquarters or other units of the firm, or location relative to the locations of competitors. *Site* factors are the absolute measures of the relevant physical characteristics of a specific parcel of land. Often included are size, shape, topography (especially slope), drainage, soil characteristics and load-bearing capacity, highway frontage, traffic counts, visibility, and ease of ingress and egress. In location searches, situation should always be determined before site issues are addressed. Within any desired general area (situation) there will be several acceptable sites from which to choose. The search always should proceed from the global scale to the local scale. It should proceed "down" the geographic hierarchy.

Intractable problems, simplified solutions.   Location searches always are greatly simplified. The number of possible locations is virtually infinite. The data demands alone are daunting. Furthermore, even if it is assumed that the ultimate locational criterion is profit, and the unrealistic assumption is made that all relevant factors are known and all variables can be objectively measured, the problem still remains insolvable in an absolute sense. The future always is uncertain, and decisions about the size, type, and location of a facility are simultaneously interactive. Also simultaneously interactive are determinations of least-cost and maximum-revenue locations. As Smith puts it,

> it would be possible to compute a market potential surface...and then to identify a total cost surface by working out for a sample of locations the total cost and then to identify a total cost of producing and distributing the required quantity of output. However, such a procedure would imply that the level of sales attainable at a given location was unrelated to production costs, whereas in reality scale might affect average cost, cost might affect price, and price might influence sales and revenue....And to start with a total cost surface and then proceed to a demand or revenue surface is equally tortuous, since both average cost and total cost can be a function of volume of sales.[6]

In spite of this intractability, locations must be and are chosen. The most commonly practiced simplifications are to (1) determine the type and size of facility before the location search, and then adjust size and product mix after experience in a location; (2) choose regions which maximize revenues; and then (3) choose localities and sites which meet requirements and provide acceptable total costs. The geographic scope of the search is always severely restricted; most areas of the world, or even of a city, are never even considered. Obviously, such simplifications are practical, but there are missed opportunities and unknown costs.

Locating a Manufacturing Plant

First situation, then site

Single-plant manufacturers rarely make explicit, proactive decisions about the location of the manufacturing facility at any scale other than the local. The broader, situational, relative location dimensions are more commonly examined by multiplant companies when locating a branch plant. The most common, and recommended, practice is to follow a hierarchical sequence, where the country or large region is first selected, subregions are evaluated and one is selected, then a set of acceptable types of places (e.g., large cities, small towns) is identified, and the most promising places within this set are chosen for local analysis. Only after the situational aspects are analyzed should there be concern about or a search for a specific site.

Location factors

There are, literally, hundreds of factors that might be taken into account when deciding where to place a manufacturing plant. A checklist helps in keeping track of them. Fortunately, relatively few factors are really important. A recent survey narrows down the many factors to those in Fig. 2.2.

The top ten overall.   Note that five of the top ten factors in Fig. 2.2 relate to personnel. The other cluster of factors relate to access to markets; this influence is reinforced by the "transportation costs" variable, since a large portion of transport costs results from shipping products to markets.

The top ten at four geographic scales.   However, this overall view is too simple because the relative importance of the factors varies according to the stage (spatial scale) of the search, as suggested in Fig. 2.3. At the national scale of search, for example, nearness to markets is most often the critical variable. At the other end of the search hierarchy,

Factor	Percent of respondents
1. Labor quality/productivity	54.7
2. Direct labor costs	43.4
3. Skilled labor availability	50.7
4. Access to markets	34.2
5. Transportation costs	32.9
6. Labor-management relations	26.3
7. State and local taxes and fees	25.0
8. Ability to attract and retain managers and professionals	25.0
9. Facility costs	22.4
10. Utility costs	21.1

SOURCE: *Survey of Corporate Real Estate Executives: Site Selection Factors, Attractive Areas, Company Plans. Premier Decision Management, Irvine, CA, October 1990.*

Figure 2.2   Top ten manufacturing site selection factors.

Rank	National		Regional		Subregional		Local	
1	Markets	102	Markets	170	Markets	91	Labor availability	161
2	Transport costs	55	Materials	54	Environmental regulations	65	Utilities	158
3	Geography of the firm	44	Transport cost	35	Materials	58	Transport availability	156
4	Materials	38	Geography of the firm	31	Labor, unions	55	Site characteristics	148
5	Environmental regulations	27	Labor costs	23	Utilities	55	Quality of life	134
6	Labor costs	26	Utilities	18	Developmental organizations	51	Environmental regulations	104
7	Quality of life	24	Developmental organizations	16	Labor costs	43	Transport costs	100
8	Utilities	22	Transport availability	15	Transport availability	43	Markets	90
9	Transport availability	20	Environmental regulations	14	Labor availability	42	Labor costs	87
10	Labor availability	20	Labor availability	14	Business climate	42	Materials	83
	Total counts	378	Total counts	390	Total counts	545	Total counts	1221

Source: H. A. Stafford, "Environmental Protection and Industrial Location," *Annals of the Association of American Geographers*, p. 235.

Figure 2.3   Ten most important location factors at four spatial scales for the interviewed firms ($N = 54$).

at the local scale, labor attributes tend to prevail. Also, the relative importance of these factors will vary among industries and firms, and over time.[7]

Transportation costs and attributes of areas. The location factors may be divided another way. One group includes all those which account for the costs (money, time, service, or worry) of getting products and/or people from place A to place B. These are the transportation costs from suppliers, and to markets, and communications costs. A second group contains all the important attributes of areas, including labor characteristics, infrastructure, governmental influences (subsidies, taxes, regulations, etc.), and quality of life.[4]

Recent studies indicate that the traditional location factors of access to markets, labor, materials, and transportation remain the most important, but productivity, taxes, and community attitudes toward business, and quality of life, are now also of major consideration.[8] Foreign firms locating plants in the United States primarily seek markets of substantial size with favorable growth characteristics, and from which they can better serve customers.[9] Conversely, there is much in the popular press about the appropriateness and effectiveness of locational incentives offered by areas, incentives which include financial and technical assistance, and employment and training services, and are most often offered by national and state governments; these are usually of little importance in the location selection process, except when they may serve as tiebreakers when the final decision is down to a few similar places.[10] Likewise, taxes are an often overrated location factor.

Making the plant location decision: situation

Mapping present and past situations. It is impossible to determine the optimum location, but it is possible to eliminate poor choices. The first step in the analysis is to map, from company experience, the locations of current facilities, including production plants, warehouses, and administration and sales offices. Then diagram all the *important* interconnections and flows, of parts, people, ideas, or money, between the various units/locations. This will provide a necessary view of the "geography of the firm." Dynamic firms might wish to map their geographies for some previous time periods, to see where they have been, and to perhaps better sense where they are going.

Next, maps should be made of market (sales) data for selected time periods, by sales territory. All mapping nowadays is best done on the computer, using some sort of geographic information system (GIS) (see under "An example using a geographic information system," later in this chapter); then analyses via map overlays, correlations, updating, and presentations are facilitated.

Maps of the future. And now a critical and difficult step: the projection of sales into the foreseeable future (one to five years) and disaggregated into subareas, such as nations, sales regions, or specific places. These forecasts are most often projections of trajectories of sales regions' histories, from the company files, and the best guesses of management as to what the future market will look like, how much and where.

Then, computer maps should be constructed showing the locations of major suppliers, including those from whom component parts are purchased, and also the locations of any requisite personnel or business services which are not widely available. Separate maps (or computer map layers) can be made for major components. If spatial changes in cost items can be anticipated, then these maps should be incorporated into "what if" scenarios in a GIS.

Why here, why not there? These maps will give insightful pictures of the spaces the firm now inhabits. Next, consider all the spaces/places in which the firm does not operate, and ask "Why not?" If the plant is in Ohio, is there merit in thinking of an Arizona facility? If the firm operates only in the United States, does Europe or Southeast Asia make any sense for sales or production? Again, why or why not? In this process, vast areas of the world will be eliminated as not being attractive. Some mistakes may be made, some good places overlooked, but managerial instincts are usually valid at this scale and eliminating much of the world makes the location search feasible; remember, there are search costs. It is, however, very beneficial to go through this phase. It stimulates new perspectives and helps people break out of old "business as usual/locations as usual" syndromes, and some exciting new places may emerge.

Needs and constraints. Make a list of all the firm's or facility's significant locational needs and constraints. If it is necessary to be in a metropolitan area for communication needs, then denote all such areas within the feasible region. If it is necessary to be within one truck hour from a major customer to meet JIT requirements, then draw a one-hour radius around the customer's facility. If it is important to be within two hours by corporate jet from headquarters, then delimit that space. If it is desirable to put the next facility within the European Community, then shade in all the countries which are now members, and also note those likely to soon join the community. And so on.

Analysis: Phases and techniques

There are three, perhaps four, phases to solving the industrial location decision problem:

1. The problem is narrowed to a manageable set of countries and regions perceived to be likely general areas for production.
2. Operations research techniques may be used to narrow the selection set even further, by eliminating places too far from some "near-optimal" situation.
3. A multicriterion decision model is implemented to short-list the one, two, or three "best" local areas from among the alternatives identified in Phase 1 (and maybe Phase 2).
4. A specific site within the selected local area is chosen.

Regarding Phase 1, a common failing in location selection research and practice is that the searches are too soon narrowed to local searches. There is a great tendency, especially within the United States among companies who do not think globally, to consciously narrow the location search too quickly, and not be aware that a decision to look at a few places is a decision to not allow other places to *ever* enter the locational calculus. However, firms are increasingly going global, and this brings with it both opportunities and concerns.

There are many operations research techniques which are of potential use in making somewhat more objective the subjective business of locating a manufacturing plant. There are techniques for dealing with continuous space, and with dynamic situations, but these are little used by actual decisionmakers. The most common practice is to operate in discrete space, and to compare in detail only a few well-chosen potential locations.

If good data are available, and the decision factors fairly precisely modeled, a variety of elegant mathematical techniques can be used to determine "optimum" locations and flows of inputs and outputs. These operations research techniques include minisum or minimax location methods, location-allocation models, and quadratic assignment techniques. Linear programming, in the form of the "transportation problem" is an example. Typically involved are an elementary "geography of the firm" with market and production point identifications (thus, a single- or multiple-plant firm in discrete space), perhaps capacity projections (an approach to the dynamics of reality via forecasting), and comparisons of costs. Both transportation costs and production costs can be incorporated in a joint objective function. There are numerous other extensions.[4,11]

In reality, linear programming and related techniques are little used in industrial location decision making. Even when they are used, experienced managers view the results as just part of the evidence. More often used are straightforward, multivariate additive comparative costs and weighted costs-and-benefits approaches. These are *substitution* models, where the pluses on one factor may be offset by negatives on another; the critical issue is to not make the location decision based on a favorite variable, but rather to simultaneously integrate several criteria. Weighted, multicriterion models also are used at the final site selection stage.[12] Of course, all the quantitative aids must be used with caution. The operations research techniques tend to look more impressive than they are. They suffer from inadequate data, suspect forecasts, and incomplete models. Since cost data often are most readily available and reliable, the operations research techniques tend to rely too heavily on them; indeed, because sales also vary with location, firms can be ruined by minimizing distribution or total costs.[13] Weighting systems are mathematically inelegant, and can be very dangerous if not handled with care. Slight changes in ranks or weights can lead to very different results. These schemes should be tested for robustness. Even more, they should be regarded as primarily valuable as mechanisms for focusing decision-making attention; the process is more important than the product.

A factor comparisons illustration

For a simulation of an actual industrial location process, begin with a decision-making *team* of five corporate executives. Endow each with some general knowledge of the firm, and also with specialized information and individual desires. Thus, all will know the firm's product range and recent history, and all will wish for the firm to be profitable, but the sales manager may emphasize nearness to markets as the most important locational factor for a new plant, the production manager may be especially concerned about easy access to materials, and the personnel manager may insist that the only good locations are areas with abundant and efficient labor available. The treasurer may make return-on-investment calculations when appropriate and be expert on tax rates, and the president may inform the team of long-range strategic plans.

Now make another realistic simplification by assuming that the new plant will be placed somewhere within a large metropolitan area in the United States. This simplification illustrates several things. Continuous space is transferred to discrete space. Many prior decisions have been made, even if unconscious; in this example, no areas outside the United States or nonmetropolitan areas can now enter the decision process. The question of locating the plant *within* a metropolitan area is deferred until the region has been selected. Thus, the illustration is of regional location decision making, midway down the scale sequence (hierarchy).

The most common next steps are for the involved decision makers to individually and collectively develop a short list of key variables upon which each potential location (metropolitan area) will be evaluated. Again, coming to a practical short list involves compromises and some effective formal or informal ways for executives to share expert opinions. Location searching can be expensive, so simplifications are necessary, but it is important to include the correct variables in the matrix. Some useful suggestions are: Do not bother to include any factors which do not vary over the space under consideration; use corporate experience and that of close competitors (analogs) to guide selections; buy relevant data from commercial vendors; explicitly, consciously, *map* out some guesses about the *spatial* patterns for the firm and for the industry five years hence; to the extent possible, use the familiar decision-making rules and procedures which are successfully used in other phases of corporate operations.

Figure 2.4 illustrates factor information for an hypothetical metals fabricator of large vessels used in fertilizer plants. The figures for each city are calculated assuming that the new plant is located in that city, and shipments are then made to every other city according to its demand, and from each of the supply cities to the city being assessed as a possible location.

Concerns for markets and sales are most explicitly expressed in the first three columns. A large local market (column 1) is desirable to have a good chance of securing a more stable customer base. However, low total product shipping costs (column 2) is a desirable attribute for sales to the entire U.S. market. The sales manager

	1	2	3	4	5	6	7	8	9	10
Metropolitan Area	Local Market Size Index	Outputs Shipping Costs (000)	Airline Passenger (000,000)	Inputs Transport (Costs 000)	Wage Costs, Annual (000)	Corporate Taxes (000)	Value Added per Employees (000)	Unemployment Rate	Air Pollution (ozone)	Total Locational $ Costs (000) (Cols. 2+4+5+6)
Atlanta	300	$400	22	$148	$2,364	$130	$41	5.4	poor	3,042
Baltimore	400	329	4	124	2,608	264	46	4.3	poor	3,325
Boston	200	456	10	200	2,482	175	43	5.1	fair	3,313
Cincinnati	1,600	299	4	188	2,590	125	49	4.4	fair	3,302
Houston	900	570	7	300	2,598	100	61	5.4	very poor	3,568
Kansas City	100	436	4	222	2,572	135	47	4.4	fair	3,365
Los Angeles	408	1,052	19	62	2,796	250	46	4.6	very poor	4,160
New Orleans	500	486	3	206	2,448	130	69	6.6	good	3,270
St. Louis	600	339	10	182	2,640	140	51	5.6	poor	3,301
Seattle	100	1,092	7	530	2,670	200	40	3.5	good	4,492

Figure 2.4 Hypothetical manufacturing costs and attributes for selected cities.

prefers a metropolitan area with a large airport (column 3) because it will have many flights to many places, which will make the connections of the sales force easier.

Production costs are illustrated in the next three columns. It would cost a modest $62,000 to get the requisite types and amounts of materials to Los Angeles; transportation costs on inputs would be many times higher if the plant were to be located in Boston (column 4). Conversely, the projected wages bill would be lower in Boston than in Los Angeles (column 5). Projected corporate taxes likewise vary over space (column 6).

There are important aspects of the labor force beyond wage costs. One argument is that it is not what is paid but what is received that is most important. Column 7, value added per employee, is an attempt to measure for an area this productivity issue. Another possible argument is that the most expensive labor item is employee turnover, and turnover is likely to be less in areas with relatively high unemployment rates (column 8). It also might be argued that a location in an environmentally clean area would enhance the recruitment and retention of scarce and mobile skilled labor (column 9). The environmental conditions might be of significance in determining the ease of getting construction and operating permits.

Comparative costs.   Given these conditions, which city is best for the new plant? One common way of combining the divergent opinions of the decision makers is to add up all the data measured on the same scale (usually interval) and in the same units (usually money). Thus can be produced the *total comparative costs* of column 10, calculated by adding columns 2, 4, 5, and 6. Now it is clear, for example, that Baltimore's other cost advantages more than compensate for Houston's lower corporate taxes. If the location decision were based solely on minimizing costs, and the summations of column 10 adequately measure all the relevant spatially variable cost factors, then Atlanta is the metropolitan area of choice.

The weighted-criteria method.   The comparison of total costs in column 10 should be insightful, but it obviously ignores some factors already deemed important. The factors which have not been reduced to monetary figures—local market size, airport size, labor productivity, labor availability, and air pollution in the current example—are not included in the comparative costs calculations. Thus, comparative costs are but one bit of information to be used in judgmental decision making.

The cities can be compared across more variables by reducing all the factors to a common scale, and evaluating them according to their (per-

Location	1 Total Locational $ Costs Index	2 Market Size Index	3 Airline Passengers Index	4 Value Added per Employee Index	5 Unemployment Rate Index	6 Air Pollution Index	7 Score (LS)
Atlanta	10	2	10	6	8	6	158
Baltimore	9	3	3	7	6	6	144
Boston	9	1	5	6	8	8	141
Cincinnati	9	10	3	7	7	8	176
Houston	8	6	7	9	8	4	158
Kansas City	9	1	3	7	7	8	140
Los Angeles	7	3	10	7	7	4	131
New Orleans	9	3	3	10	10	10	165
St. Louis	9	4	5	7	8	6	154
Seattle	6	1	7	6	5	10	109
Weights =	10	4	1	3	2	1	

[10 = Best or Most Desirable]

$LS = \Sigma \ W_i X_i$

where W_i = the weight for factor i, and
where X_i = the value (index number) of factor i

Figure 2.5   Indexed and weighted criteria for selected cities.

ceived) significance. Figure 2.5 illustrates the process for the example data. The metropolitan areas were indexed on a 10-point scale on each of the factors, with 10 being best or most desirable. Obviously, any scaling could be used, and many more factors can be added.

It also is obvious that not all the variables are of equal importance in locating the plant. They must be differentially weighted. Here again, decision makers must rely on their collective experiences and analogs to somehow determine the appropriate weights. For the current illustration, air pollution is determined to have a weight of 1, and total locational costs to be 10 times more important in guiding the location decision. Figures 2.5 and 2.6 show the "location score" results for a sample of cities. Assuming faith in all the data and judgments that went into the process, Cincinnati is the place to put the new plant. Of course, the result could be different if the weights are different. If the value added per employee index is weighted at 6 instead of 3, then New Orleans is virtually tied with Cincinnati on the composite location score. If New Orleans were deemed the most likely area for future market growth, then it might be the best choice. Likewise, Houston, Atlanta, or St. Louis might be viable candidates. Conversely, it would take a lot of faith in Seattle's future to seriously consider locating there; it appears to be beyond the *spatial margin* in this example. Since this stage can easily be put on a computer, it is easy to play "what if" games to assess the results of different weights—i.e., to perform sensitivity analyses—and then to incorporate the results into a wider decision-making framework. In this illustration the data were managed, and sometimes generated, and the maps produced within a GIS.

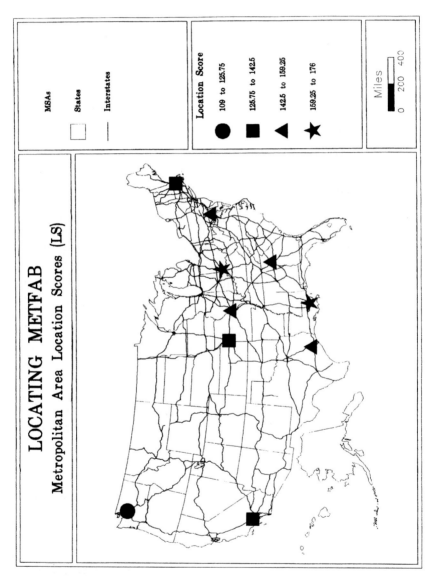

Figure 2.6   Locating METFAB—Metropolitan area location scores.

Making the location decision: Site selection

After narrowing the list to a preferred place, it is now necessary to literally get "on the ground" so as to select a specific site for the manufacturing facility. Choosing a particular plot of territory for the facility comes near the end of the location selection process, but it may be the most time-consuming phase. Only a few sites can be assessed in detail.

This phase has two components; one is the selection of a community or neighborhood within the local area, and the other is the identification of suitable buildings and/or parcels of land. Of course, the two are not mutually exclusive; the availability of suitable premises often influences the selection of neighborhood, and a strong desire to be in a particular neighborhood could occasion modifications in preferred site attributes.

Information sources.   Detailed information is necessary for both phases. The list of data to be obtained will be somewhat unique for each firm, plant, and time. Excellent sources are state and regional economic development offices; Chambers of Commerce; local planning agencies; and utility companies, especially those with economic development divisions. Ask to see the data and maps for each attribute of interest, and combinations of attributes via map layers— e.g., a map showing major transportation routes, industrially zoned land, and available industrial buildings—and be receptive to suggestions of additional considerations. Especially good sources of information and insights on community characteristics, labor supply, costs, and customers, and on area "business climates," are other manufacturers already operating in the areas of interest. They usually are remarkably cooperative.

Neighborhoods.   Sites reside in neighborhoods. The attractiveness of a specific site is conditioned by the attributes of its immediate surroundings. Of course, some of the community/neighborhood characteristics will already be known since they were part of the data inputs when making the subregion selection from among the short list of feasible regions. Information will already have been collected, for example, on the area's transportation facilities, labor supply and costs, utilities, business climate, local governments, and overall quality of life. Additional information on local social and demographic conditions is useful. For example, if the labor pool needed is concentrated in a particular section of the city, then that section may be an area where the site search should be concentrated. However, variables which do not exhibit significant spatial variability across the metropolitan area need not be considered.

Site factors. Additional information to be collected is very site-specific. The particular needs of your operation will dictate the list; also, decisions regarding buying versus leasing, or a new site versus an existing building will influence the criteria and the number of potential sites that can be evaluated. In general, the critical site considerations are:

1. *Availability of land or buildings.*
    a.   Availability of land parcels of sufficient size (the parcel should be five or more times larger than the physical size of the plant, to allow for parking, sidings, and expansion) and/or
    b.   Availability of buildings of suitable size and configuration
2. *Zoning.* These land use regulations are supposed to protect incompatible uses from each other by imposing spatial separations. Ideally, industrial users are protected from residential and commercial users, as well as vice-versa. A manufacturer wants to be protected from future restrictions caused by residences subsequently moving too near the facility—these might result from allegations of noise pollution, for example. Unfortunately, many zoning systems are "progressively inclusive": that is, "lower-level" uses (e.g., manufacturing) are excluded from "higher-level" (e.g., residential) areas, but the reverse does not hold; this is changing, but slowly. The exact zoning categories vary from community to community, and there may be hundreds of variants within a single metropolitan area. Those around sites of interest must be examined in detail, but in general there are a few major categories into which most classifications fall: heavy industry, light industry, and mixed (manufacturing and other commercial users). *Heavy* and *light* are broad, often ill-defined, and not very descriptive terms, but they are widely used; "heavy" industries are typically those which operate on a large scale, and may have some acute environmental concerns, in terms of noise, water or air pollution, or the disposal of solid wastes. "Light" industry is typically thought of as relatively "clean" and thus has less difficulty in finding a neighborhood welcome. Mixed areas obviously present both opportunities and risks. If a desirable parcel of land is identified, but the current zoning is not appropriate, it is possible to petition for a change. However, obtaining a zoning change usually is more trouble than it is worth; furthermore, you may not be well protected in the future by being in a jurisdiction which is too ready to bend the rules for the next potential users. For clues to future zoning, examine community master plans and land use plans.
3. *Neighbors.* Zoning may be sufficient to keep the facility separated from undesirable neighboring land uses, but it is a good idea to check out the nearby areas. Conversely, it may be desirable to be near certain other types of activities. Although industry tends to

be scattered across urban areas, specific types of industry show spatial concentrations. This "strongly suggests the importance of *external economies* in the locational decision of manufacturing enterprises."[14] Most large areas have directories showing the locations of industry; ask at the local planning office, utility, or Chamber of Commerce.

4. *Industrial parks.* These are organized areas of fifty to several hundred acres which provide all physical services and simplified zoning and permitting.

> The typical industrial park today is characterized by three features: (a) a comprehensive plan for the development of the park with enforceable restrictions that control land use, building-to-land ratios, construction and even landscaping; (b) availability of all utilities, roads and services that meet the particular need and varied requirements of park residents; and (c) supervision and management that can enforce regulations and guarantee the maintenance of the esthetic and efficient environment.[15]

If a facility is compatible ("heavy" or "noxious" plants often are excluded), and the park is otherwise well located, then such an organized area may simultaneously address the zoning and neighbors' concerns; also, an industrial park already will have received its environmental approvals.

5. *Labor accessibility.* Labor is both geographically mobile and sensitive to the friction of distance (commuting time). Workers with cars willingly commute 5 or even 10 miles to the job, but much less frequently 20 or more miles. Workers without cars are dependent on public transportation, and are thus much more restricted. Although labor forces are scattered throughout a city, there typically are concentrations relatively near the place of work. Other things being equal, workers are more likely to know about and choose jobs near home, and they are also more likely to select residences with at least one eye on the journey to work. The lessons for site selection are:

   a. Be aware that the odds of attracting and retaining the desired workforce can be enhanced by an accessible location.

   b. Determine if the workforce will have any special spatial needs or preferences—e.g., on a public bus route, or in a suburban setting near an expressway exit.

   c. Use this knowledge in delimiting local search areas.

6. *Transportation access.* The issue of the accessibility of labor via car or public transportation has been noted. There also is the obvious problem of getting materials into the plant and product shipped out. Since most manufactured goods move by truck, the local network of highways is of great interest. In addition to the

existence of "industrial-grade" highways, attention must be given to access—e.g., the locations of interchanges—and to congestion, which will vary through the day. Traffic counts can be obtained from the regional planning agency. Operations which need or want to use rail or water modes are, of course, more restricted in their locational choices.

7. *Visibility.* For some operations it is desirable to increase local public awareness by being visible from heavily trafficked streets and highways. If this is important, a drive-by is the best way to assess the ease with which the site can be seen and the number of people likely to see it.

8. *Utilities.* For the majority of manufacturing plants (which are small operations), the sewer, water, gas, and electric services to most industrial parks and other industrially zoned areas are adequate. Of themselves, then, they are not of great locational concern. However, if there are special needs, such as being able to tie into a large sewer or a high-pressure gas line, then the sites with relatively easy access to such services probably are the only ones to consider seriously. The local utility companies can provide detailed information. It also may be necessary to consider access to dump sites or hauling services for solid industrial waste; it will certainly be necessary if the operation produces hazardous waste.

9. *Topography, soils, and foundations.* It is easier and cheaper to build on level land which is underlain with soils and bedrock which allow for the rapid construction of heavy buildings. Day-to-day operation of the plant is easier on large, level parcels of land. If the site is otherwise attractive, then it may warrant spending money to compensate for physical imperfections, but some sites may simply be too difficult.

10. *Environmental audits and permits.* Owners are responsible for any mandated cleanup of pollution caused by themselves *and by previous occupants.* It is *essential* that an *environmental audit* be performed *before* property is purchased.[16] The audit should be performed by a qualified, independent commercial environmental consulting firm, and should be paid for by the seller. Ask the local and regional environmental protection agencies and planning agencies for advice.

A separate environmental concern is the obtaining of environmental *permits.* The environmental protection laws and the enforcement organizations (federal, state, and local environmental protection agencies) require permits and provisions for the safe discharge of wastes into the air and water. The specifics vary by industry and by locality. It is necessary that the firm's environmental experts confer with state, regional, and local environmental protection personnel. Research has shown that the main problem most locating businesses have with environmental regulations is not the costs of compliance, but rather the time required to obtain permits.[17] Therefore, the permitting process should be started by the firm as early as possible in the location search.

11. *Access to business services.* It may be desirable to locate near frequently used services, such as printers and repair shops. Nearby restaurants can be useful for entertaining clients and for employees.

12. *Taxes.* The importance of taxes is the most hotly debated topic in all of industrial location. Taxes get much media attention, and businesspersons express concern, but objective research indicates that taxes are of relatively little importance in locational decision making. At the national and regional location search scales, taxes should be, and are, of little importance. At the local scale, however, they may make some difference. If location A is just across the municipal or state boundary from location B, and has the same access to markets, materials, labor, etc., then significantly lower taxes in location A *might* be a legitimate decision factor. On the other hand, it might be wise to gladly pay higher taxes if commensurate services are provided by local governments. Some good rules are;

   a. View taxes as costs *and* benefits.

   b. Evaluate them as unemotionally as possible.

   c. Evaluate *all* taxes: state, local, corporate, personal, income, property, etc. Tax burdens have a tendency to even out among different areas when all things are considered.

   d. On average, high-to-low *variations* among places in state and local taxes are less than 1 percent of sales, so if other locational differentials exceed this figure, taxes are not worth much concern.

13. *Cost of land.* Industrial land costs vary across metropolitan areas. In part this is because of special situations and attributes, but more generally the spatial variations are functions of highway, especially expressway, access; closer land is usually worth more, other things being equal. Obviously, land can be a significant up-front cost, and the capital is tied up for many years. It pays to shop wisely. However, again it is important not to let this very tangible cost drive the location decision too completely. As should by now be abundantly clear, there are many factors which must be considered. Even regarding land, keep in mind that:

   a. Land costs are a small proportion of total fixed investments.

   b. Land and building costs (purchase or rental) together are a small proportion of operating costs.

   c. Land becomes a permanent asset, and is likely to appreciate over time. A good price on the right site is good business. At the same time, a "good price" on the wrong place is no bargain.

Final site selection and judgmental decision making. By now much information has been collected and many opinions have been formed and exchanged. All that remains is for the location decision team to make a recommendation on the site for production. In many cases consensus is reached easily, but sometimes it is necessary to more formally summarize what is known about the sites which made the final short list. This process also has the merit of focusing debate. This step does not have to be especially complex in terms of data or methods of analysis; a simple chart such as Fig. 2.7 might suffice to guide debate.

   Figure 2.7 has two sections. The first section lists the constraints; either a site has the attribute or it does not. Thus, potential sites 4 and 5 each fail on one critical dimension and are eliminated from further consideration. (Of course, the rejection of sites because of "attribute failure" probably has occurred earlier in the narrowing down process as well.) For site 3 there is debate as to whether adequate electric service is a go/no-go attribute, or a variable in the

Constraints of Potential Sites					
Factor	**1**	**2**	**3**	**4**	**5**
Zoning	Y	Y	Y	Y	Y
Parcel size	Y	Y	Y	Y	Y
Environmental	Y	Y	Y	Y	N
Sewer service	Y	Y	Y	N	Y
Gas service	Y	Y	*	Y	Y

Y = Adequate
N = Not adequate
* = Possible

Ratings and Weightings of Potential Sites				
Factor	**1**	**2**	**3**	**Factor weight**
Labor accessibility	4	2	1	2
Transportation access	5	3	1	3
Business services	5	2	2	2
Neighborhood	3	4	5	2
Taxes	1	2	4	1
Cost of land	1	2	4	1
Total scores	41	29	26	

5 = Excellent
1 = Poor

Figure 2.7 Final manufacturing site selection: factors, constraints, and weights.

sense that what the site now lacks can be rectified by spending some money. The decision is to leave it in for the next stage.

Section B lists the site selection variables deemed important. Each site is rated for each variable on a scale of 5 (excellent) to 1 (poor). Each variable is weighted by significance. Some arithmetic produces the total scores. If the team members agree that the process has thrown up the best site, then the investment should be made at site 1.

This last stage can seem deceptively simple. Actually, it points up the sophistication of the location decision process. It is a high-stakes process which is part "science" and part "art." It demands sophisticated judgmental decision making from the members of the team, at all stages of the process, from the global to the local. Because location decisions are relatively rare events in the lives of most firms, managers typically are little experienced in making such decisions. Thus, it might be a good idea to hire a professional consultant. A consultant can guide the firm through the international to local process, and, as importantly, help adjudicate the inevitable debates. However, it is important that the firm's management *not* abdicate to the consultant. The firm's management must be *involved.* They must provide much of the information, and their judgments are critical. Whether or not consultants are employed, it is true that involved managements are usually happier with the final location selection; and locational happiness also can be important in operating productivity and efficiency.

Locating a Retail Outlet

In general, the procedures useful in deciding on the best location for a store are much like those for choosing the place to put a plant. However, there often are noticeable differences of emphasis between the location search for a manufacturing facility and the selection of a location for a retail outlet. Compared to manufacturing, retail location puts supreme emphasis on *access-to-market* considerations, at all scales of the search, and heavy emphasis on the geography of the firm. Again, there is the division between situation and site.

Situation

The first locational decision is whether or not the new store will be in a new region or will in-fill in areas where the firm already has outlets. If current stores are in northwestern Europe and the firm is headquartered in Sweden, is the establishment of retail outlets in Canada a good idea? If the firm now operates in the mid-Atlantic states, does a move into Ohio make sense? Firms may wish to move into *new regions* because:

1. The new regions offer especially attractive sales potentials.

2. The regions in which the firm currently operates are saturated, so outward expansion is the only way to grow.

3. It is important to be in the new region, to complete a service network, or for image, or to gain a foothold for anticipated future markets, or to preempt the locational moves of competitors.

Conversely, even if one or more of these lures is present, firms may be reluctant to risk ventures into new territories because:

1. The company is not well known in the new territory.

2. Communication lines are stretched.

3. Supply lines are stretched.

4. Managing a more spatially spread organization is more difficult.

5. New cultural customs may have to be learned.

6. New financial (e.g., currency exchange, taxes) and regulatory systems (e.g., permits) have to be incorporated.

7. Costs of building the minimum number of stores to carry overheads, advertising, and so forth, or of building a sufficiently large magnet store, are prohibitive.

Locational analyses must help with the decisions of (1) selecting no new places or (2) expanding in regions in which the company currently operates or (3) going into new regions. Locational analyses also should help acquisition decisions, since existing locations are inherited and must be integrated into the firm's spatial fabric. Expansion by acquisition can be attractive because it is quick, and the stores have strong images. However, the acquired stores come with locations which may not be the places desired by the acquiring company.[18]

Thus, the *situation* phase of retail location first involves the strategic decision of whether the next outlet will extend the geography of the firm or will in-fill the existing spatial pattern. Chain-store retailers tend to have locational phases; a period of expansion into new territories may be followed by a period of more local concentration. Fill-in stores tend to be profitable more quickly. However, even if the decision is to expand into new territories, the *spatial increment* rule is prudent; expansion adjacent to existing sales territories has advantages for supplying and managing the network and for consumer name recognition.

Situation and site: Trade area analysis

Retail success is a function of two things: good management and good location. The probability of choosing a good location for a store is enhanced by knowing its favorable trade area characteristics. A *trade area* is the region surrounding a store from which comes a large proportion of its customers. The best way of determining what kinds of trade areas "work" is to match the relative success of existing stores against the explanatory characteristics of the market areas; commonly used criteria include:

Population size

Population change history

Per capita income

Property values

Special cultural and ethnic attributes

Age distributions

Market size

Competitors

The methods used to determine the correlates of success may range from experienced observers comparing notes, to simple cross-tabulated tables, to sophisticated multivariate statistical models. Once the location analysts are comfortable with the selection criteria, the logical search procedure is to look for market areas with growth potentials in the desired characteristics. Conversely, the same information can be used to evaluate the expansion or even continuation of existing stores. If a store in a good market area is doing poorly, it makes sense to change management or product lines rather than close the stores.

Trade area analysis can be reduced to the ability to reasonably measure three elements:

1. *Extent of the trade area.* A well-delimited trade area is the foundation of retail location evaluation, especially at the local scale. Trade areas can be delimited in several ways, including customer spotting; geometrically apportioning spaces to competing stores; applying known drawing-power distances, usually from experiences with "sister stores"; and judgment (see under "Trade area delimitation" below).

2. *Business available.* Within the trade area, how much money is available to be spent? How big is the retail pie? The potential dollars available for the type of retail operation of concern usually are determined by the demographics of the area. Common measures are population, income (e.g., median family income), change rates, and age structure.

3. *Share.* How much business can a store at a specific location be expected to get, given the competition? The key elements are the

numbers, characteristics, and strengths of the competitors. Also, retail synergism is important. A critical consideration is that a new store in an area generate new revenues for the firm and not transfers from the firm's existing stores.

The concepts of retail location are straightforward. It is the measurements which are difficult.[19]

Trade area delimitation

This discussion of retail location rests on the assumptions that (1) the *customers* travel to the place of business, rather than the goods moving to the consumers' locations, and (2) this *friction of distance* means that the probability of a customer patronizing a store is directly related to the distance to the store, so that (3) there is a *distance decay function,* where most of a store's business comes from nearby potential customers, and (4) thus stores have higher sales the closer they are to their customers. Since the question is the location of retail outlets, and these outlets are market-oriented, it is necessary to analyze *spatial markets.*

Existing trade areas are shown on maps as the areas surrounding the locations of current company stores and of major competitors. There are many ways to do this; two simple methods will illustrate the basic concepts. The first procedure exploits the concept that customers are more likely to patronize the nearest store rather than a more distant comparable outlet. These "spatial monopoly" trade areas can be drawn by connecting perpendicular bisectors of the straight lines joining the centers (towns, shopping malls, or stores) in the set (see Fig. 2.8). The boundary lines could be refined to reflect the differential attractions (pulling power) of the centers or physical barriers via a gravity model; the more attractive or accessible centers would have the larger trade areas (not shown on Fig. 2.8).

Another way to delimit trade areas is to plot the known locations of customers of already operating stores; this is known as *customer spotting,* and is accomplished using credit card or other address lists, or by license plate spotting in parking lots, or other methods to generate address files. For a given store, the spatial scatter can be examined for density and spread. Critical contour lines can be plotted to summarize the key elements of the pattern; in Fig. 2.8, the line within which 50 percent of the customers live is shown.

An example using a geographic information system

A *geographic information system* (GIS) presents the modern and best way for retailers to manage and analyze spatial data.[20] Only the most

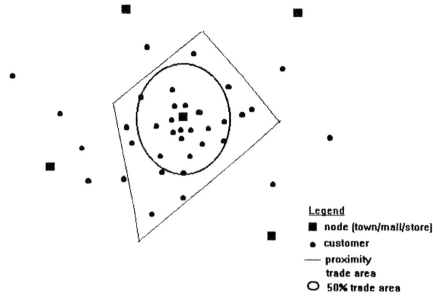

Legend
■ node (town/mall/store)
● customer
—— proximity
   trade area
○ 50% trade area

Figure 2.8   Trade area delimitation.

progressive retailers are now using a GIS, but the trend will accelerate. Retailers operate in especially dynamic spatial environments, and
they need the tools to stay current and successful. The database management, spatial analysis, and graphics capabilities of a GIS are well
suited to retail location selection. The following hypothetical example
uses ATLAS*GIS (Strategic Mapping, Inc., San Jose, California), one
of several suitable commercial software packages.

The initial problem is to decide the approximate location for a new,
large widgets supply outlet somewhere within the Gotham region.
The first step is to spatially subdivide the region. Here, the zip code
areas of Gotham are used. Relevant attribute data for each area are
then put into the database; total population, median family income,
daytime population, employment by occupation type, and retail sales
are typical examples. These data can be collected from many sources,
such as the censuses of population, retail trade, or manufacturing. If
the data are reported for a different geography than that being used
for the primary analysis, the GIS can transpose from one to another;
for the current example, census tract data can be apportioned to zip
code areas. Already formatted data can be purchased from commercial vendors. Proprietary data can be entered by hand.

Using appropriate models, either constructed by the firm based on
experience, or purchased from a vendor, the attribute data can be
weighted, factored, and combined to produce new decision variables.
Figure 2.9 shows such an index variable, market size by zip code,

ZIP_Codes

NewStoresA

NewStoresB

Stores

Market Size

67.11575

Miles

0   2   4

Figure 2.9   Gotham retail location example: Markets and sites.

where the market is defined by the type of goods to be sold by the
store being located. Market size might have been calculated, for
example, by multiplying total retail sales per zip code area (such data
can be purchased) by the regional proportion spent on widget-type
goods. It could have been further refined by factoring in specific
demographic or employment characteristics of each area, and by
making forecasts about future levels. (Used in this illustration are
hypothetical data from the ATLAS*GIS tutorial.)

The three existing stores operated by competitors are superimposed
on the market map. Given this information, in this simple example,
where should the firm just coming into the region locate its first store?

Somehow, the newcomer has tentatively chosen two store sizes,
large (the A stores) and moderate (the B stores), and four neighborhoods for further analysis. The three existing stores and the four
locations being considered also are shown in Fig. 2.9. Trade area and
market share analyses are now conducted to assess the relative merits of each tentative locality. Based on experience, and recognizing
the economic and social geographies of Gotham, it is decided that the
relevant trade area, where 80 percent of the store's business is
expected to come from, for an A store is defined by a circle with a

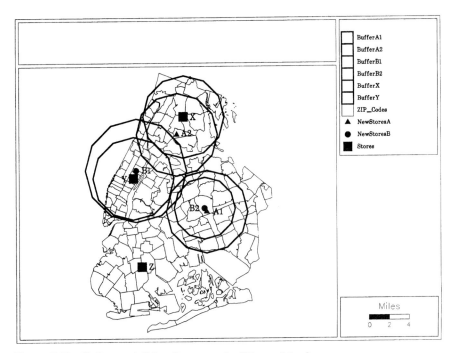

Figure 2.10   Gotham retail location example: Sites and trade areas.

radius of 4 miles from the store. The trade area radius for a smaller B store is 3 miles. However, because of retail synergism, the radius for a B store adjacent to a competitor's store is 5 miles (e.g., B1). Each of the competing stores has a trade area radius of 4 miles.

A trade area is drawn around each potential location, and markets from the zip code areas within each are apportioned. The six relevant trade areas are shown simultaneously in Fig. 2.10. The resulting total anticipated market for the trade areas, given the assumptions *and ignoring competition,* are: A1 = 918,000; A2 = 1,064,000; B1 = 678,000; B2 = 509,000.

Now account for the competition. A store at potential location B1 would have to share its market area with an existing competitor. As the trade areas around the northern (X) and western (Y) competitors indicate, a store at location A2 would have to share most of its primary market area. Thus, the adjusted expected markets for the four potential locations are: A1 = 918,000; A2 = 639,000; B1 = 678,000; B2 = 509,000.

The next step is to determine if *any* of these locations can be expected to generate sufficient revenues to justify the establishment and operation of a retail outlet. If it is assumed that all four of these

possibilities are viable, the best bet if the firm is confident it will
have a "killer" store may be A1. However, if competition is taken
more seriously, A1 looks most promising. If an attractive deal cannot
be made for a specific site in this locality, then it might be prudent to
look in the vicinity of B1; and so on. Of course, the above illustrate
just some of the fundamentals of the location search. More variables,
different assumptions, and more potential locations to be analyzed
are all likely in real situations.

Site selection

Now that the target market area has been determined, and with it
the approximate site, it is necessary to choose and purchase (or lease
or option) a specific parcel of land or building within a shopping district. The key site considerations include:

1. *Accessibility.* How easy is it to get to the site by car? By public
   transportation? How much foot traffic goes past the site? Is parking adequate?

2. *Visibility.* How easy is it to see the site from the street or road?
   From within the shopping center?

3. *Neighbors.* What other retailers are nearby? Are they complementary types which it is desirable to be near so as to catch business
   when customers are on multipurpose shopping trips? Are they
   competitors to be near so as to build volumes by providing com-
   parison-shopping opportunities? Or are they competitors to avoid,
   so as to preserve market share? More generally, is the immediate
   area attractive or unattractive to shoppers?

4. *Zoning.* Is the area zoned for business? Is the category exclusively
   for compatible uses, or is it for mixed uses?

5. *Infrastructure.* Are the utilities, parking lots, and roads in place
   and adequate?

6. *Freestanding or shopping center.* Should the store be freestanding
   or in an organized shopping center or mall? What have been the
   company's experiences? How do competitors operate? Why?

7. *Costs.* What are the rents?

As in the case of site selection for a manufacturing facility, final
retail selection debates can be focused by the use of comparable factor weights and rankings for each of the potential sites on the short
list. See Fig. 2.11 for an example. Here, site 2 should be the first
choice for negotiations. Again, the data management, spatial analyses, and mapping for this final phase can be handled within a GIS;

Potential Sites					
Factor	1	2	3	4	Weights
Zoning	5	5	3	5	8
Utilities	3	5	4	5	6
Access	4	5	5	1	10
Size	2	4	5	3	8
Cost	5	2	4	1	3
Neighbors	4	5	3	2	9
Visibility	5	4	4	5	5
Total scores	190	223	197	150	
*5 = Excellent 1 = Poor					

Figure 2.11   Final retail site selection: Weighted factors.

especially useful at this scale are the digitized computer street maps available from the U.S. Census Bureau (the TIGER line files) and enhanced versions from commercial vendors.

The locational analysis/site evaluation does not end with the opening of the store. Areas change. Merchandise offerings may need adjustment to reflect changing demographics or changed competition. New markets open, and new market segments are born, presenting new opportunities, but also new challenges for the optimum organization of the geography of the firm. All of these can be enhanced by explicit recognition of the importance of spatial analyses, and of the efficient management of spatial data.

Like good medicine, good location selection for manufacturing and retailing is part science and part art. It is the exercise of informed good judgment.

Acknowledgments

This chapter has benefited from the inspiration, related research, and comments of Dr. Risto Laulajainen, Department of Human and Economic Geography, Gothenburg University Business and Law School, Gothenburg, Sweden. Dr. John Haake, Director of Market Research, Fay's Incorporated, Liverpool, New York, provided critical insights and commentary, especially regarding retail location selection.

References

1. Fassl, J., "High-Tech Systems Key to Site Choice," *Food Engineering,* March 1989, pp. 115–118.

2. Schmenner, R. W., "Look Beyond the Obvious in Plant Location," *Harvard Business Review,* January–February 1979, pp. 126–132.
3. Townsend, R., *Further Up the Organization,* Harper & Row, New York, 1984, p. 79.
4. Stafford, H. A., *Principles of Industrial Facility Location,* Conway Publications, Atlanta, 1980.
5. Jones, K., and J. Simmons, *Location, Location, Location: Analyzing the Retail Environment,* Methuen, Toronto, 1987.
6. Smith, D. M., *Industrial Location: An Economic Geographical Analysis,* Wiley, New York, 1981, p. 278.
7. Calzonetti, F., and R. Walker, "Searching for New Manufacturing Plant Locations: A Study of Location Decisions in Central Appalachia," *Regional Studies,* vol. 24, no. 1, 1989, pp. 15–30.
8. Blair, J. P., and R. Premus, "Major Factors in Industrial Location," *Economic Development Quarterly,* vol. 1, no. 1, 1987, pp. 72–85.
9. Haigh, R. W., *Investment Strategies and the Plant-Location Decision: Foreign Companies in the United States,* Praeger, New York, 1989.
10. Lyne, J., "Incentives Are Important, Executives Say, but Business Concerns Drive the Location Process," *Site Selection and Industrial Development,* April 1992, pp. 282–294.
11. Love, R. F., J. G. Morris, and G. O. Wesolowsky, *Facilities Location: Models & Methods,* North-Holland, New York, 1988.
12. Mills, N. L., "A Multi-Phase Approach to Site Selection," *Site Selection and Industrial Development,* June 1989, pp. 769–773.
13. Rand, G. K., "Methodological Choices in Depot Location Studies," *Operations Research Quarterly,* vol. 27, no. 1, 1976, pp. 241–249.
14. Struyk, R. J., and F. J. James, *Intrametropolitan Industrial Location,* D. C. Heath, Lexington, MA, 1975, p. 16.
15. Hunker, H. L., *Industrial Development,* D. C. Heath, Lexington, MA, 1974, p. 179.
16. Barbi, G. J., "Should a Site Audit be Performed? The Universal Answer is Yes," *Site Selection and Industrial Development,* December 1990, pp. 1397–1403.
17. Stafford, H. A., "Environmental Protection and Industrial Location," *Annals of the Association of American Geographers,* vol. 75, 1985, pp. 227–240.
18. Laulajainen, R., "Chain Store Expansion in National Space," *Geografiska Annaler,* 70B(2), 1988, pp. 293–299.
19. Haake, J., personal correspondence, 1992.
20. Goodchild, M. F., "Guest Commentary: Geographic Information Systems," *Journal of Retailing,* Spring 1991, pp. 3–15.

Further Reading

There is a large literature, both theoretical and applied, on commercial location selection. Five book suggestions are:

Ghosh, A., and S. L. McLafferty: *Location Strategies for Retail and Service Firms,* D. C. Heath, Lexington, MA, 1987.
Jones, K., and J. Simmons: *The Retail Environment,* Routledge, London and New York, 1990.
Schmenner, R. W.: *Making Business Location Decisions,* Prentice-Hall, Englewood Cliffs, NJ, 1982.
Stafford, H. A.: *Principles of Industrial Facility Location,* Conway Publications, Atlanta, 1980.
Watts, H. D.: *Industrial Geography,* Wiley, New York, 1987.

A useful journal, published six times each year, is *Site Selection and Industrial Development,* published in Atlanta by Conway Data.

3

Strategic Master Site Planning

N. L. Hannon
Vice Chairman, The Leawood Group, Ltd.
Leawood, Kansas

W. Wrennall
President, The Leawood Group, Ltd.
Leawood, Kansas

A *master site plan* is a set of documents and drawings that defines the purpose, intentions, and designs for a contiguous land area and the physical facilities thereon. Where relevant, it may have information on the political, geographical, cultural, and physical environment. A *strategic master site plan* includes sufficient historical background to provide context for the current situation and describes the strategic framework that gave rise to the plan.

A well-designed site derives from and supports the high-level strategies of a business. It allows for progressive, orderly, and economical expansion while providing smooth product flows and enhanced communication. The strategic master site plan supports operations in meeting market requirements.

As discussed in Chap. 1, "Facilities Design for World Class Competition," the physical facilities (land, buildings, equipment, and infrastructure) provide the capacity for operations. They are among the most costly of a firm's investments and the most difficult to change.

A poorly designed site increases costs in many ways. It can be difficult or impossible to reconfigure. It may lead the firm toward an inappropriate business strategy. A site that does not support market requirements offers a competitive advantage to other suppliers.

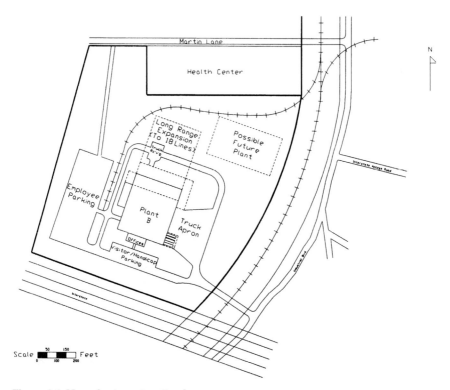

Figure 3.1 New plant master site plan.

Master site planning is the highest-level facility planning which translates directly to physical configuration. It is, therefore, one of the most important tasks in facility design.

A simplified example of a site plan drawing is given in Fig. 3.1. Typically the master site plan will show:

- The site plot plan with boundaries, access, and egress points
- Easements on and close to the site
- Buildings and other structures
- The site circulation system: e.g., roads, railroads, waterways
- Site features such as lakes, streams, hills, and settling ponds
- Utilities such as water towers, electric transformers, gas tanks, gasoline storage tanks, sewerage system, and water, steam, gas, telephone, and power lines
- Truck and car parking areas

A strategic master site plan is a master site plan which supports the strategic intent of the company. The strategic master site-planning process described in this chapter will:

- Stimulate thoughts and actions to provide long-range capacity to support the firm's strategic operating plan
- Provide for rapid introduction of new products
- Facilitate incorporation of new technology
- Phase further expansion to a well-planned saturated site
- Signal when additional sites will be needed
- Anticipate and provide for scenario options
- Accommodate changes in manufacturing/operating strategy

The work of site planning often falls to an engineering branch of the organization. A plant or facilities engineering department usually has most of the information and many of the required skills. But such a functional department rarely has the breadth of knowledge and experience required for such a key strategic task. Interpretation of facts, development of strategy, and development of site scenarios should come from a broad-based group. The master site-planning group should include people from almost every part of the business. For an effective plan, top management must play an active role in developing strategies and scenarios. Often their participation is through a steering committee of from three to ten managers who make major decisions. They typically guide a planning team.

A *master site-planning team* should perform the work. This team should represent all major functions such as marketing, finance, production, product development, and manufacturing engineering.

We recommend that plant and facilities engineers, perhaps with the help of consultants, orchestrate the master site planning. This group, referred to hereafter as *the team,* normally consists of middle managers and professional staff. It occasionally includes representatives from the shop floor. The team should work on-site. This is necessary to check details and grasp intangibles that they may otherwise miss.

The *strategic master site-planning process* consists of five primary tasks:

- Information acquisition
- Strategic integration
- Site scenario development
- Site plan design

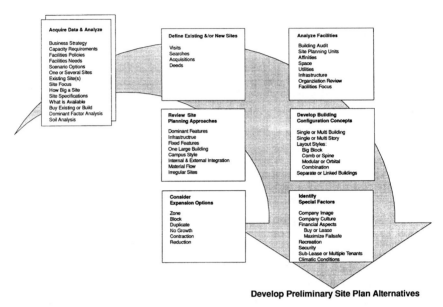

Figure 3.2 Master site-planning structure.

■ Master site plan document

Each primary task has numerous subtasks. Their nature, depth of analysis, and level of detail vary from one project to the next. Figure 3.2 illustrates a task planning structure to adapt as required. The figure does not show task sequence.

Information Acquisition

This major block of site-planning tasks will consume about two-thirds of the project's time. It is the logical foundation for the design and the reason for the large capital investment that will be required. The planning cost is typically 5 to 7 percent of the capital investment. The operation cost consequences of the planning are enormous and the costs of correcting inadequate or incorrect planning can be disastrous to a firm's competitiveness and profitability. These facts go largely unrecognized when financial planning budgets assume supremacy.

Information acquisition assembles a wide variety of data, facts, and opinions. The planners organize, summarize, and digest this information. Experience over many such projects helps to decide what information is necessary and where it is likely to be easily obtained. The team then develops a list of *findings,* significant facts and summaries that lead to preliminary conclusions.

Historical information

Land has a life way beyond the current site use. Historical information provides insight into the past use and the impact of the structures and processes that were part of previous site plans.

The current firm's history will influence future attitudes, approaches, and events. Historical information can be used to:

- Predict growth and trends
- Identify potential environmental issues
- Uncover hidden site features
- Forecast space needs
- Understand corporate culture and traditions

A brief history of the site or sites and their facilities will provide the planners with the reasons for past plans. This is important with older sites which may have altering demographic patterns, or where environmental concerns have been changing. It can also be instructive to note how occupancies have changed over the years and how this might impact future site development plans.

Existing Facilities and Conditions

"As-built" drawings alone do not provide a sound basis for a site development plan. An assessment of the existing facilities, buildings, utilities, services, physical infrastructure, along with records and drawings updates, confirms the planning baseline.

Review and assessment of those off-site entities which can impact the on-site activities is also necessary. Typical examples are:

- The coming conversion of the current two-lane road adjoining the site to four-lane with relatively limited access
- The new interstate interchange, which will force relocation of a site entrance and a major change of traffic flow on the site
- The proposed erection of a new high-tension power line along the perimeter of the site, which could interfere with the nearby R&D electronics laboratories' operations
- The granting of planning permission to build private housing adjoining the plant site
- Changes in neighboring industrial site occupancy
- The extension of a nearby airport and introduction of different flight paths

Space data

For site analysis and planning, collection of information on amounts of space by class is recommended. Typical classes are:

- Under-roof, footprint, and total
- Employee and visitor parking
- Truck parking, turnarounds, and aprons
- Outside storage
- Special service areas including power stations, transformers, water towers, cooling or sludge ponds, and waste treatment
- Recreation areas
- Setbacks, buffer zones, landscaping, green areas, unusable corners, roads, rail, barge docks, helipads
- Dedicated expansion areas
- Vacant land

A review of the age and existing conditions of buildings, utilities, plant, and equipment with the time and cost of restoration to current standards brings realism to the starting point of the planning project. The following gives guidance as to the areas for audit and review:

- Buildings
- Fixed handling equipment
- Utilities
- Services
- Topography
- Soil
- Climate conditions
- Easements

Local and regional context

Before embarking on a site plan or replan, review how the current and intended site uses fit into the local and regional environment. Changes in local attitudes, area development, city and county government, and neighbors can have a significant impact. A map such as the one of the Illinois R&D Corridor shown in Fig. 3.3 illustrates the point. Investment in facilities in the wrong location leads to ghost towns. On the other hand you may strategically want to avoid such a high concentration of similar businesses.

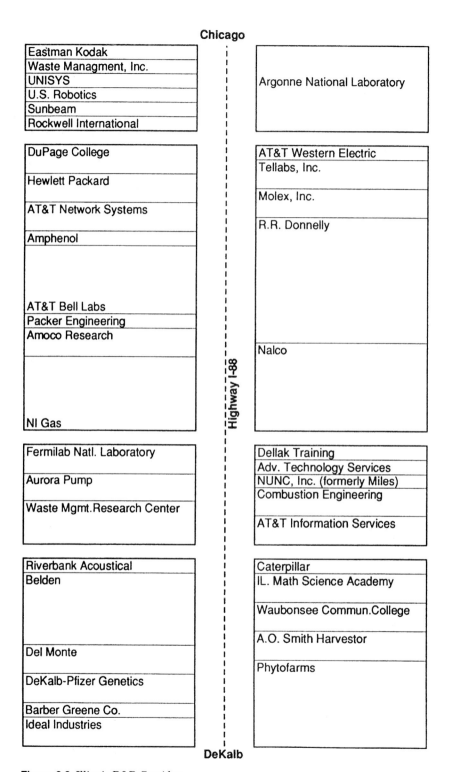

Chicago

Eastman Kodak	Argonne National Laboratory
Waste Managment, Inc.	
UNISYS	
U.S. Robotics	
Sunbeam	
Rockwell International	

DuPage College — AT&T Western Electric / Tellabs, Inc. / Molex, Inc. / R.R. Donnelly / Nalco

Hewlett Packard

AT&T Network Systems

Amphenol

AT&T Bell Labs
Packer Engineering
Amoco Research

NI Gas

Fermilab Natl. Laboratory — Dellak Training / Adv. Technology Services / NUNC, Inc. (formerly Miles) / Combustion Engineering / AT&T Information Services

Aurora Pump

Waste Mgmt.Research Center

Riverbank Acoustical — Caterpillar / IL. Math Science Academy / Waubonsee Commun.College / A.O. Smith Harvestor / Phytofarms
Belden

Del Monte

DeKalb-Pfizer Genetics

Barber Greene Co.
Ideal Industries

Highway I-88

DeKalb

Figure 3.3 Illinois R&D Corridor.

Dominant factor analysis

Dominant factor analysis summarizes site physical and nonphysical features. It visualizes the site and aids analysis of strengths, weaknesses, and opportunities.

To use this analysis gainfully, it is important for the facilities planner to:

- Have an in-depth knowledge of the site
- Identify the key or dominant factors on the site which will affect long-range planning
- Be able to communicate to others the importance and relevance of these factors

Review the list of typical factors given in the *dominant factors checklist* shown in Fig. 3.4. These are divided into four groups:

- *Physical features—external.* Those outside the site which could affect operations on site
- *Physical features—internal.* The dominant features on the site which could affect the site planning
- *Intangible factors—external.* Outside factors such as zoning, local government, fire protection availability, which will affect site operations
- *Intangible factors—internal.* Factors such as appearance, image, safety, personnel considerations, management philosophy

By the time the *dominant factor analysis* shown in Fig. 3.5 has been prepared, many local contacts have been made, and management can be given a clear understanding of the scope of the site-planning project. Management can also see the limitations and dominant features of the site and their impact on planning for its future use.

Strategic Integration

Fundamental changes to facilities often require significant time, resources, and special skills. The configuration of facilities, and relative marketing strategies and market needs, are among the determinants of long-term business success. This makes facilities an important element of business strategy.

Such factors also imply that facilities configuration should derive from a strategy plan. This contrasts with the more common situation in which a firm's strategy largely derives from an ill-configured facility.

Business and financial strategies are beyond the scope of this book. However, there are several references at the close of Chap. 1.

This is a checklist of factors to use as a reference when making a Dominant Factors Analysis. These are typical only; there may well be others which are relevant in particular situations.

PHYSICAL FEATURES - EXTERNAL (off site)

- Proximity of highways, roads, streets, overpasses
- Possibility of upgrading, widening adjoining roads or streets, and future impact on the site
- Site access limitation
- Availability to the site of power, water, gas, sewers, roads, rail and related easements
- Airport flight path problems
- Neighbors, adjoining land
- Storm water run-off to or from adjoining properties
- Proximity of river, creek, bridges; flood plain situation
- Prevailing winds

PHYSICAL FEATURES - INTERNAL (on site)

- Topography: lay-of-the-land; high and low points; natural drainage
- Soil conditions; mining activities
- Marshy areas, creeks, sink holes, ponds - can they be built on?
- Existing buildings, size, condition
- Existing roads, foundations
- Existing utilities, sub-stations, water tower, etc.
- Road access, railroad interferences
- Easements on site for power, water, gas, sewers
- Railroad spurs
- Rights of way, abandoned railroad tracks
- Wells
- Pumping stations
- Waste water treatment, pollution control

INTANGIBLE FACTORS - EXTERNAL (off site)

- Zoning, building restrictions
- Water and mineral rights
- Easement restrictions
- Type of local (city, county, state) governments
- Police and fire protection
- Historical, archaeological
- Community relationships; "power" groups

INTANGIBLE FACTORS - INTERNAL (on site)

- Appearance, landscaping, image
- Personnel considerations - benefits, recreation
- Managerial philosophy
- Security
- Safety

Figure 3.4 Dominant factors checklist.

Manufacturing and operations strategy planning is a topic discussed in Chap. 1. In this chapter, we assume that a manufacturing or operations strategy is substantially formulated and explicitly stated. Such plans are important inputs for master site planning. If they are not developed, the master site-planning project must initiate this activity.

Description: New Mfg. Plant / 25 Acre Site	Proj. No. 12345
Company: XYZ Corp.	Prep By: PAH
Location: Sun City Page: of: 1 1	Date: 15-Aug

			Existing Year — Now		Future Year — +20	
			Impact	Control	Impact	Control
PHYSICAL FEATURES	**EXTERNAL**	Herrick Ave. borders south side of site. Four lanes now, six lanes later?	E	X	A	X
		River Road borders east side of site. Will become busier.	E	X	A	X
		High tension power line (11KV) crosses site NE to SW.	A		I	
		Sun River borders north side of site.	I	X	I	X
		Open land to west for sale as probable light industrial site.	I		I	
		No access to site within 200' of Herrick / River Road crossroads.	E	X	E	X
		No access to site within 200' of bridge over Sun River.	E	X	E	X
		City water and sewers on south side of Herrick.	I	X	I	X
		No history of flooding.	U		U	
		Municipal Airport 4 miles west. City center is east on Herrick.	O	X	O	X
	ON SITE	Topographic and soil conditions mainly OK. Some rock NW corner.	O		O	
		Sixty foot (60') easement with 11KV power line crosses site NE to SW. (a)	A		I	
		Marshy area with natural spring in NE corner.	I		I	
		Gas main in 25' easement parallels Herrick.	E	X	E	X
		Minimum building setbacks 75' from street and 25' from property lines.	E	X	E	X
		(OK to use setbacks for roads, access, parking)				
		High point of site in NW corner.	I		I	
INTANGIBLE FACTORS	**EXTERNAL**	Zoning and county building restrictions.	E		E	
		Easements on site for gas and power.	E		E	
		Sound abatement restrictions in the area.	O	X	O	X
		Pollution control becoming more stringent.	O	X	I	X
		Traffic on Herrick will increase.	I	X	E	X
		Traffic on River Road will increase.	O	X	I	X
		County fire and police protection (may change to city later). (b)	E		E	
		County government now. Probably will be annexed by city later.	O	X	O	X
		Not in major airport flight path.	U		U	
	ON SITE	Building / site appearance and image.	E		E	
		Building and parking security.	E		E	
		Need to provide outdoor lunch area.	I		I	
		Personnel considerations - recreation area and changing rooms.	I		I	
		Signage to all important areas.	E		E	
		Management limitations on long term buildings/site growth.	E		I	

Impact on Facilities Planning: A-Very High,   E - High,   I-Medium,   O-Minor,   U-Minimal
Control: Mark an "X" if beyond control of company
Notes:
(a) Preliminary discussions indicate power line may be moved to west side of site.
(b) Assurances from county and city to respond to fire emergencies.

Figure 3.5 Dominant factor analysis.

Site focus will already have occurred, leading to the site mission. The secondary focus will be within the site. Figure 3.6 illustrates how focus can result in clusters on the site and indicate candidates for transfer from the site when the need arises.

During strategic integration, *scenarios* help planners understand the external environment at the enterprise level. Such scenarios

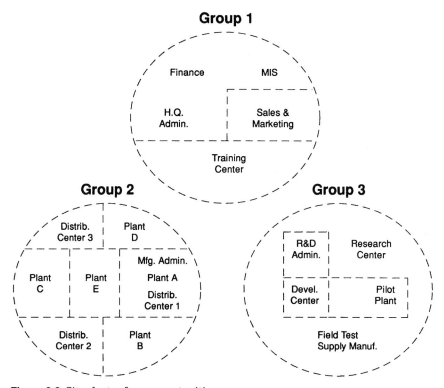

Figure 3.6 Site cluster focus opportunities.

describe factors such as markets, world economic developments, and new product technologies.

Out of the strategy formulation, the master site-planning team will need at least the following:

1. Enterprise scenarios
2. Site mission statement(s)
3. Manufacturing/operations strategy summary

Site Scenario Development

Scenarios are concise, qualitative statements of future events which may impact an organization's performance. They are particularly useful for strategic thinking in uncertain environments.

Master site planning usually requires scenarios at two levels. During strategy formulation, *enterprise-level scenarios* assist top managers in looking at future events and developing appropriate strategies.

After the team has gathered information and formulated strategy, it must develop *site scenarios.* While site scenarios encompass strategic scenarios, they concentrate on events that impact the particular site.

Long-range master site planning developed and was relatively successful in the comparatively stable 1950s and 1960s. Long-range planning was traditionally based on forecasts. Since the early 1970s forecasting errors have become more frequent and occasionally of dramatic and unprecedented magnitude.

Forecasts are not always wrong. They are often reasonably accurate, and therein lies the danger. Forecasters usually assume that tomorrow's world will be very much like today's.

Many of the facilities of the 1970s did not match capacity demands. During a period of rapid growth in the United States in the 1970s it was common to build excess capacity and grow into it. Product life was relatively long; processes and markets were stable, and competition was not very intense. To cope with uncertainty, managers bought better forecasting techniques and hired more forecasters. Today many forces work against a forecast. The future is no longer stable and trends are discontinuous. Rather than try to make better forecasts, it is necessary to accept uncertainty, try to understand it, and make it part of our reasoning.

Uncertainty today is not just an occasional temporary deviation from predictability; it is a structural feature of the business environment. The method used to think about and plan for facilities must be appropriate to such an environment. Scenario planning is such a method. Scenarios help managers structure uncertainty when (1) they are based on a sound analysis of true reality, and (2) they change the managers' assumptions and compel them to reorganize their mental model of reality. This process entails much more than simply designing good scenarios. A willingness to face uncertainty and to understand the forces that drive it requires a significant transformation.

The greatest danger in turbulent times is not the turbulence; it is to act with yesterday's logic, according to Peter Drucker.

Scenario planning is a strategic tool for senior management. To develop scenarios, first recognize:

- *Predetermined elements*—Those events that have already occurred, but whose consequences have not yet happened

- *Uncertainties*—Those events which may happen in the future

Emphasizing only obvious uncertainties leads to:

First-generation scenarios: These present raw uncertainties but are not a basis on which to exercise judgment. Their goal is not action, but understanding. They are needed to develop the second-generation scenarios.

Second-generation, or decision, scenarios:

- The most likely scenarios
- The "challenge" scenarios

Decision scenarios acknowledge uncertainty. They aim at structuring and understanding it, not by criss-crossing variables and producing hundreds of outcomes, but by creating a few alternative and internally consistent pathways into the future. Decision scenarios describe different worlds, not just different outcomes in the same world. Scenarios coalesce thousands of design decisions and possibilities into a few. Four is a practical number of scenarios to consider. This allows management to make rational and consistent decisions.

Scenarios should focus less on predicting outcomes, more on understanding the forces that will eventually compel an outcome; less on figures, more on insight.

Scenarios deal with the two worlds of *facts* and *perceptions*. Their purpose is to gather and transform information of strategic significance into fresh perceptions. It is most important in scenario design that:

- The scenarios be realistic
- They respond to managers' deepest concerns
- There is a willingness to face uncertainty
- The scenarios are protective, anticipating and understanding risk
- The scenarios are entrepreneurial, discovering strategic options of which there was no previous awareness

Strategic scenario options as illustrated in Fig. 3.7 combine:

- *Strategic vision*—A clear structured view of what you want your company to be
- *Focused scenarios*—Fresh perceptions in understanding the uncertainties
- *Option creation*—Considering all possible options for implementing the strategic vision and scenarios for the company's future

For strategic facilities and site master planning use the most likely strategic scenarios option as the planning base. Then identify the most logical "what ifs" or alternative scenarios and develop facilities planning options to meet these.

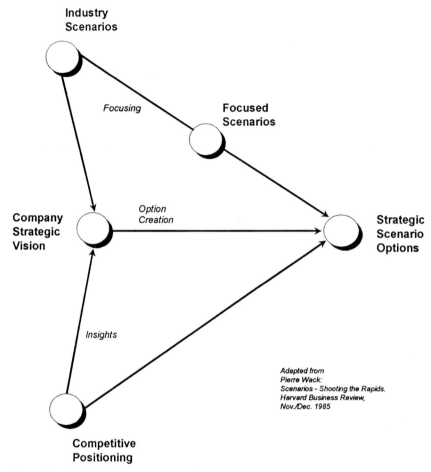

Figure 3.7 Developing strategic scenario options.

Site Mission

The *site mission* states the purpose of each site with goals and objectives. To be useful to the facility planner, the following points need to be covered:

- Why do we have the site?
- Existing facilities use
- Building upgrading or replacement plans
- Ecological, meteorological, and environmental considerations
- Desired site image

- Resale factors
- Who will move when site is supersaturated?
- What does the nonspecific mission require of the site?
- Methods of meeting long-range challenges

The mission statement gives the planning team clues on how to interpret corporate strategy and provide for it. For one site the mission and the derived facilities reflect the corporate and marketing strategy; for multiple sites the missions are an important component of operations strategy.

Key Issues

Layout styles

There are many different layout styles that can be used for planning building sites. Seven basic styles are:

- Big block
- Comb
- Spine
- Modular
- Orbital
- Campus
- Combined

The "big block" style is the concentration of all under-roof activities in one building. The "comb" through "campus" styles are illustrated in Fig. 3.8. There may be combinations of the separate styles on one site.

Which style to adopt may be predetermined on an existing site. Where it is feasible to choose, the preference may be a stated management one or it may emerge from the logic of the design process.

Single or multistory buildings

Decisions regarding single or multistory will depend on:

- Types of occupancy and operations
- The organization and layout focus
- Zoning ordinances
- Topography of the site
- Materials flow and handling (Is gravity flow important?)

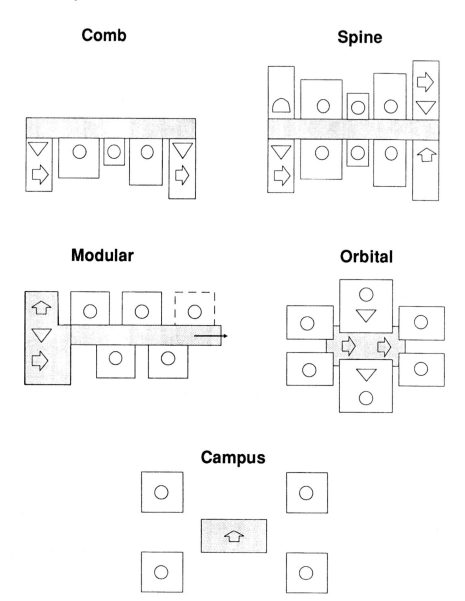

Figure 3.8 Layout styles.

One or many buildings

Whether to have a single, several, or many buildings can depend on:

- Characteristics of the processes
- X affinities (the need to keep things apart), and hazards

- Properties of materials—dusty, reactive, radioactive, dangerous, toxic
- Noisy, vibrating processes versus sensitive, clean activities
- Any fundamental basis for segregation or separation such as:

  Wet from dry

  Hot from cold

  High and low security

  High- and low-value materials

- Regulations that demand physical separation
- Areas that must be kept away from the general public
- Different environmental requirements
- Organizational reasons for separation by company, division, market, product
- Customer requirements
- Special utilities
- Special waste disposal needs, e.g., for hazardous waste

Multiple buildings can be *separate* or *linked.* The connecting of buildings is desirable in cold or wet climates or with "clean" processes or products. Flexible buildings accommodate future changes and match short use or layout life with long building life.

Expansion options

Of the five recognized expansion options, three—*zone, block, and duplicate*—are illustrated in Fig. 3.9. The others are:

- *Static.* Fill the site initially. No expansion is contemplated.
- *Transfers.* Make space provision for activities from other sites or from acquisition.

Special factors

Company culture and image.   Are there any special aspects of company culture and image that should be reflected by the site plan, such as:

- Show the best side of the facilities and operations to the general public
- Project a special image with frontage treatments for manufacturer of daily-use consumer products

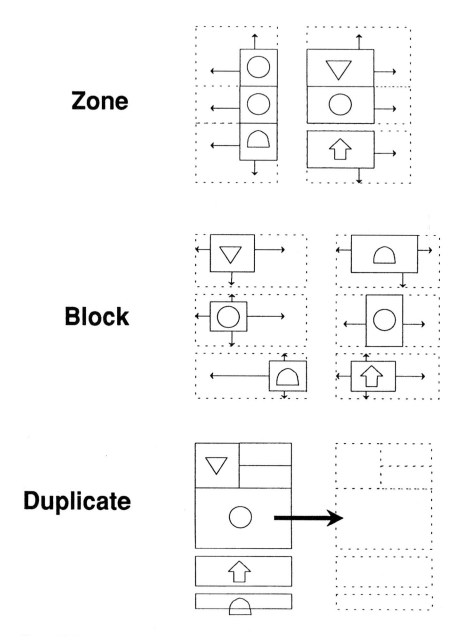

Zone

Block

Duplicate

Figure 3.9 Expansion options.

- Provide buildings and environment to attract and maintain high-quality personnel

The facilities represent the company image to customers, employees, suppliers, and local authorities.

Recreation. Consider company policy on providing playing fields, clubhouse, fitness facilities, locker and changing rooms, etc. Will an outside lunch/picnic area be needed? *Note:* If a precedent is established when there is plenty of room on the site, consider the impact if some of the recreation areas need to be used later for production or other purposes.

Financial aspects. Consider with finance representatives the major questions of company financial policy:

- Should you buy or lease the existing or "to-be-built" property?
- Should you maximize "fail safe"? Consider long-term resale value if you might ever need to sell all or part of the property. The more specialized the building, the more difficult it will be to sell.

Security. How can the site and buildings be adequately secured to meet company needs? Are there any special security requirements—e.g., computer centers, laboratories, potentially hazardous or confidential process areas, valuable materials?

Sublease or multiple tenants. Do you plan to overbuild initially and sublease temporarily until you need space? What about separate access, security, etc.?

If you acquire an existing building, will there be more space than you need? Should you sublease, have single or multiple tenants? Or should you be a tenant in someone else's building?

Climatic conditions. Consider impact on:

- Arrangement on-site
- Under-roof/open area activity decisions
- Special weather extreme needs
- Open or closed docks and which direction they face considering winter prevailing winds
- Drainage and storm water runoff
- Construction needs, regulations, and timing
- Utility consumption

Site Space

Land-to-building ratios

Land-to-building ratios are used to determine what size of site is needed to support a given requirement of buildings. For an existing site, determine how much of the site is covered by buildings—that is, the under-roof footprint. Then, find out how much is yard or open space and how effectively it is being used for:

- Employee, visitor, and truck parking
- Storage areas
- Service support utilities
- Green areas, setbacks
- Roads, rail, etc.
- Vacant land

Normalize this open space usage to determine what it should be to support the under-roof footprint. Use these normalized open space proportions related to under-roof projected needs for developing long-range total saturated-site needs.

A table of land-to-building ratios published by the U.S. Department of Commerce is shown in Fig. 3.10. This shows that for a 1:1 ratio 2 acres would be required. This would be for 1 acre under-roof and 1 acre

According to the size (footprint) of structure and land-to-building ratio

Size of Structure (footprint)		Acres Required for Selected Land-to-Building Ratios									
s.f.	acres	1 to 1	2 to 1	3 to 1	4 to 1	5 to 1	6 to 1	7 to 1	8 to 1	9 to 1	10 to 1
5,000	.11	.23	.34	.46	.57	.69	.80	.92	1.03	1.15	1.26
10,000	.23	.46	.69	.92	1.15	1.38	1.61	1.84	2.07	2.29	2.52
20,000	.46	.92	1.38	1.84	2.29	2.75	3.21	3.67	4.13	4.59	5.05
30,000	.69	1.38	2.07	2.75	3.44	4.13	4.82	5.51	6.20	6.89	7.57
40,000	.92	1.84	2.75	3.67	4.59	5.51	6.43	7.35	8.26	9.18	10.10
50,000	1.15	2.29	3.44	4.59	5.74	6.89	8.03	9.18	10.33	11.48	12.63
60,000	1.38	2.75	4.13	5.51	6.89	8.26	9.64	11.02	12.40	13.77	15.15
70,000	1.61	3.21	4.82	6.43	8.03	9.64	11.25	12.85	14.46	16.07	17.68
80,000	1.84	3.67	5.51	7.35	9.18	11.02	12.86	14.69	16.53	18.36	20.20
90,000	2.07	4.13	6.20	8.26	10.33	12.40	14.46	16.53	18.59	20.66	22.73
100,000	2.30	4.59	6.89	9.18	11.48	13.77	16.07	18.36	20.66	22.96	25.25
200,000	4.59	9.18	13.77	18.37	22.96	27.55	32.14	36.73	41.32	45.91	50.50
300,000	6.89	13.77	20.66	27.55	34.43	41.32	48.21	55.10	61.98	68.87	75.76
400,000	9.18	18.37	27.55	36.73	45.91	55.10	64.28	73.46	82.64	91.83	101.01
500,000	11.48	22.96	34.45	45.91	57.39	68.87	80.35	91.83	103.30	114.78	126.26
1,000,000	22.96	45.91	68.87	91.83	114.78	137.74	160.70	183.65	206.61	229.57	252.52

43,560 s.f. per acre   Source: Office of Area Development (U.S. Dept. of Commerce)

Figure 3.10 Site requirements.

for all other supporting land. A 4:1 ratio would have 5 acres site total with 1 acre under-roof supported by 4 acres for everything else.

Some local authorities decree the minimum land-to-building ratio. Others regulate car parking space per employee and the maximum number of employees for the site.

The analysis described above can give a good indication of how much property would be needed to support a given footprint of buildings if the site were full. Typically a simple warehouse operation can be as low as a 1:1 ratio. But large, complex, multibuilding sites may require 4:1, 5:1, or more at saturation. When full, industrial facilities tend to fall in the 2:1 to 3.5:1 range. If a high "image" level is required, calling for major setbacks from the road and extensive beautification, then 3:1 to 5:1 may be required at saturation.

These land-to-building ratios are approximate only. Much will depend on the shape of the property, what kind of easements there are, access to and egress from the site, soil conditions, and other factors.

The ratios are based on under-roof *footprint* only. Land to total multistory square feet under-roof is a completely different type of ratio and needs to be used for site-planning purposes with caution.

Development of facilities needs—Space

In site planning, space is classified as *under-roof* or *outside* (*yard*). Space is determined for:

Under-roof operations, including:

- Manufacturing
- Warehouse
- Support
- Offices
- Laboratories
- Other

Outside or yard areas, including:

- Employee and visitor parking
- Truck aprons and parking
- Outside storage
- Cooling ponds
- Waste treatment areas
- Roads and rails
- Nonusable space
- Other

Description: Site Planning	Proj. No. 12345
Company: XYZ Corporation	Prep By: PAH
Location: Corp. Wide	Page: of: 1 1 Date: 17-Apr

ID	Facilities	Year Ready	Add	Delete	Cumul.	Mfg.	%	Storage	%	Support	%	Offices	%	R&D	%	Units	Sales	Ratio
1	Existing Plant				247,640	116,400	47.0%	47,050	19.0%	44,575	18.0%	29,715	12.0%	9,900	4.0%		27,000	9.17
2	R&D Extension	+1	4,500			0		0		0		0		4,500				
					252,140	116,400	46.2%	47,050	18.7%	44,575	17.7%	29,715	11.8%	14,400	5.7%		29,800	8.46
3	Mfg. Extension	+2	22,000			16,200		3,000		2,400		400		0				
					274,140	132,600	48.4%	50,050	18.3%	46,975	17.1%	30,115	11.0%	14,400	5.3%		32,700	8.38
4	Corp.Office Exten.	+3	6,400			0		0		1,400		5,000		0				
					280,540	132,600	47.3%	50,050	17.8%	48,375	17.2%	35,115	12.5%	14,400	5.1%		36,000	7.79
5	Distrib. Extension	+5	8,000			0		6,000		1,200		800		0				
					288,540	132,600	46.0%	56,050	19.4%	49,575	17.2%	35,915	12.4%	14,400	5.0%		43,500	6.63
6	Old Paint Shop	+5		3,800		(2,800)		(600)		(400)		0		0				
					284,740	129,800	45.6%	55,450	19.5%	49,175	17.3%	35,915	12.6%	14,400	5.1%		43,500	6.55
7	Coating Facility	+5	2,800			2,000		500		300		0		0				
					287,540	131,800	45.6%	55,950	19.5%	49,475	17.2%	35,915	12.5%	14,400	5.0%		43,500	6.61
	Rough Projection (total space)	+10	132,460			62,700		23,850		22,725		16,585		6,600				
					420,000	194,500	46.3%	79,800	19.0%	72,200	17.2%	52,500	12.5%	21,000	5.0%		70,000	6.00
	Rough Projection (total space)	+15	145,000			67,100		27,600		25,000		18,100		7,200				
					565,000	261,600	46.3%	107,400	19.0%	97,200	17.2%	70,600	12.5%	28,200	5.0%		113,000	5.00

NOTES:
Sales in constant annual $000
Ratio = Cumulative space / constant annual sales ($000)

Figure 3.11 Space balance.

These will be in terms necessary to meet the growth projections for site-planning purposes only. The methods for calculating space requirements are discussed in detail in other chapters.

Space balance

The *space balance analysis,* example in Fig. 3.11 shows how the buildings and other facilities on the site could grow to meet future needs. The analysis shows each addition, removal, or demolition as it is made, and the proportion of space under each classification. Each addition is related to the projected business growth.

In the early projected growth time frame, additions are shown year by year. This allows finance and corporate management to determine projected capital needs in two-, three-, and five-year business plans.

Projected space additions are also related to business growth in the same and future time frames. For long-range master site and facilities planning it is often more relevant to use volumes as indicators for new facilities, rather than calendar years.

In industrial and commercial organizations the provision of new sites and/or buildings can be done fairly quickly (one to three years). In government facilities such as the large national laboratories it may take four to eight years from planning start to implementation. In these cases planning horizons of ten, twenty, or thirty years, or even longer, should be the norm.

Site-Planning Approaches

The following eight approaches are useful for site planning:

- *Dominant features.* Identify the dominant external and internal site features which will tend to force the master site planning into a certain mode or direction. Examples of features that constrain the plan are lakes, streams, hills, swamps, sinkholes, and residential areas.

- *Infrastructure.* Plan the basic road (and rail) circulation system for the site. Plan the main utilities runs. Within this infrastructure fill in the pieces—buildings, structures, parking lots, outside storage areas, etc.

- *Fixed features.* Identify and locate the most fixed items, the ones that are least likely to be moved once they are located, such as power plant, water tower, foundry, plating, effluent treatment, main utilities runs, rail spurs. Then fill in the rest of the buildings and parking areas around these.

- *One large main building.* House virtually all under-roof activities in one main building.

- *Campus style.* House operations on-site in a number of separate buildings with or without walkway interlinks.

- *Internal and external integration.* Ensure that access and egress to and from the site are logical: road, rail, water, air, materials, personnel services. Road entrances must not be too close to crossroads, cloverleafs, T and Y junctions, bridge entrances, and one-way traffic limitations. You may need to provide buffer zones of trees, open land, berms, levees, etc. (on your property), if there are adjoining residential or other special areas.

  Is your site part of an industrial park? If so, there may be many restrictions on what you can do.

  Review the long-term purpose of the locality. Will you fit in without future problems? Will you fit with long-term city growth plans? What impact will there be on and of neighbors?

- *Material flow.* Plan the buildings and facilities on the site based on optimized material flow. Common basic site flow patterns are shown in Fig. 3.12.

- *Dealing with irregular sites.* It is usually better to accommodate features of an irregular site rather than change them. But don't build in long-term operating or expansion problems.

  Roads can be closed or moved; railroads can be redirected; rivers and streams can be diverted; lakes and swamps can be filled.

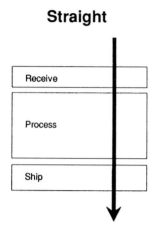

Straight

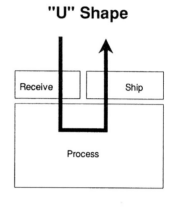

"U" Shape

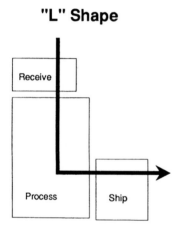

"L" Shape

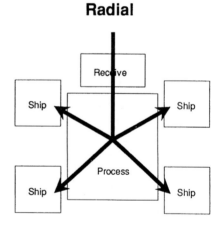

Radial

Figure 3.12 Basic site flow patterns.

Land may be acquired, sold, donated, or exchanged to make the site a more regular shape.

Site Space-Planning Procedure

This is the conversion of the business strategy, site missions and objectives, policies, and other site data into input for the site-planning process.

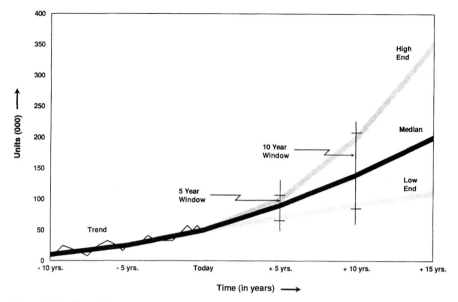

Figure 3.13 Growth projections.

Growth planning base

Site plans are required to provide for additional or new capacity or for consolidation of existing capacity. Projections can be expressed in constant sales dollars, units produced, or another parameter that will give a realistic forecast of what growth to expect.

An example is shown in Fig. 3.13. This gives the planner an indication of a widening planning "window." A base scenario may plan for the median growth projection and most likely constraints. Alternative scenarios can utilize the "what-ifs" of the high and low projections.

Infrastructure

Identify the basic road, rail, water, and air circulation system for the site (there may be little, if any, in place on a new site):

- Main points of access and egress
- Road/rail interferences, particularly if rail has to be crossed at traffic peaks or shift changes
- Any particular problem areas such as roads narrowing between buildings, heavy traffic crossing roads between buildings

- Site split by public roads with operations both sides and intraplant traffic across the roads

Identify also the main utilities distribution structure. Develop the infrastructure for circulation and utilities distribution to site saturation level, that is, as if the site were full. Then show whatever partial implementation is needed to meet the growth stages of facilities on the site.

The planning process

The planning structure given in Fig. 3.2 provides an outline of how to proceed. The steps to take in developing the master site plan are now reviewed.

Capacity requirements

Determine current and proposed capacity needs. Relate these to facilities, equipment, and support requirements, balancing with shift and other working arrangements. Consider any planned or possible changes resulting from the introduction or enhancement of just-in-time, cellular operations, time compression, or other world class manufacturing techniques that can have major impact on facilities needs.

Facilities needs

Determine capacity requirements in terms of major items of manufacturing and support equipment only. Do this in sufficient depth to enable short-, medium-, and longer-range facilities needs to be determined. Relate them to overall building requirements in gross terms of projected amounts of manufacturing, storage/warehousing, support services, offices, and R&D.

Site saturation

Determine what the site could look like if fully occupied. This is known as *site saturation planning*. It can help avoid making mistakes in locating short-term planned facilities which might preclude effective long-term site development.

Development of site layout options

The development of site layout plans is essentially a large-scale layout project as described in other chapters. In this case whole buildings or major sections of these, parking areas, truck turnarounds, site entrances, etc., are identified as the *site-planning units* (SPUs) rather than internal operating and support areas within a building.

The site-planning process is in two parts:

1. *Develop a site saturation plan*—What the site would look like when full, with buildings generally located and the physical infrastructure (roads, utilities, etc.) fully developed.

2. *Develop the tactical implementation plan* covering the next two to five years, showing buildings and infrastructure to meet this, all within the scope of the site saturation plan.

The planning steps to be taken are:

- Identify SPUs for under-roof facilities and outside or yard activities.

- Determine the affinities between the SPUs, considering material or transport movement and nonflow relationships.

- Prepare a *configuration diagram* illustrating the affinities between the SPUs. Modify the diagram to indicate a long road with multiple access points to the site, rather than constraining the diagram to a single-point site entry. This is shown in Fig. 3.14. Also shown are focus opportunities.

Diagram 1
Area 16 (outside road) shown as single hub

Diagram 2
Area 16 shown as multi-point access road
Clear indication of focus clusters and sub-clusters

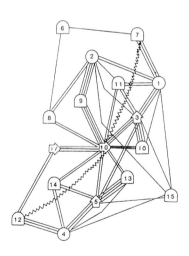
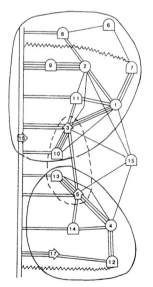

Figure 3.14 Affinity configuration diagrams.

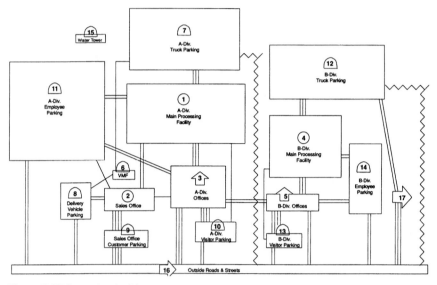

Figure 3.15 Layout primitive.

- Develop space requirements for each of the SPUs.
- Construct a *layout primitive* by "overlaying" the space to scale on the configuration diagram as illustrated in Fig. 3.15.
- Prepare alternative site space plans using the layout primitive as a base and constraining it because of the following influences:

  - Dominant factors analysis
  - Existing buildings and infrastructure
  - Easements
  - Soil analyses
  - Organization strategy
  - Facilities focus study
  - Fixed features on the site
  - Layout style to adopt
  - Expansion options
  - Special factors (e.g., image, financial)

Develop a few sound space plans from the above procedure for evaluation and selection of a preferred option, as described in Chap. 11.

Phased Implementation

Develop an implementation program

The next step is to prepare a phased implementation plan to reach the preferred site saturation layout option. This should be done in tactical planning detail for the next two to five years, in more general terms for the longer-range growth steps.

Site-planning financial needs

Work closely with finance management to decide what level of financial information should be included in the strategic master site plan document. Capital projects will be identified, cash needs determined, and so forth. Finance will advise on corporate policy on whether to do all planning in current-year dollars or to use some form of inflation parameter.

Alternative scenarios—The "what ifs"

Review the alternative scenarios options previously identified. Determine what impacts these could have on the base scenario and develop the appropriate strategic site layout, facilities planning, implementation, and financial options.

Master schedule

An example of a *master schedule* identifying all known and possible facilities planning projects shown in Fig. 3.16 gives the big picture. It may be more realistic to relate future projects to strategic levels of business activity rather than dates. For example, "When we reach $200 million in sales we will require x facilities; when we reach $500 million, y facilities."

Master Site Plan Documentation

Strategic master site-planning outputs will be site- and company-specific but should include task outputs under the following headings:

- Introduction
- Missions, goals, and objectives
- Site history
- Strategies and scenarios
- Existing facilities and conditions
- Planning analysis and process

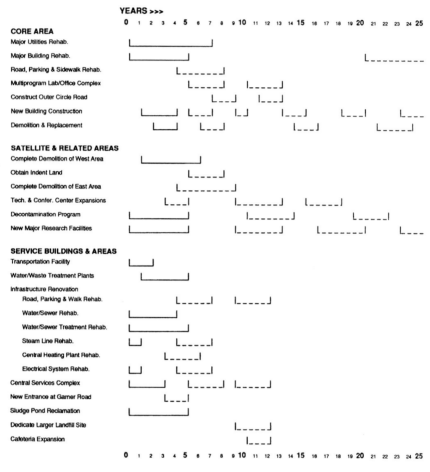

Figure 3.16 Long-range project schedule.

- The long-range facilities master plan
- Implementation plan
- Support data

Summary

Having a fully developed strategic master site plan that supports the strategic business plan is necessary, be it for a 3-acre site for a small service group or a 1,500-square-mile national test site. A lot of effort goes into making a good one. The need for looking long-range when planning sites and facilities can make the facilities planner the catalyst who sparks the strategic business planners into opening their minds to longer-range opportunities.

The approved strategic master site plan becomes a major planning tool for the company. It should be reviewed with the business plan each year (or more often if necessary) and modified or updated as appropriate. It should be considered as a "living document" that all can use as reference.

The facilities represent the physical operating ability of a company to meet its strategic business intent.

Further Reading

Durand, Jacques: "A New Method for Constructing Scenarios," *Futures,* 1972, December 1972, pp. 325–330.

Hodson, William K. (ed. in chief): *Maynard's Industrial Engineering Handbook,* 4th ed., McGraw-Hill, New York, 1992.

Lynch, K.: *Site Planning,* MIT Press, Cambridge, MA, 1992.

O'Connor, R: "Planning under Uncertainty: *Multiple Scenarios and Contingency Planning,* The Conference Board, New York, 1978.

Salvendy, G: *Handbook of Industrial Engineering,* 2d ed., Wiley, New York, 1992.

Tompkins, J. A., and J. A. White: *Facilities Planning,* Wiley, New York, 1984.

Wack, P.: "Scenarios: Unchartered Waters Ahead," *Harvard Business Review,* September–October 1985, pp. 73–89.

Wack, P.: "Scenarios: Shooting the Rapids," *Harvard Business Review,* November–December 1985, pp. 139–150.

4

Factory Layout
and Design

William Wrennall
President, The Leawood Group Ltd.
Leawood, Kansas

Introduction

Factory layout is the planning, designing, and physical arrangement of processing and support areas within a manufacturing facility. The goal is to create a factory that meets company and operating strategies. A good layout will optimize the use of resources while satisfying other criteria such as image, control, and quality, plus many other factors.

Due to these many factors, the task of plant layout can quickly become complex. This chapter guides the planner and clarifies the plant layout process. The procedure also shows how to design layouts that support world class manufacturing strategies. New strategies will have a significant impact upon space requirements and activities focus.

In this chapter, we will answer the following questions:

- What are the major steps required to create a good plant layout?

- What data must be collected and analyzed?

- What are the issues to resolve?

- What decisions need to be made?

- How do you quickly but accurately develop space requirements?

- How should space-planning units (SPUs) relate to each other?

- How do designers optimize resource use?

■ What are the ways to select the best layout developed?

In the early 1980s the first news began to filter out of Japan about the manufacturing system called *just-in-time* (JIT). Since then, JIT has matured. It is one of several manufacturing strategies emphasizing inventory reduction and people-technology integration. Manufacturers may use other names, such as *world class manufacturing* (WCM). Such strategies usually have the following elements:

JIT production	Scientific maintenance
Rapid setup	JIT production control
Group technology	Focused factories
Supplier network	Capacity reserve
JIT accounting	Total quality
The team approach	Work simplification

These strategies affect plant layout. The plant layout design has, in turn, a profound effect on the success of the strategy.
   WCM layouts feature:

■ Workcells

■ Linked production

■ Focused factories

■ Kanban stockpoints

■ Point-of-use delivery

■ JIT material handling

■ Reduced space

Workcells

Manufacturing *workcells* are small, self-contained work units that typically build a single product or group of similar products. They typically employ 2 to 12 persons. Ideally, they contain all the equipment to manufacture complete products.
   Workcells balance tasks, encourage teamwork, improve quality, and respond quickly to customer requirements. While workcells are simple in concept, they are often difficult to design and implement. Effective operation requires unconventional approaches to supervision and management.

Linked production

In JIT and world class factories, each process has close ties to upstream and downstream processes. They are physically close and have small inventory buffers between them. The *linked production* reduces space requirements and increases response time. It also requires that the designer have a thorough understanding of products, processes, and work flow.

Focused factories

Wickham Skinner of Harvard first presented *focused factory* concepts more than twenty years ago. These factories limit products, processes, and markets to a manageable range. They concentrate on customers with compatible requirements: products of a similar nature and processes they can master. Focused factories simplify work flow, facilities, and infrastructure. Designers using focus concepts optimize an entire manufacturing system for a specific set of manufacturing tasks.

Kanban stockpoints

Where *Kanban* controls production, the system requires *stockpoints*. They are often next to producing work centers. These stockpoints are, frequently, the only significant *work-in-process* (WIP). Large staging areas and WIP warehouses disappear with their elaborate tracking systems and the people who operate them. Shortages, miscounts, and concerns about inventory accuracy also disappear.

Point-of-use delivery

With JIT/WCM strategies, the traditional and adversarial vendor relationships move toward cooperation. With changes in attitude come changes in supplier deliveries, quality, and transactions. The supplier may require direct access to production areas. This bypasses the usual receiving, inspection, and warehouse functions. Deliveries are small quantities of material brought to those who immediately require it. That is *point-of-use delivery.*

JIT material handling

Conventional manufacturing moves large loads of material infrequently. WCM strategies require small loads of material with frequent or continuous movement. These movements are often more direct and over shorter distances. Handling systems therefore change. Fork trucks and automatic guided vehicles (AGVs) give way to hand trucks and simple conveyors. The paths of movement also become paths of communication as workers talk with each other about future requirements and common problems.

Reduced space

Many features translate directly to reduced space. The dramatic inventory reductions that occur also reduce space. JIT/WCM factories may often require only 40 to 50 percent of the space used by conventional layouts.

The Strategic Framework

Manufacturing strategy is the underlying philosophy of a manufacturing system. It manifests itself as:

- The pattern of management decisions over time
- The range and grouping of products
- Types of support systems
- Selection and arrangement of equipment
- Employees and their attitudes

A manufacturing strategy should be explicit, consistent, and well thought out. More often it is implicit, inconsistent, and haphazard.

Regardless of management's strategic sophistication, the designer of a factory layout should know what strategy the firm will most likely follow. The layout can then reflect and support that strategy.

Focused manufacturing limits activity in an organization to a manageable and consistent set of tasks. These tasks directly support the firm's marketing strategy. This concentrates expertise and promotes superior performance, although in a narrow range. It is not uncommon to see inappropriate manufacturing focus built into plant layouts. A usual case is where a functional layout is producing high- or medium-volume products. An example is a manufacturer of minicomputers. Such a company with five basic models assembling 3 to 10 of each per day could use a large functional assembly area. A preferred layout for a new plant would be a small assembly line for each basic model. Prototypes and replacement parts require a functional area.

Most manufacturing organizations (and other organizations as well) display a strong bias for the functional mode. The reasons for this bias are unclear but some possibilities are:

- Functional layouts are often easier to design.
- Accounting systems do not penalize the high inventory required by functional layouts.
- Financial policies emphasize high equipment utilization and, in theory, favor functional layouts.
- Engineers favor high-tech, costly, and large-scale equipment that demands high utilization. This also favors a functional layout.

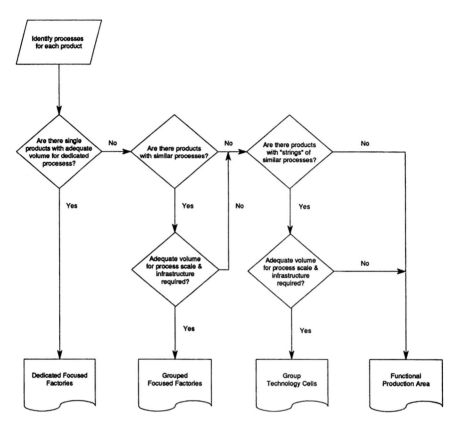

Figure 4.1 Manufacturing focus algorithm.

Manufacturers misuse the functional mode most often. However, any mode has the potential for misapplication. In one situation, a Detroit-style assembly line builds massive off-road vehicles at 1.5 per day. The results are poor.

The layout designer can achieve an optimum degree and mix of manufacturing focus by using the algorithm given in Fig. 4.1. The *product*-focused modes, especially line and continuous, offer many advantages in quality, low inventory, and efficiency. The *process*-focused *functional* mode is most frequently misapplied. For these reasons, the algorithm starts with a pure *product* focus and line production. It then backs away through the *Toyota, cellular, and functional* modes toward an acceptable alternative.

Start with an operation process chart for each distinct product. Any industrial engineering handbook or text shows the conventions for constructing these charts. Line the charts up, side by side, as shown in Fig. 4.2.

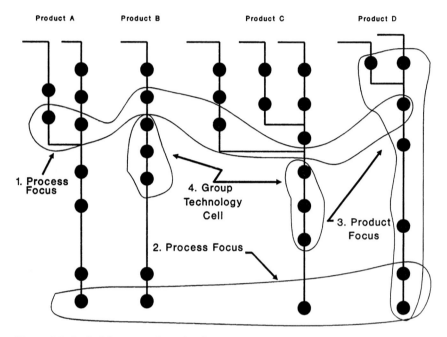

Figure 4.2 Optimizing manufacturing focus.

Each operation on a product is a *product operation* (PO). In the typical situation, a single PO requires too little time and equipment for a dedicated workstation or department. The layout designer must somehow group various POs.

One method groups all POs that require the same equipment or type of process. Envelopes 1 and 2 illustrate this. Such grouping provides a pure process focus.

Alternatively, a designer might group all operations required by a single product. Locating these in a single product department or workstation gives a pure product focus. This is illustrated in Envelope 3.

In Step 2, the designer examines each product for a trial product-focused grouping. There are, we believe, only two valid reasons for rejecting a pure product focus:

1. The available processes have a large product volume that cannot economically process a single product. For example, a small turned pin requires only 0.25 machine hours per week for the expected product volume.

2. Some element of the infrastructure is large-scale. It cannot effectively serve a single low-volume product. For example, a highly skilled electronic technician calibrates a circuit board. Production only requires the technician's skills about two hours per week.

If a trial product passes both tests, it should have its own manufacturing area and possibly its own *plant-within-a-plant* (PWP). Remove such products from further consideration.

In Step 3, the designer examines each remaining product, then identifies subsets of the manufacturing operations (*strings*) that are common between two or more products. When such a string occurs, the designer tests it for adequate volume vis-à-vis process and infrastructure scale. If passed, such operation strings become *group technology* (GT) cells as in Envelope 4.

Step 4 collects all remaining operations into functional areas.

The identified areas become layout cells. Such layout cells may be GT cells, functional cells, or product cells. Figure 4.1 charts the procedure step by step.

The focus criteria selected should be consistent with corporate goals, market strategy, manufacturing processes, and infrastructure. Developing such focus is one of the most important elements of manufacturing strategy.

Focus is our best means to reduce manufacturing complexity and direct technical and knowledge resources at customer demands. By doing this, manufacturing becomes an important part of the product mix rather than a hapless supplier of commodities.

The Layout Project Structure

A structure to follow for any factory layout project includes:

The layout life cycle as a framework

The layout process

A generic project plan

The layout life cycle

The layout process proceeds from the general to the particular, from a general framework or structure to the location of each piece of equipment in each work place. In this way it is much easier to get a sound and logical arrangement of *space-planning units* (SPUs) or blocks of space. That is, get the layout sound in principle first. Next, increase the detail within the approved and sound macro layout, until the layout is complete.

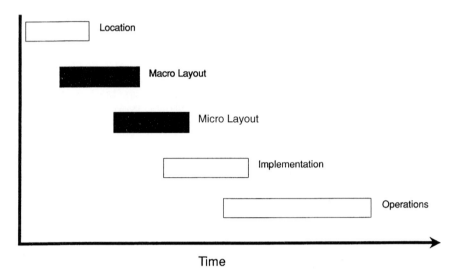

Figure 4.3 Layout project life cycle.

A structured approach to layout projects

In our structured layout-planning approach, we recommend phasing the facilities project. Figure 4.3 illustrates a five-phase plan. The five phases lead to an organized method for developing a layout.

In the *phase diagram* time is represented horizontally. The time for each phase is the *phase block length.* The lengths given are for purposes of demonstration and do not represent any particular project.

Phase overlap shows that phases can begin before the completion of the prior phase. Phases cannot be completed before the completion of the prior phase. The greater the phase overlap, the shorter is the time for completion of the project. This is desirable to improve productivity of layout planning.

Phase I: Location.   This is where the project is defined and the layout planners are oriented to the project. The following questions need to be answered:

- Which site is to be considered?
- What building is to be used?
- What is the scope of the project?
- What is the schedule?
- What are the tasks and deliverables?
- Who will staff the project?

Phase II: Macro layout.   This is the main planning phase, where business plans, strategy, focus, and so forth are integrated to develop macro layouts.

The *macro layout,* sometimes called a *block layout,* consists of the arrangement of blocks of space—SPUs, or cells—in a regular shape.

In the macro layout the SPU outlines only are given. These outlines should not be interpreted as walls. They may or may not be, depending on the refined shape after Phase III and structural building requirements.

The enveloped shape may be that of an existing building, floor, or area for the layout. It may be conceptual and free from structural or size constraints. This may be the basis for a future building design.

Phase III: Micro layout.   This phase is for the layout in detail. Only when the layout is accepted at the macro level should the designer begin to place equipment in each SPU. In this phase, each SPU becomes a semi-independent layout design.

The blocks, or SPUs, from the macro layout are now given operational meaning. Equipment, furniture, utilities, building features, aisles, and material locations are determined.

Phase IV.   Phase IV is where the physical arrangement occurs. The layout plans are extended into an implementation plan which is then executed. The result is the physical layout.

Phase V: Operations.   Phase V is when the layout design is tested. Operations start-up occurs in this phase and the plant is transformed to a production unit.

In Fig. 4.4, planning costs and benefits are added to the layout life-cycle phase diagrams.

Good layout planning leads to good layouts and efficient operations. Retrofits or afterthought corrections are also avoided.

The layout life-cycle cost curve shows the expenditure rate or cost at each phase and the strategic impact. Note that in the early phases, cost is low but strategic impact is high. These early phases thus have an important long-term effect on operations and largely decide profitability. *The early phases are not the place to economize.* In the later phases—micro layout, implementation, and operation—costs peak, yet the activities during these phases have little strategic impact.

The Generic Project Plan

Factory layout projects transform a manufacturing strategy into a physical capability. The project structure given has the following steps:

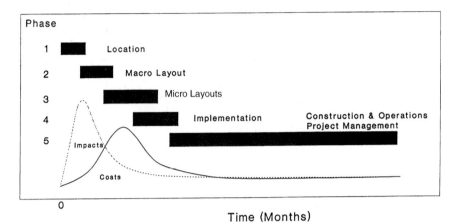

Figure 4.4 Layout planning phases (costs and impacts).

Definition and data collection

Strategic analysis

The layout process

Integration

Detail

Implementation

To manage a layout project, it is convenient to extend the above steps into a network of activities and tasks. The network is illustrated as a *layout and material-handling generic project plan* in Fig. 4.5.

The tasks and their outputs or deliverables are organized into the phase groups given in the phase diagram shown as Figs. 4.3 and 4.4.

The *generic project plan* (GPP) defines a set of tasks and sequences that guide a layout project, large or small, simple or complex.

To arrive at a layout, the designer must accomplish each task in some manner. The formality of analysis, the specific techniques, rigor, creative insight, resources expended, and time required will vary from project to project.

The first task, Task 2.01, is to scope and schedule the project. For a major new facility this may require a detailed schedule using computer-aided project management. For a small plant or department, the scope and schedule may exist only in the mind of the designer.

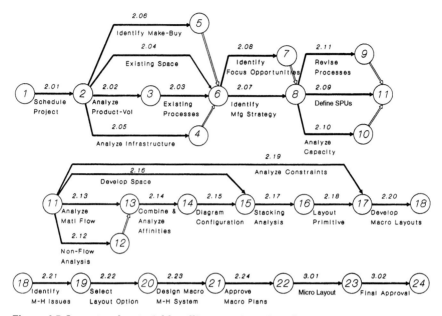

Figure 4.5 Layout and material-handling generic project plan.

Information acquisition

In this part of the project, Tasks 2.03 to 2.06, the designer or project team embarks on a series of tasks which gather information, organize it, and develop preliminary conclusions. These include:

- Documenting the existing processes
- Cataloging of current space usage
- Identifying infrastructure requirements
- Developing make-buy policies

When these tasks are completed, the project team has a thorough understanding of the current and past situation. It is then in a position to develop a strategic base for the layout.

Strategic definition

Tasks 2.07 and 2.08 develop the strategic foundation for the layout. The focus analysis identifies opportunities for focused manufacturing and a policy for focus. The manufacturing strategy development produces a policy-level summary that guides the layout team in their work. This manufacturing strategy summary also will guide operating people as they make day-to-day decisions. The summary should address the broad areas of process, facilities, and infrastructure.

Process refers to the equipment and operations that work directly on the product to add value. The summary recommends an appropriate scale for equipment. It makes general recommendations on technology types, ease of setup, and ranges of product capability. It should also address quality capability.

Infrastructure supports the manufacturing process. It includes production control, personnel, accounting, maintenance, and other functional areas of the business. These infrastructural activities are often the most dissonant, for many reasons beyond the scope of this chapter.

Facilities are the physical infrastructure. Buildings, utilities, roads, and land support the process. These, too, should fit with the other elements of the manufacturing system. For example, a strategy that depends on cellular manufacturing and new product flexibility requires a layout, building, and electrical system that can easily change. This may be to accommodate new products, new cells, and changed volumes.

The factory layout should incorporate and support the manufacturing strategy of the firm. If the layout fails to do so, it will induce, at the very least, considerable stress. At worst, the manufacturing system may fail to meet market requirements and risk failure of the entire firm.

Layout design

The GPP Tasks 2.09 to 2.11 translate existing facts, existing knowledge, and agreed-on strategies into a factory layout. These tasks include the definition of SPUs that establish the organization and function of the layout. The designers analyze capacity requirements and review the processes.

Next, quantify material flows, calculate space requirements, and identify constraints. Incorporating this information in progressive diagrams will lead to macro-layout options.

Management then selects from the macro layouts or synthesizes additional macro layouts from the options presented. With an approved macro layout, proceed with designing material-handling systems. The design and approval of detailed layouts for each cell of the macro layout can now follow.

The GPP builds on experience with techniques developed by Muther, James, and others. It adds improved data analysis, strategic considerations, techniques that organize constraints, and evaluation of design options.

Layout elements

Every layout has four basic elements and four derived elements. The basic elements are:

1. SPUs
2. Affinities
3. Space
4. Constraints

The derived elements are:

1. Configuration diagram
2. Layout primitive
3. Macro-layout options
4. Micro layouts

Figure 4.6 shows how the basic elements interact to give the derived elements. These lead to the development of the plant layouts.

The first layout element and basic layout block is the *cell, or space-planning unit (SPU)*. In this chapter we use the two terms interchangeably. Most projects require 15 to 25 SPUs. If more appear necessary, the project scope may be too big. In such cases reduce the project scope or divide the project into two subprojects.

An SPU may be:

- Product focused
- Functional department
- Storage area
- Building feature

A fluorescent light assembly layout is an example of product focus. A powder paint unit is a functional department. A tool crib is a storage area, and a loading dock is a building feature.

Ten ways to develop SPUs

Explore the following sources when developing SPUs:

1. Use of existing SPUs
2. Operation charts
3. Organization charts
4. Group technology
5. Technology forecasting

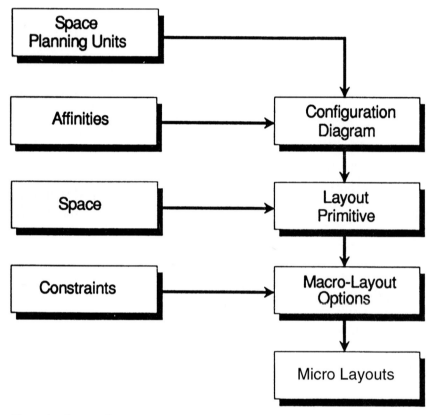

Figure 4.6 Layout elements.

6. Research and development
7. Infrastructure
8. Company policy
9. Codes and regulations
10. Company strategy

The example given in Fig. 4.7 shows a useful way to record the results of SPU definitions in summary form. Each SPU has a discrete identity number. The SPU description includes areas that are included and excluded. A source listing on the right-hand side of the form also gives the primary focus for each SPU.

Description:	Plant Layout	Proj.#	1154
Company:	USPS	Date:	7-1
Location:	GMF	By:	KB

page 1 of 4

No.	Name	Inclusions	Exclusions	Sources										Primary Focus				
				Exist	Oper.	Org.	GT	Tech	R&D	Infra	Polc.	Code	Strat.	Proc.	Prod.	Other	None	Note
1	Receiving Platform - Dock	Dock handling, sorting of inbound mail	PAU/BMAU Dispatch areas	X										X				
2	Dispatch Platform - Dock	Dock handling Dispatch staging area of outbound mail	PAU/BMAU Receiving areas	X										X				
3	Initial Sort	Sawtooth, slide and manual sorting (mostly sack sorting)	Dispatch sorting	X										X				
4	PAU/BMAU	Office, staging, scales and dedicated dock area (presort and bulk mail only)	Other receiving and dispatch areas	X										X				
5	Opening & Prep	Opening pouches/sacks Opening letters/trays	010 Originating Mail	X										X				
6	Cull/ Face/ Cancel (AFCS/ISS)	010 (DPRCS and AFSM) Equipment, support, WIP, sweep, staging, controls	Mail that does not need to be culled, faced and cancelled	X										X				
7	MLOCR (all ISS)	Equipment, support, WIP, sweep, staging, controls (machinable letters)	Flats and parcels Non-machinable letters	X										X				
8	BCS	Equipment, support, WIP, sweep, staging, controls (9D barcoded letters)	Flats and parcels Non-9D barcoded letters	X										X				
9	DBCS	Equipment, support, WIP, sweep, staging, controls (9D barcoded ABC mail)	Flats and parcels Non-9D barcoded letters	X			X							X				

Figure 4.7 SPU definition summary.

How to classify SPUs

SPUs are classified by purpose or the use of the space they represent.

In this process we use an extended version of the American Society of Mechanical Engineering (ASME) process charting symbols. The conventions are given in Fig. 4.8, Strategic Facilities Planning (SFP) Conventions.

- The circle or operation symbol represents an operation area or space such as product final assembly.

- A right-hand pointing arrow transport symbol represents a shipping dock.

- The square inspection symbol indicates a test or inspection area.

- A storage activity symbol is the inverted triangle or hopper.

- The D, temporary storage symbol, represents work-in-process, or set-down space.

Other symbols are formed by rotating two of the ASME symbols.

- An upward-pointing arrow symbol identifies office space.

- A letter D rotated $-90°$ is a service symbol.

The next step in the layout process is to determine affinities.

Basic Planning Symbols & Colors

Process Chart Symbols & Action*	Symbols Extended to Identify Activities/Cells & Areas	Color Ident. **	Black & White **
◯ Operation	◯ Primary-Form-Treat.	Green	\\\\\\\\\\\\
	◯ Secondary-Assembly-Packaging	Red	|||||||||||||||
⬱ Transportation Circulation	⬱ Entrances, Exits, Docks	Orange Yellow	:::::::::::::::
▽ Storage	▽ Storage Activities/Areas	Orange Yellow	:::::::::::::::
D Delay	D Set-down or Hold Areas	Orange Yellow	:::::::::::::::
☐ Inspection	☐ Inspect, Test, Check Areas	Blue	═══════
Adapted from *A.S.M.E. Stand.	◸ Service & Support Activities/Areas	Blue	═══════
**I.M.M.S. Stand.	⌂ Offices or Building Features	Brown (Gray)	#########

For Product Focus use separate colors on an uncolored layout to identify all activities/cells related to specific product lines or families.

Site Planning Colors

Site/Land Occupancy	Color Ident.	Black & White
Underroof Land Areas (or Production Space)	Purple	/////////
Green Areas, Grass, Beautification, Setback, Open Areas	Green	\\\\\\\\\
Water, Pond, Stream, Service Outbuildings Utility Lines	Blue	═══════
Outdoor Storage	Orange Yellow	::::::::::::::
Office Buildings, Laboratories, Administration, and Personnel Service Buildings	Brown	#########
Property Lines and Easements	Red	═══ ∙∙ ═══
Building Outlines	Black	─────
Personnel Roads and Car Parking	Uncolored or Gray	#####
Rail, Truck Roads, Truck Parking	Yellow or Gray	::::::::#####

Ratings

Rating			Proximity	Color Code **	Evaluating Results	Rating
A	4	////	Absolute	Red	Almost Perfect (Excellent)	A/4
E	3	///	Exceptional	Orange Yellow	Exceptionally Good (Very Good)	E/3
I	2	//	Important	Green	Important (Good)	I/2
O	1	/	Ordinary	Blue	Ordinary (Fair)	O/1
U	0		Unimportant	Uncolored	Unimportant (Poor)	U/0
X		⋏	Not Near	Brown	Not Acceptable (Unsatisfactory)	X/?
XX		⋏⋏	Apart	Black		

Figure 4.8 Strategic facilities planning (SFP) conventions.

How to develop affinities

Affinities are the degree of attraction between cell or SPU pairs that lead to the layout configuration. They represent a requirement for proximity or closeness between a pair of SPUs, cells, or activities in a layout.

Affinities may be positive, negative, or neutral. Positive affinities indicate attraction or closeness required. Negative affinities indicate that separation is desirable or necessary. This requirement arises from *material flow and nonflow factors.*

The steps used in developing affinities are:

- Calibrate material flows.
- Determine nonflow affinities.
- Combine flows and nonflow affinities.
- Evaluate and review affinity proportions.

Calibration of material flows

Material flow calibrations follow the material flow analyses. Analysis methods are covered in depth in Chap. 6.

Material flow intensities between SPU pairs are converted into affinities based on flow only by:

- Transferring material flow data from a from-to chart to a ranked bar chart. An example is given in Fig. 4.9.
- Calibrating the flows into an AIOU format.

The chart will typically indicate a Pareto distribution. There will be an important few high-intensity flow paths and many "trivial" low-intensity flows. The few high-intensity flows will have a significant effect on the layout design. The other flows modify the layout.

Developing nonflow affinities

Material flow is a major determinant of layout design. *Nonflow affinities* also are an important consideration for most layouts.

Some reasons for nonflow affinities are:

- Environmental constraints and concerns
- Process similarities
- Shared equipment
- Shared supervision or management
- Shared workforce

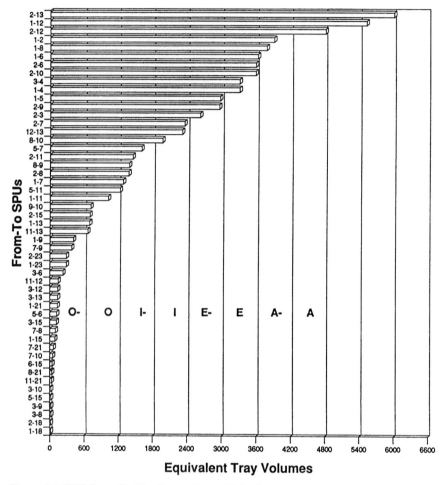

Figure 4.9 SPU flow affinities for medium cellular layout.

- Product quality enhancement
- Utility distribution
- Security and hazard concerns
- Fitting the manufacturing/corporate strategy
- Codes and regulations
- Company image
- Communication between cells

Each project will reveal other unique causes that create affinities.

The vowel convention scale used for calibrating material flow is used to rate the nonflow affinities. The scale is extended for negative affinities.

A   is **A**bsolute proximity

E   is **E**special proximity

I   is **I**mportant proximity

O   is **O**rdinary proximity

U   is **U**nimportant proximity

X   indicates that proximity is not desirable

XX   indicates that separation is necessary

Because the determination of nonflow affinities is a matter of judgment, it is useful to develop *benchmark affinities*. These can be a basis for judging all other affinities.

A manufacturing example of benchmark affinities, with typical reasons or bases, is:

Affinities	SPU pairs	Basis for proximity
A	Coil storage and shear, clean, and paint	Similar handling problems; avoids contamination
E	Primary and secondary painting	Share materials, share equipment, similar skills
I	Subassembly and final assembly	Control
O	Wire department and stores	Convenience
U	Harness assembly and tool room	Infrequent contact
X	Welding and solvent storage	Fire hazard

Material flow ratings are the result of quantitative analysis but nonflow ratings are subjective. The following are sources of informed opinion for developing nonflow data:

- Planners

- Managers

- Supervisors

- Employees

Considerable difficulty can occur in collecting the necessary input from each source. The three most common methods for collecting this data are interviews, surveys, and consensus meetings. Each method has its own advantages and disadvantages.

Interviews have the benefit of face-to-face interaction. However, interviews can be lengthy and time-consuming. Surveys are a quick means of gathering data using survey forms. This method provides input from people who are unavailable for interviews. Surveys result in few responses and questionable results. Individuals who respond know their own jobs very well. That is no guarantee of their judgments of affinities.

The most reliable method is the consensus meeting. The participative approach fosters cooperation and provides additional information.

A chart such as the one shown in Fig. 4.10 is a useful way to record affinities. The procedure for entering the data is as follows:

1. Record the project identification data in the top project block.
2. Fill in the reasons or bases in the basis box, at the top right. These are the reasons for the affinities chosen earlier.
3. List the SPUs down the left-hand column.
4. Record the affinity in the appropriate intersect.
5. Record the basis or bases for choosing the affinity rating.

Figure 4.11 shows the intersects on an enlarged scale for SPUs 1 to 4. The procedure for filling in the diamond that overlaps SPUs 1 and 2, "Incoming Platform" and "Letter Process A," is:

■ Place the affinity rating between SPUs 1 and 2 in the top half of the diamond. Here, it is E.

■ Fill in the bases for this affinity in the bottom half of the diamond. Here, numbers 1 and 4.

Developing combined affinities

Combined affinities are the result of combining or merging flow ratings with nonflow affinities. The procedure is to:

1. Determine the flow-to-nonflow ratio
2. Multiply flow rating by weight or ratio
3. Multiply nonflow ratings by weight
4. Sum flow and nonflow scores
5. Place scores on ranked bar chart
6. Calibrate total affinities

The flow-to-nonflow ratio will vary from project to project, and from industry to industry. The designer will decide the "weight," or

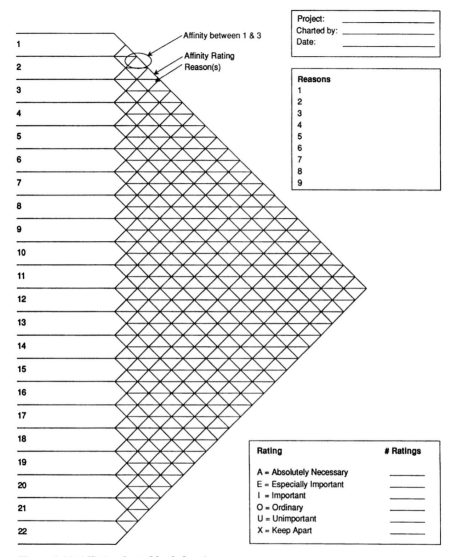

Figure 4.10 Affinity chart (blank form).

relative importance, in each case. Typical ratios of flow-to-nonflow affinities range from 1:1 to 2:1.

Applying the selected weighting to the flow and nonflow affinities, Steps 2 through 4, provides a *merged affinity*. This is done by building a matrix of flow and nonflow ratings. Multiply the weight by the numerical value of the ratings, then add the two products.

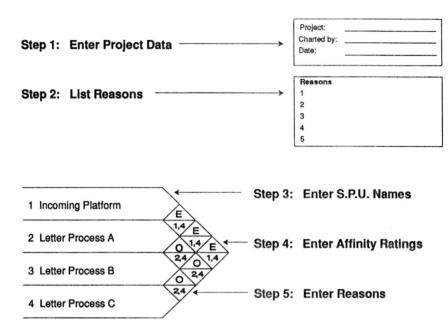

Figure 4.11 How to complete an affinity chart.

The matrix in Fig. 4.12 uses a 2:1 flow-to-nonflow ratio. Note that X values flag negative closeness. The numerical value of -1 for an X cannot be subtracted from a flow value of A (4) to give a resulting affinity of E $(4 - 1 = 2)$. The issue has to be dealt with in a nonarithmetical way. It may be a fire wall or curtain or process change.

An affinity combination example which shows the process is shown in Fig. 4.13. In Fig. 4.14 an example of combined affinities from an actual project is shown.

Checking affinity quality

An affinity frequency distribution should display:

- A few A ratings
- Progressively larger numbers of E's, I's, and O's
- A few X's and XX's

The distribution of the combined affinity ratings should be within certain limits. A typical distribution for a process-focused layout is:

		Non-Flow										
		4	3-1/2	3	2-1/2	2	1-1/2	1	1/2	0	-1	-2
Flow		A	A-	E	E-	I	I-	O	O-	U	X	XX
8	A	12	11-1/2	11	10-1/2	10	9-1/2	9	8-1/2	8	•	•
7	A-	11	10-1/2	10	9-1/2	9	8-1/2	8	7-1/2	7		
6	E	10	9-1/2	9	8-1/2	8	7-1/2	7	6-1/2	6		
5	E-	9	8-1/2	8	7-1/2	7	6-1/2	6	5-1/2	5		
4	I	8	7-1/2	7	6-1/2	6	5-1/2	5	4-1/2	4		
3	I-	7	6-1/2	6	5-1/2	5	4-1/2	4	3-1/2	3		
2	O	6	5-1/2	5	4-1/2	4	3-1/2	3	2-1/2	2		
1	O-	5	4-1/2	4	3-1/2	3	2-1/2	2	1-1/2	1		
0	U	4	3-1/2	3	2-1/2	2	1-1/2	1	1/2	0	-1	-2

Note: X values cannot be subtracted from flow values

Figure 4.12 Flow/nonflow matrix.

Affinity rating	Percent
A	1–3
E	2–5
I	3–8
O	5–15
U	20–85
X	0–10

With a product-focused layout there will probably be fewer total affinities and fewer high-level affinities (A's and E's). This is because of simpler material flows and communication lines in a product-focused layout.

In a functional or process-focused layout there are more lines of communication and heavier, more complex material flows between major departments. The result is a higher number of total and high-level affinities.

If a nontypical affinity distribution occurs, it will be necessary to review the ratings.

Configuration Diagrams

These are the derived elements from cells and affinities. An example is given in Fig. 4.15.

Description:	New Layout	Project:	1162
Company:	Wautomi GMF	Date:	7-2
Location:	New Facility	By:	KB

From	To	Material Flow			Non-Flow			Total Score	Comb. Rating
		Rating	Wt.	Score	Rating	Wt.	Score		
5	6	4	.66	2.6	4	.34	1.4	4.0	A
1	5	4	.66	2.6	3	.34	1.0	3.6	A
9	10	3	.66	2.0	3	.34	1.0	3.0	E
1	7	3	.66	2.0	2	.34	0.7	2.7	E
6	7	2	.66	1.3	4	.34	1.4	2.7	E
4	6	3	.66	2.0	1	.34	0.3	2.3	E
1	4	3	.66	2.0	1	.34	0.3	2.3	E
1	3	3	.66	2.0	1	.34	0.3	2.3	E
2	6	3	.66	2.0	1	.34	0.3	2.3	E
3	6	2	.66	1.3	3	.34	1.0	2.3	E
1	2	2	.66	1.3	3	.34	1.0	2.3	E
1	8	2	.66	1.3	3	.34	1.0	2.3	E
6	8	2	.66	1.3	2	.34	0.7	2.0	I
1	6	2	.66	1.3	2	.34	0.7	2.0	I
15	16	1	.66	0.7	2	.34	0.7	1.4	I
9	11	1	.66	0.7	2	.34	0.7	1.4	I
1	10	2	.66	1.3	0	.34	0.0	1.3	I
7	8	2	.66	1.3	0	.34	0.0	1.3	I
11	16	2	.66	1.3	0	.34	0.0	1.3	I
7	9	2	.66	1.3	0	.34	0.0	1.3	I
5	9	2	.66	1.3	0	.34	0.0	1.3	I
3	9	1	.66	0.7	1	.34	0.3	1.0	O
2	9	1	.66	0.7	1	.34	0.3	1.0	O
4	9	1	.66	0.7	1	.34	0.3	1.0	O
1	11	1	.66	0.7	1	.34	0.3	1.0	O
6	11	1	.66	0.7	1	.34	0.3	1.0	O
5	8	1	.66	0.7	1	.34	0.3	1.0	O
4	5	1	.66	0.7	0	.34	0.0	0.7	O
2	3	1	.66	0.7	0	.34	0.0	0.7	O
2	4	1	.66	0.7	0	.34	0.0	0.7	O
2	5	1	.66	0.7	0	.34	0.0	0.7	O
9	13	1	.66	0.7	0	.34	0.0	0.7	O
14	15	1	.66	0.7	0	.34	0.0	0.7	O

Note: List continues with over 100 other "U" ratings.
        Xs are handled on an individual basis.

Figure 4.13 Affinity combination.

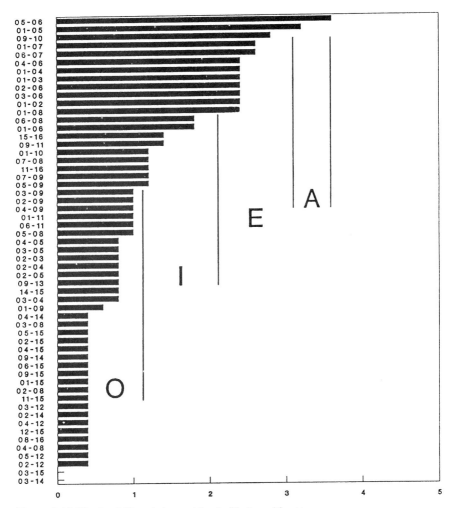

Figure 4.14 Wautomi Mountain combined affinity calibration.

Configuration diagrams are similar to "bubble" diagrams used by architects. The diagram is a nodal representation of a layout or a layout without space. It is a logical step in a structured layout process.

The source of information is the combined rating affinity chart (Fig. 4.16). The procedure for developing the configuration diagram is given in Fig. 4.17.

The number of lines indicate the affinity between two SPUs in the configuration diagram. The conventions are as illustrated earlier in Fig. 4.8.

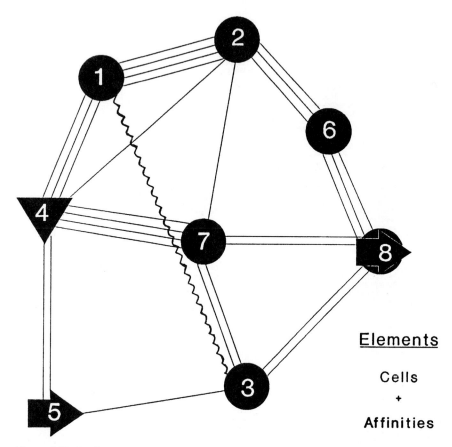

Figure 4.15 Configuration diagram.

Consider the rating lines as rubber bands, with the A rating giving a strong pull like four bands. SPUs, cells, or activities with A ratings are close together. Those with E (3), I (2), or O (1) ratings would have lesser pulls and are drawn apart. Connect X or negative ratings with wiggly lines representing a spring pushing them apart. For X affinities use one wiggly line and use two similar lines for XX affinities.

Figure 4.17 presents the four steps in developing the configuration diagram. Begin with placing the A's and E's. The A's have four lines between them, the E's have three lines. The next step is to add the I's. The diagram is then redrawn. A good way to do this is to look for hubs and terminals, rearranging and adding O's and X's. The diagram is rearranged for the best fit.

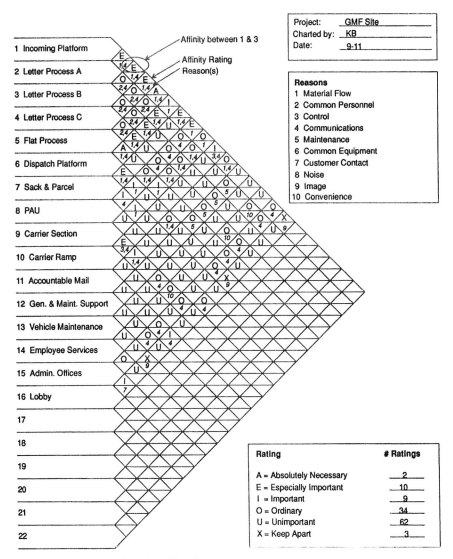

Figure 4.16 Affinity chart (combined).

SPU pairs are identified by number and activity symbol. Single or multiple lines connect the SPU symbols.

Developing Space

Space is one of the four basic elements of every layout. The space on existing layouts is a matter of record, but the space for new layouts

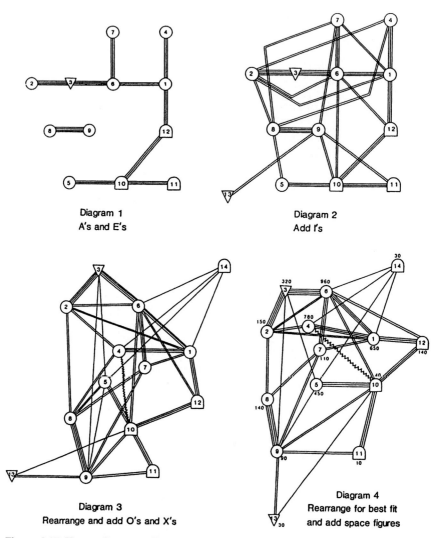

Diagram 1
A's and E's

Diagram 2
Add I's

Diagram 3
Rearrange and add O's and X's

Diagram 4
Rearrange for best fit
and add space figures

Figure 4.17 How to diagram affinities.

has to be determined.

The designer of a new layout first develops the cells and their affinities, then calculates space. Developing space is Task 2.16 on the GPP.

Six methods are available for computing space:

- Elemental calculation
- Transformation

- Standard data
- Ratio forecasting
- Visual estimating
- Proportioning

Each has advantages and disadvantages. These methods give the designer a toolbox from which to select the most appropriate method or methods. Each SPU may require a different method for establishing the space needed.

The accuracy level required for space calculations varies widely. The level of effort, credibility, and time horizon also vary for each method.

Elemental calculation

Figure 4.18 illustrates the approach to *elemental calculation.*

In this method, collect the space for each process, item of equipment, or aisle that will be provided. The total will be the SPU space. Use elemental calculations for small areas with few items or where a few items dominate space requirements. Use it also where space standards do not exist.

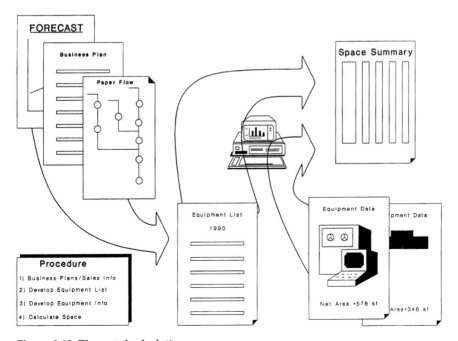

Figure 4.18 Elemental calculation.

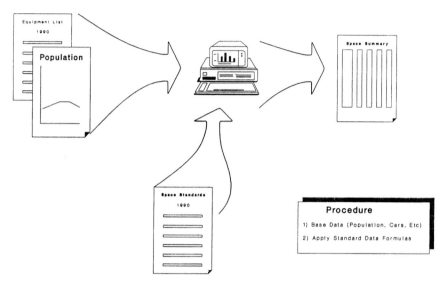

Figure 4.19 Standard data calculation.

Elemental calculation requires a complete list of equipment and other needs. For large or complex layouts it is laborious.

The accuracy of elemental calculation reflects the accuracy of the equipment and furniture schedule. The computed space can be quite accurate when based on an accurate and stable list. Inaccuracy will result from a functional manager's inflated opinion of departmental importance.

Standard data

Standard data calculation is a close cousin to elemental calculation. With the standard data method, the designer takes a known unit such as persons, automobiles, or product. Existing space information is applied to calculate the space. The procedure is illustrated in Fig. 4.19.

Standard data calculation is the method of choice for large firms with many rearrangements. Once the standard data are complete and approved, space calculations are fast and accurate. We recommend the following procedure to develop new space standards:

1. Identify complaints/symptoms of nonstandardization.
2. State goals/objectives.
3. Survey current situation and reconfirm goals/objectives.
4. Develop data using regression analysis and synthesis.
5. Gain approval.

This method, like elemental calculation, reflects the accuracy of the data. When the data are accurate and the standards are valid, the space calculation is highly accurate.

Where standards are nonexistent, their development can be difficult and time-consuming. In situations where furniture and equipment are nonstandard, use the elemental method.

Transformation

Transformation is the most versatile of the six methods, particularly for macro and site-level layouts. With an experienced layout designer, the method is reasonably accurate. It is simple and does not require complete furniture, equipment, or personnel schedules.

How to use the method is illustrated in Fig. 4.20. To estimate the space for a particular layout cell, the analyst begins with a comparable existing space. Next adjust or *normalize* the existing space to account for abnormal current conditions. The analyst will review business forecasts, operating plans, and task definitions. If this will have a space effect, then adjust the space to allow for the anticipated changes in activity levels.

Visual estimating

Visual estimating, Fig. 4.21, illustrates the use of layout templates and the visual experience of the designer to estimate space. The

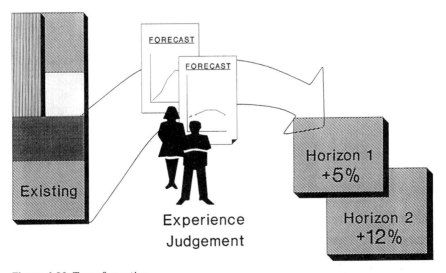

Figure 4.20 Transformation.

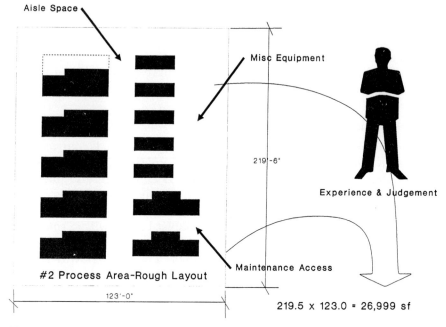

Aisle Space

Misc Equipment

219'-6"

Experience & Judgement

Maintenance Access

#2 Process Area-Rough Layout

123'-0"

219.5 x 123.0 = 26,999 sf

Figure 4.21 Visual estimation.

method is fast and has moderate accuracy and high credibility. The designer does not require a complete furniture and equipment schedule. However, visual estimating must be done only by highly experienced designers.

The approach uses a set of scale templates based on a preliminary furniture and equipment schedule as the starting point. Arrange the templates on a layout grid with space for aisles, circulation, and miscellaneous equipment. The gross space occupied is the estimate for that particular layout cell.

Note that the prototype layout does not consider affinities or constraints. It visualizes the space required.

Proportioning

Proportioning, illustrated in Fig. 4.22, applies the principle of proportioning space use. Classes of space often require a constant proportion. Standards, past practice, experience, or existing layouts may provide such proportions. If so, the designer can use them to calculate space.

With known proportions, the method is fast, easy, and accurate. Without known and tested proportions, only the experienced designer should use this method.

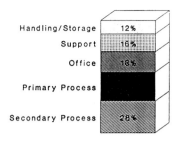 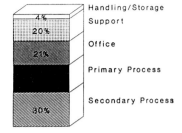

Standard Proportions   New Facility

Figure 4.22 Proportioning.

Ratio forecasting

Ratio forecasting has the lowest near-term accuracy of the six methods but has the best long-term accuracy. Its primary use is for site planning and long-range planning. In stable industries it can produce fair accuracy on twenty- to forty-year forecasts.

Begin the process with historical data as visualized in Fig. 4.23. The space is classified and measured by class at several past-time intervals. These intervals vary but typically are five to twenty years. Develop ratios by relating the space in each class to some parameter such as sales dollars or site population. The basis parameter should be stable over the historical period and expected to be stable during the forecast period.

The ratios of space class to basis parameter often have clear long-term trends. The analyst chooses a set of ratios and projects them. Such projection may use visual or statistical measures.

The ratios for future horizons are used to indicate the space required at that horizon. For example:

- The ratio of units produced per square foot of a factory in ten years is projected to be 120 per day. The forecasted daily sales for the product in ten years is 2.4 million.
- The estimated production space will be sales per day in ten years divided by the number of units produced per day per square foot.
- That is 2.4 million divided by 120 = 200,000 square feet.

Hybrid methods

These are combinations of methods. When computing space, the analyst or layout designer normally selects a method for each layout cell. Sometimes several methods can combine for good results.

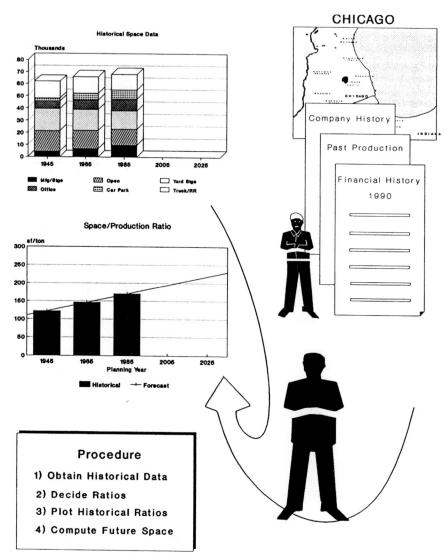

Figure 4.23 Ratio forecast.

The analyst also may use more than one method as a cross check. Space calculations for several SPUs using transformation can then be cross-checked using elemental calculation.

Selecting a method

Several factors affect the designer's choice of method for determining space. Among these are:

- Accuracy
- Computation effort
- Experience of designer
- Credibility to users

Figure 4.24 compares the six methods on each of these dimensions. Large or complex layouts justify higher levels of effort than small and simple layouts. Detail or micro layouts require more effort and greater accuracy than macro or site layouts. Experienced designers may use visual estimating and transformation when the less experienced should use elemental or standard data calculation.

Users and managers often consider transformation an unreliable method. In the hands of an experienced designer, however, it is often far more accurate than the forecasts on which elemental or standard data calculations are based.

Apart from credibility, true accuracy is also important. Factors which affect the required accuracy are:

Methods	Accuracy	Effort	Planning Horizon	Input Data Complete	Experience Req'd	Credibility
Elemental	High	Very High	Short	Very	Low	High
Standard Data	High	High	Short	Very	Low	High
Visual Estimate	Mod.	Mod.	Med.	Mod.	High	High
Transformation	Mod.	Low	Med.	Mod.	High	Mod.
Proportioning	Mod.	Low	Long	No	Mod.	Mod.
Ratio Forecast	Low	Mod.	Very Long	No	Mod.	Mod.

Figure 4.24 Computing space—selecting a method.

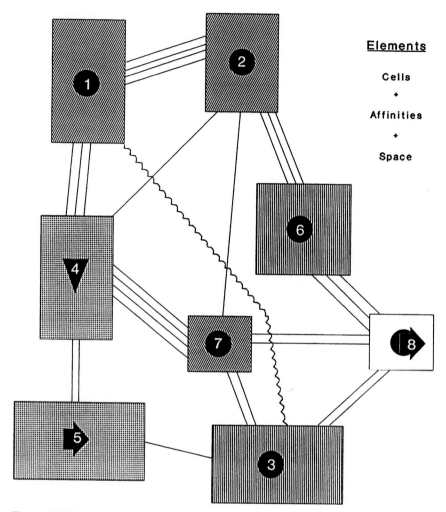

Figure 4.25 Layout primitive.

- Design phase
- Equipment type
- Business plan uncertainty
- Consequences of over- or underestimating

Typically, a design requires less accuracy in the site and macro phases of layout. Two reasons account for this. First, random errors compensate. Bias may cause all SPU sizes to be either over or under the true space requirement. The second reason is less sensitivity to error. During the detail layout in the next phase, refinement corrects small errors.

The uncertainty of input parameters may overshadow the accuracy of some methods. The resulting pseudo-accuracy brings comfort to some but does not improve the uncertainty of the business plan or environment. In an unstable business environment, accuracy is fleeting and flexibility is important.

Layout Primitive

Adding space to the configuration diagram leads to the derived element *layout primitive* (Fig. 4.25). This does require adjustment to the configuration diagram to accommodate the space blocks. The block sizes are to scale but normally do not assume the eventual shape of the SPU. If the aspect ratio (length to width) is known, then it can be shown as required in the final layout. For example, a phosphate wash process size and shape will be known.

The layout primitive is an exploded view of a layout. This leads us into the constraint process of developing the macro layout.

Constraints

Every layout project is subject to some constraints. At this stage of the layout process we accommodate the physical constraints such as space available, shape of space, doorways, and elevators.

The constraint summary example in Fig. 4.26 provides a lead into the *constraining process*. This is the step that *transforms the layout primitive into macro, or block, layout options.*

This is where in the process the facility designer can introduce "secondary clusters" or "post-focus." There is an opportunity to divide or combine SPUs. For example, a warehouse may be split into two parts. One part can store purchased parts and the other part can hold finished-goods packaging material.

Macro Layouts

The outputs from the constraint process are macro, or block, layout options. Layout options are different, sound designs. They provide managers with choices. The selection of a preferred option tests the extent to which the designer has interpreted the strategic company intent. It also releases additional information. An example of a macro, or block, layout is shown in Fig. 4.27.

Management approval of a block layout provides the input for the detail layouts.

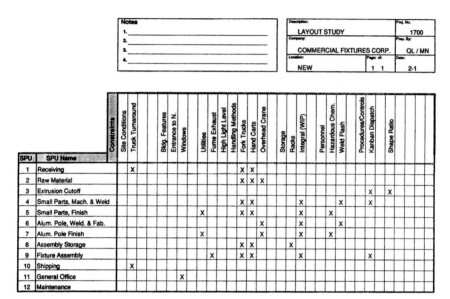

Notes
1. _____
2. _____
3. _____
4. _____

Description: LAYOUT STUDY	Proj. No. 1700
Company: COMMERCIAL FIXTURES CORP.	Prep. By: QL / MN
Location: NEW	Page of: 1 1	Date: 2-1

SPU	SPU Name	Site Conditions	Truck Turnaround	Bldg. Features	Entrance to N.	Windows	Utilities	Fume Exhaust	High Light Level	Handling Methods	Fork Trucks	Hand Carts	Overhead Crane	Storage	Racks	Integral (WIP)	Personnel	Hazardous Chem.	Weld Flash	Procedures/Controls	Kanban Dispatch	Shape Ratio
1	Receiving	X						X	X													
2	Raw Material							X	X	X												
3	Extrusion Cutoff																			X	X	
4	Small Parts, Mach. & Weld							X	X								X		X	X		
5	Small Parts, Finish					X		X	X								X	X				
6	Alum. Pole, Weld. & Fab.												X				X		X			
7	Alum. Pole Finish					X							X				X	X				
8	Assembly Storage							X	X					X								
9	Fixture Assembly					X		X	X								X			X		
10	Shipping	X																				
11	General Office					X																
12	Maintenance																					

Figure 4.26 Constraint summary.

Cell Layouts

When details of the individual equipment locations are shown in each block of space or cell, a cell layout results. An example is given in Fig. 4.28.

The procedures for generating cellular layouts are beyond the scope of this chapter.

Implementation

The next step is to implement the layout design. The layout may be the rearrangement of an existing plant or a layout for a new building. In the first case emphasis will be on meeting customer needs during the rearrangement.

For a layout in a new building, there is a need to integrate the plant layout with construction and installation of utilities. To aid in this process a *physical infrastructure checklist* is given in Fig. 4.29.

Operations

The selected plant layout design leads into the implementation plan. This provides the steps to provide the physical capacity. Plant operations will follow. The result of the design and implementation process

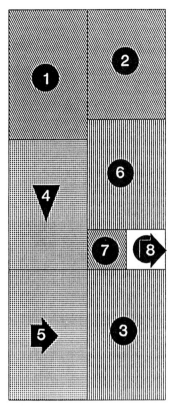

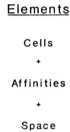

Elements

Cells

+

Affinities

+

Space

Figure 4.27 Block layout.

determines the capability of the plant to achieve the objectives set in Phase I of the project.

Summary

At the macro level, layouts using WCM principles achieve material-handling reductions of 80 to 90 percent over conventional layouts. But measurable material-handling cost reduction is perhaps the least of the advantages. The strategic gain comes from the reduction in factory throughput time, typically from days to hours or minutes.

World class layouts improve communication and teamwork while improving work flow and reducing inventory. In doing this they also support the efforts of total quality management. The factory layout can be a good focus point for introducing JIT and WCM principles.

Vanderbilt Sash Subcells

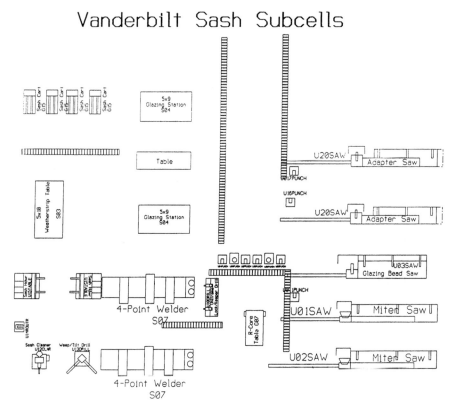

Figure 4.28 Cell layout.

For multiple sites or sites with multiple buildings, focused factories offer significant advantages over large plants with many products, processes, and markets. Studies have confirmed that, in general, a 50 percent reduction in the product range brings a 20 percent improvement in overall costs.

WCM and JIT concepts are simple enough. Execution, however, can be very difficult. This is particularly true for organizations that have evolved along functional lines.

The procedures presented in this chapter provide a range of structured approaches to help with the design of WCM facilities. The focus algorithm identifies opportunities for focused factories at the site and building level as well as at the plant and workcell level. The generic project plan structures the macro-layout process at the plant level.

PROJECT		
Description:		Proj No:
Company:		Date:
Plant:	Page: Of:	By:

Utilities

Electrical:
- ☐ Outdoor Substation
- ☐ Indoor Substation(s)
- ☐ Switchgear Room
- ☐ Motor Control Center
- ☐ Uninterruptible Power
- ☐ 115v, 1-ph
- ☐ 208/120v, 3-ph
- ☐ 230v, 1-ph
- ☐ 230v, 3-ph
- ☐ 460v, 3-ph
- ☐ 460/277v, 3-ph
- ☐ Other_____
- ☐ _____
- ☐ _____

Lighting:
- ☐ High Intensity
- ☐ Low Intensity
- ☐ Natural Light
- ☐ Other_____
- ☐ _____

Water/Sewer:
- ☐ Drinking Water
- ☐ Process Water
- ☐ Cooling Tower
- ☐ Water Treatment
- ☐ Wastewater Treatment
- ☐ _____
- ☐ _____
- ☐ _____

Steam:
- ☐ Boiler Room
- ☐ Distribution System

Fuel:
- ☐ Fuel Gas Storage
- ☐ Fuel Gas Mix Plant
- ☐ Oil Storage
- ☐ Coal Yard

Compressed Gas:
- ☐ Air Compressor
- ☐ Surge Tanks
- ☐ Distribution System
- ☐ Oxygen Storage
- ☐ CO2/NO2/Ar/He/Xe
- ☐ Other:_____
- ☐ _____
- ☐ _____

Other Utilities:
- ☐ Vacuum
- ☐ Dust Collection
- ☐ _____
- ☐ _____

Transportation

External:
- ☐ Rail Siding
- ☐ Rail Dock
- ☐ Truck Docks
- ☐ Truck Turnaround
- ☐ Truck Parking
- ☐ Truck Scale
- ☐ Truck Maintenance
- ☐ Road Access/Egress
- ☐ Airport/Heliport
- ☐ Off Site Storage

Internal:
- ☐ Pedestrian Aisles
- ☐ Pallet Truck Aisles
- ☐ Tractor/Fork Trk Aisles
- ☐ Tractor/Fork Trk Repair
- ☐ Battery Charge
- ☐ Overhead Crane
- ☐ Freight Elevators
- ☐ Scale
- ☐ Other:_____
- ☐ _____
- ☐ _____

Maintenance

Shops:
- ☐ Electrical/Electronic
- ☐ Mechanical
- ☐ Carpenter
- ☐ Tool & Die
- ☐ Yard Maintenance
- ☐ Other:_____

Storage:
- ☐ Service Parts
- ☐ Vehicles/Equipment
- ☐ Obsolete Equipment
- ☐ Other:_____

Administration

- ☐ Visitor/Reception
- ☐ Conference Rooms
- ☐ Customer Service
- ☐ Display Room
- ☐ Training Center
- ☐ Photo/Artwork

Storage:
- ☐ Archives/Tapes
- ☐ Secure Storage
- ☐ Catalogue/Display
- ☐ Microfilm/Microfiche
- ☐ Surplus Furniture/Equip.
- ☐ Other:_____
- ☐ _____
- ☐ _____

Human Resources

Administration:
- ☐ Personnel Office
- ☐ Employment Office
- ☐ Credit Union
- ☐ Library
- ☐ Training Facilities

Food Service:
- ☐ Cafeteria
- ☐ Vending Area
- ☐ Lounge/Break Room
- ☐ Executive Dining Room

Health:
- ☐ Excercise Room
- ☐ Jogging Track

Personnel:
- ☐ Toilets
- ☐ Washrooms/Showers
- ☐ Locker Room
- ☐ Water Fountains
- ☐ Public Telephones
- ☐ Employee Parking
- ☐ _____

Other:
- ☐ _____
- ☐ _____

Health And Safety

- ☐ Medical Center
- ☐ Disaster Alarm System
- ☐ Fire Egress
- ☐ Sprinkler Controls
- ☐ Fire Station
- ☐ Tornado/Fallout Shelter
- ☐ Other:_____
- ☐ _____

Production

- ☐ Production Offices
- ☐ Quality Laboratories
- ☐ Gage Calibration
- ☐ Tool Crib
- ☐ Formulation Labs
- ☐ Clean Room
- ☐ Other:_____

Storage:
- ☐ Packaging Materials
- ☐ Lubricants/Cutting Oil
- ☐ Discrepant Material
- ☐ Incidentals
- ☐ Other:_____
- ☐ _____

Notes:

The Leawood Group (c) 1988 92371

Figure 4.29 Physical infrastructure checklist.

Further Reading

Hodson, W. K.: *Maynard's Industrial Engineering Handbook,* 4th ed., McGraw-Hill, New York, 1992.

Lee, Quarterman: "Layout for the Just-In-Time Plant," *National Plant Engineering and Maintenance Conference Proceedings,* Chicago, IL, March 25 1986.

Lee, Quarterman: "Manufacturing Focus—A Comprehensive View," *Manufacturing Strategy, Technical & Medical Publishers* (1990, ed. Voss), Chapman & Hall Scientific, London, June 1990.

Lee, Quarterman, and M. Niedenthal: "Layout for the Just-In-Time Plant," *National Plant Engineering and Maintenance Conference Proceedings,* Chicago, IL, March 16, 1987.

Lee, Quarterman, and W. Wrennall: "Manufacturing Focus for Strategic Advantage," *IIE Integrated Systems Conference and Society for Integrated Manufacturing Conference Proceedings,* San Antonio, TX, October 28–31, 1990.

Muther, Richard: *Systematic Layout Planning,* 2d ed., CBI Publishing, Boston, 1973.

Salvendy, Gavriel: *Handbook of Industrial Engineering,* 2d ed., Wiley-Interscience, New York, 1992.

Wrennall, W.: "Productivity of Capital: Some Benefits from Facilities Planning," *The Journal of Methods-Time Measurement,* vol. XI, no. 3, Fairlawn, NJ, 1986, pp. 2–6.

Wrennall, W.: "Facility Layout—The Key to Workflow, Cashflow and Profit," *Manufacturing Technology International,* Sterling Publications, London, 1988.

Wrennall, W.: "Productivity Strategies for the 1990's," *Operations Management Association (OMA) Conference Proceedings,* Warwick, England, June 26–27, 1990.

Wrennall, W., and M. McCormick: "A Step Beyond Computer Aided Layout," *Industrial Engineering,* Vol. 17, no. 5, May 1987, pp. 924–934.

5

Material Movement
and Storage

William Wrennall
President, The Leawood Group, Ltd.
Leawood, Kansas

Quarterman Lee
Vice President, The Leawood Group, Ltd.
Leawood, Kansas

Material movement dominates the design of many facilities. Movement adds cost, time, and complexity to manufacturing and distribution; it adds no value until the final move, delivery to the customer. The priority for facility designers, therefore, is to reduce material movement.

The minimizing of material movement requires an effective layout based on a sound manufacturing strategy. The chapter on factory design and layout explains affinities and focus. These are powerful tools which can reduce both the amount and complexity of handling.

Layout affects material flow in three ways. First, the *space-planning unit* (SPU) or cell definition sets a pattern for overall material flows. Figure 5.1 illustrates how *production mode* impacts both the intensity and complexity of intercell flows. Second, the arrangement of cells in the layout can increase or decrease particular route distances. Finally, the arrangement of cells sets a large-scale flow pattern (or nonpattern in some cases).

Material flow analysis (MFA) examines the movement of materials over time. It helps develop affinities for the layout and evaluation of layout options, and is basic to the design of material-handling systems. Unlike other reasons for affinities, material flow is tangible

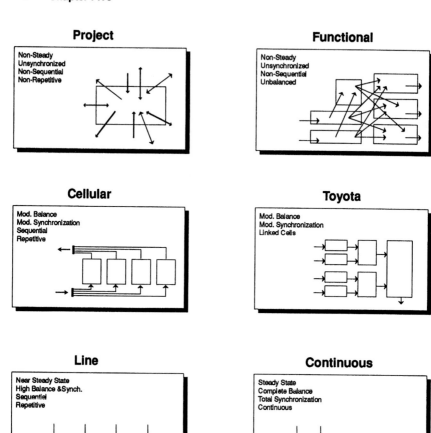

Figure 5.1 Material flow patterns.

and measurable. The use of quantitative methods adds rigor to the facility layout planning process.

After rigorous analysis and simplification, the remaining and necessary material moves are economical and timely.

Material Flow Analysis

The objectives of MFA are to:

- Compute affinities based on material flow
- Evaluate layout options
- Assist handling-system design

Figure 5.2 Material varieties.

A macro layout of 30 cells has 435 possible material flow routes. Most facilities have many materials, processes, handling methods, and multiple movement paths with flows in both directions. Figure 5.2 illustrates some of the possible variety. Then there is the variety of handling equipment, containers, route structures, and control methods. Analysis and the subsequent development of handling systems can be complex and difficult.

This chapter explains how to bring order and structure to the process. The MFA steps, shown in Fig. 5.3 are:

1. Classify materials.

2. Define flow unit.

3. Determine data source.

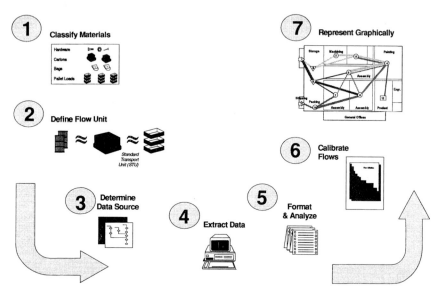

Figure 5.3   Material flow analysis.

4. Extract data.
5. Format and analyze.
6. Calibrate flows.
7. Represent graphically.

These seven steps comprise Task 2.13 on the generic project plan (GPP) of Chap. 4, "Factory Layout and Design." They give the layout planner an understanding of the material flows in the facility. The calibrated flows are used to develop affinity ratings. These initial steps are also the basis for subsequent evaluation of layout options and material-handling system design.

Step 1—Classify materials

Most manufacturing and warehouse operations have large varieties of products and materials. Situations with 10,000 or more distinct items are not unusual. To analyze flow or design a material-handling system around so many individual items is not practical. Classification reduces materials to a manageable number of items. Classes are then the basis for determining flow rates, containers, and handling equipment.

Description:	New Layout	Project:	93115
Company:	XYZ Corp.	Date:	9-Mar.
Location:	New Facility	By:	QL

| Material Class | | Classification Criteria | | Typical |
Description	Class	Physical	Other	Examples
1 Electronic Assemblies - Small	A1	12" max. dimension		Drives, Fan Modules
2 Electronic Assemblies - Med.	A2	24" max. dimension May be painted		Printers, Micro Engines
3 Electronic Assemblies - Large	A3	Painted		Standard Engines
4 Printed Circuit Boards	E1	Printed circuit boards with components	ESD critical	PCBS
5 Board Level Components	E2	May be high value Fragile leads	ESD critical	Proms, Capacitors, Resistors
6 Cables	E3	Long and flexible		Power leads, Computer cables
7 Sheet Metal Large	M1	Larger than 18" May be painted		Cabinets, Doors
8 Mechanical Components	M2	Standard shapes and sizes		Brackets, Slides
9 Fasteners	M3	1.5" max. length Low risk		Screws, Nuts, Washers
10 Packing Materials	P1	Flat panels		Cartons
11 Other	Z1	Misc. items (not classified elsewhere)		

NOTES: Physical characteristics include size, weight, shape, risk, condition, etc.
Other criteria includes quantity, timing, special control, etc.

Figure 5.4   Material classification summary.

The initial classifications stratify materials for common handling methods and container design. Weight, size, shape, stackability, and special features are defining criteria. Figure 5.4 shows a material classification based on handling characteristics.

In addition, similarity in product, process sequence, or raw material is a basis for grouping items that move over the same routes.

Step 2—Identify flow units

Material flow is measured in units of material over a unit of time, and the analyst chooses appropriate units for both parameters. The time unit is usually a matter of convenience and depends largely on data availability. Typical examples are cases/hour, tons/day, pallets/week.

Selection of the material flow unit is more problematic. Where only one type of material moves, the selection is straightforward—for example, the bushel for a grain mill. But few facilities have only a single material or material type. A wide variety of size, shape, weight, and other handling characteristics must be considered, as illustrated in Fig. 5.2. For example, integrated circuits (ICs) are tiny, delicate, expensive, and highly sensitive to electrostatic discharge (ESD). The operations that use ICs also use large metal cabinets. Between these extremes is a wide range of diverse items to move.

Items with the same size may have different handling requirements and costs. A resistor and an IC are very close in size. But resistors move in bulk, in ordinary containers, and without special precautions. The individual IC is sensitive to ESD. It requires an enclosed, conductive, and expensive container. It may have a special tube or bag to further protect it. Humans may only touch it if they wear a grounded wrist strap and a conductive smock.

Individual items or materials are seldom handled separately. Most items are in boxes, tote boxes, cartons, bundles, bales, or other containers. These, then, are what need to be handled. But layout design requires a standard unit of flow. This is the *equivalent flow unit* (EFU). An EFU should be:

- Applicable to all materials and routes
- Easily visualized by the users
- Independent of the handling method

The EFU should account for weight, bulk, shape, fragility, value, special conditions, and other factors:

Weight is a common unit for most materials and is usually available in a central database.

Bulk, or more precisely, density, relates weight and size. Overall dimensions determine bulk density.

Shape impacts handling difficulty. Compact, regular shapes such as boxes stack and handle most easily. Round and irregular shapes stack with difficulty. Long, thin shapes (high aspect ratio) are difficult to handle.

Fragility—Objects may be relatively insensitive or highly sensitive to damage. This influences handling difficulty and expense. One hundred pounds of cannonball and one hundred pounds of Dresden china present very different handling tasks.

Value for a wide range of objects and materials has little influence. But high-value or special-security items such as drugs require protection from theft, damage, or loss.

Special conditions that affect material-handling difficulty and expense are stickiness, temperature, slipperiness, hazard, and ESD sensitivity.

As material moves through the production system, the handling effort, as measured by EFUs, can change drastically. For example:

- Bulk flour may be received and loaves of bread or small packets of flour may be shipped.
- Bundles of lumber may be received and furniture shipped.
- Bulk liquids and gases may be received but pharmaceutical tablets in small-dosage packs are shipped.
- Uranium ore is received and enriched uranium is shipped.
- Logs of wood are received; paper rolls are shipped.
- Bulk chemicals are received; fabric and garments are shipped.

What seems a minor change in the item sometimes brings a dramatic change in the EFUs. Figure 5.5 is a schematic flow diagram that illustrates changes in flow intensity as the material is processed for a four-drawer filing cabinet. In Fig. 5.6, a "river" diagram illustrates flow for an entire plant and all products. Notice how flow intensity increases after painting and decreases after assembly. Painted sheet metal parts are easily damaged and difficult to handle. Once assembled and packaged, the units become protected, compact, and stackable and their flow in EFUs decreases dramatically for the same quantity and weight.

When deciding on an EFU, convenience and familiarity often take precedence over accuracy. The primary purpose of this analysis is to rate flow affinities into one of four categories. Accuracy of the order of plus or minus 20 percent is therefore sufficient. For this level of accuracy, use the following procedure:

- Review potential data sources.
- Interview production and support personnel.
- Observe operations.
- Define the EFU.

    Some examples of EFUs are:

- Pallets
- Paper rolls
- Tote boxes
- Letters
- Tons of steel
- Computer cabinets

Equivalent Unit = Equiv. Pallet (E.P.)

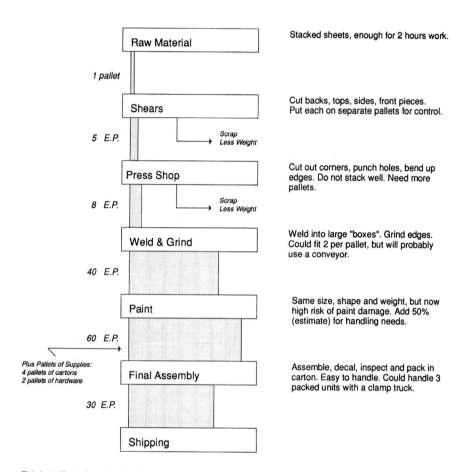

Raw Material — Stacked sheets, enough for 2 hours work.

1 pallet

Shears — Cut backs, tops, sides, front pieces. Put each on separate pallets for control.

5 E.P. — Scrap Less Weight

Press Shop — Cut out corners, punch holes, bend up edges. Do not stack well. Need more pallets.

8 E.P. — Scrap Less Weight

Weld & Grind — Weld into large "boxes". Grind edges. Could fit 2 per pallet, but will probably use a conveyor.

40 E.P.

Paint — Same size, shape and weight, but now high risk of paint damage. Add 50% (estimate) for handling needs.

60 E.P.

Plus Pallets of Supplies:
4 pallets of cartons
2 pallets of hardware

Final Assembly — Assemble, decal, inspect and pack in carton. Easy to handle. Could handle 3 packed units with a clamp truck.

30 E.P.

Shipping

This is an illustration showing the process of making a 4-drawer filing cabinet. The illustration shows the rate of flow combined with the degree of handling difficulty. The "equivalent pallets" shown here do NOT suggest that the handling must be done using pallets or fork trucks.

Figure 5.5   Equivalent unit flow analysis.

The analysis now requires factors to convert all materials into the EFU. Conversion may come from experience, work measurement, or benchmarking.

An example from an aircraft engine overhaul facility is given in Fig. 5.7. The graph converts item size to EFUs. Pallets and pallet-size containers were the most commonly moved items and the basis for most records. The equivalent pallet was, therefore, the most sen-

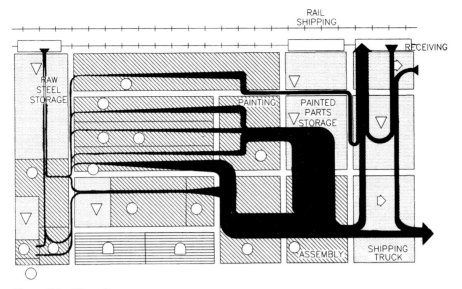

Figure 5.6   River diagram.

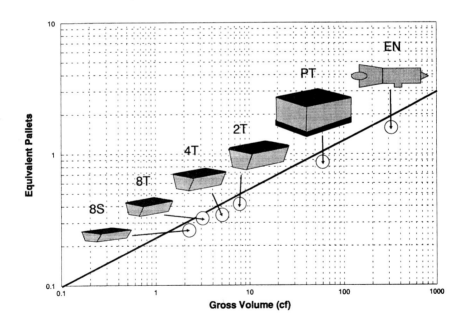

Figure 5.7   Equivalent flow unit.

sible EFU. The pallet container labeled "PT" is 1.0 EFU on the vertical scale. Its volume is 60 cubic feet on the horizontal scale.

In this system, volume is the basis for EFU conversion. Several tote pans of different sizes are labeled "2T," "4T," "8T," and "8S." An assembled jet engine has a volume of about 290 cubic feet and an EFU value of 1.6 equivalent pallets. Note that the relationship between volume and EFU is logarithmic rather than linear. This correlation is not unusual. Jet engines are 4.8 times the bulk of a pallet. On dollies they require only a small tow tractor or forklift to move. The cost and effort is about 1.6 times that for moving a pallet load.

Additional factors can affect the logarithmic volume relationship. This accounts for differences in density, shape, or other handling modifiers.

Work standards can be used as conversion factors. The time and cost of moving representative items is calculated and compared. These items become benchmarks for all other items in the facility. Or, the data might be the basis for a graphical relationship similar to the one illustrated in Fig. 5.7.

Step 3—Determine data source

In this step, the analyst decides on the source for flow data. The sources are many and varied. A good selection often makes the difference between a difficult or an easy analysis.

Almost every facility is unique with respect to the material data source. Products, volumes, and mix vary; practices are diverse, as are recording methods. Accuracy may be good, suspect, or demonstrably poor, and individuals who control data sources may be cooperative or protective.

This diversity necessitates extensive interviews with staff people who collect and compile the data. Here are some possible data sources:

- Process charts
- Routing sheets
- Material requirements planning (MRP) database
- Routing database
- Direct observation
- Handling records
- Work sampling
- Schedule estimates
- Informed opinions

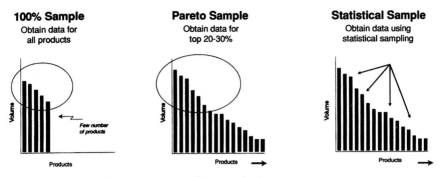

Figure 5.8 Data selection for material flow analysis.

When selecting the data source, the analyst must also decide on the range of included items. Use all items if their number is small or when computerized records make it feasible to do so. When a few products represent the largest volumes and are representative of others, use the data from the top 20 to 30 percent.

Where groups of products have similar processes and flows, a representative item might portray an entire group. When the product mix is very large and diverse, random sampling may be appropriate. Figure 5.8 illustrates data selection guidelines.

Process charts map the sequence of processes graphically; routing sheets often have much the same information in text form. With either source, examine each operation and determine in which SPU that operation will occur. This determines the route. From the product-volume analysis or other information, determine the raw flow. Then convert raw flow units to EFUs, as illustrated in Fig. 5.9. Use this procedure directly if there are only a few products. Use also where processes and flows are similar and a single item represents a larger product group. For large numbers of items, use process charts with a random sample.

Most or all of the necessary information may exist in the databases of MRP and other production and scheduling information systems. It may be necessary to change the data to a format suitable for flow analysis.

Normally, the analyst will download the information to a personal computer (PC). Spreadsheets and database managers can then manipulate it. The success of this approach depends greatly on the knowledge, resourcefulness, and cooperation of those who manage the large system as well as on the analyst's skill with PC spreadsheets and databases.

Material-handling records or direct observation is a good source of data. If material is moved by fork truck, for example, and a central

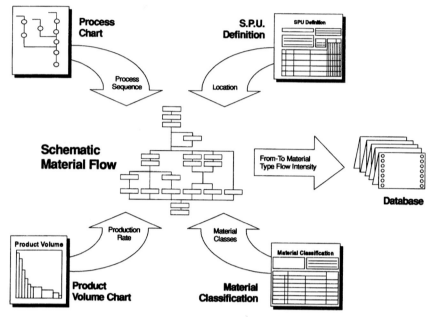

Figure 5.9   Flow data from process charts.

fork-truck pool keeps good records of pickups and deliveries, these records contain the necessary information.

In direct observation, the observer follows products through various moves and operations. In this way both process and material flow information are gathered simultaneously. The "from-to" chart of Fig. 5.10 documents flows obtained by direct observation.

Several sources may be necessary to capture all flows. For example, an MRP database may contain flows for production items but not for scrap, maintenance, trash, or empty containers. These ancillary items are often significant and sometimes dominant.

Step 4—Acquire the data

Having decided on a data source, it is now necessary to acquire the data. For computer-based data, this will usually be accessed by information services. Other data sources may require considerable clerical effort. Data derived from direct observations or work sampling may require weeks to collect and process.

Step 5—Format and analyze the data

Manual methods can suffice for the entire MFA. However, PCs reduce the time and effort and often allow more rigorous analysis.

Description:	New Layout	Project:	1234
Company:	Commercial Fixtures	Date:	1-Aug.
Location:	New Facility	By:	JMN

TO

FROM	1 Receiving	2 Raw Matl.Storage	3 Pole Plant	4 Pole Finish	5 Pole Ship	6 Fixture Fab.	7 Fixture Finish	8 WIP Storage	9 Assbly.Storage	10 KanBan Stor.	11 Fixture Assbly	12 Fixture Ship	13 R&D Shop	14 Gen.Offices	15 Chemical Storage	Totals
1 Receiving		120	12						7						6	145
2 Raw Matl.Storage			2			105				2						109
3 Pole Plant	1			12												13
4 Pole Finish					12											12
5 Pole Ship																0
6 Fixture Fab.		1					84	5		2	1		1			94
7 Fixture Finish						1		4		70	14		1			90
8 WIP Storage							4			1	3					8
9 Assbly.Storage	1									4	2					7
10 KanBan Storage									1		75					76
11 Fixture Assbly.						1			1	1		42				45
12 Fixture Ship																0
13 R&D Shop						1	1									2
14 Gen.Offices																0
15 Chemical Storage	1			2			3									6
Totals:	3	121	14	14	12	108	92	9	9	80	95	42	2	0	6	607

NOTE: Basis is equivalent tote containers in thousands (000) per year

Figure 5.10   From-to chart.

PCs are particularly valuable for facilities with a wide product mix, process focus, and a complex process sequence.

Spreadsheet programs such as Lotus 1-2-3* and Excel† are suitable for most analyses. Database programs such as Dbase‡ and Paradox§ are sometimes better than spreadsheets because of their reporting and subtotal capabilities.

With computerized analysis, data can be entered as the project progresses. Initial data may come from downloaded information or manual collection and consist of product information such as names and part numbers. It may also have annual volume, weights, and routing. The analyst should consider ancillary uses for the database as well. It may assist later for handling system development or in the determination of storage areas. It might be part of a group technology (GT) study for cell design.

*Lotus 1-2-3 is a trademark of Lotus Development Corporation.
†Excel is the trademark of Microsoft Corporation.
‡Dbase is the trademark of Ashton-Tate.
§Paradox is the trademark of Borland International.

Material Flow Report

Field:

1	2	3	4	5	6	7	8
S.P.U.		**DESCRIPTION**					
From	To	From	To	Type	EFU	Pcs/Day	FIU/Day
3	6	Letter Process A	Dispatch Platform	ADC1	1	107	107
3	6	Letter Process A	Dispatch Platform	ADC3	1	10.6	10.6
3	6	Letter Process A	Dispatch Platform	DES1	1	78.6	78.6
3	6	Letter Process A	Dispatch Platform	DES3	1	21	21
3	6	Letter Process A	Dispatch Platform	ORI1	1	110	110
3	6	Letter Process A	Dispatch Platform	TRN1	1	54	54
					Subtotal:	381.2	381.2
3	9	Letter Process A	Carrier Section	DES1	1	9	9
3	9	Letter Process A	Carrier Section	DES3	1	2.3	2.3
3	9	Letter Process A	Carrier Section	TRN1	1	6	6
					Subtotal:	17.3	17.3

NOTE: Report continues for all flow routes.

Figure 5.11 Material flow report.

Figure 5.11 is an example of a Dbase used for the layout of a mail-processing facility. Data came from a series of schematic material flow charts, in turn derived from process charts, cell definitions, and a product volume (PV) analysis, as shown in Fig. 5.9.

Fields 1 and 2 are SPU numbers which define the flow path for that entry. Fields 3 and 4 are descriptors corresponding to the cell numbers, these descriptors enhancing readability. They lend themselves to easy entry using mass update. Field 5 is a type code. All mail with the same type uses the same process and follows the same route. Field 6 stands for EFUs per piece.

Field 7 is the daily volume in pieces and field 8 is the number of EFUs per day. The database manager divides field 7 by 6 and subtotals for each route and size. These subtotals are the basis for subsequent affinity ratings and for material-handling system design.

Step 6—Calibrate flows

This step includes the calculation of material flow from each route origin to each destination. It also includes conversion of calculated flows to a step-function calibration for use in layout planning. The affinity rating system can be alpha or numeric. Here we use the vowel rating convention AEIO. A neutral value, U, indicates no material flow. Affinities may exist for reasons other than flow. These nonflow reasons are discussed in the previous chapter. Chapter 4, "Factory Layout and Design," also explains how to merge flow and nonflow affinity ratings.

Transfer the total EFU flow rate for each route to a spreadsheet. Spreadsheets often have better graphing capabilities than database managers. In addition, the calculations for combining flow and nonflow affinities are easier in a spreadsheet format.

For the calibration, rank the flow rates on a bar chart as shown in Fig. 5.12. The breakpoints are a matter of judgment and should be

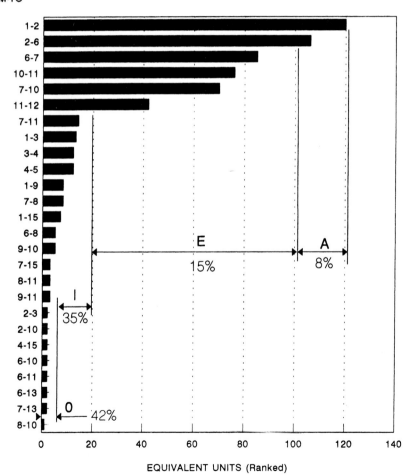

Figure 5.12 Commercial fixtures material flow analysis.

made near natural breaks. Experience from a range of projects suggests that the following proportions are a useful guideline:

A   5 to 10%
E   10 to 20%
I   20 to 40%
O   40 to 80%

An A affinity between two SPUs or cells signifies that the proximity requirement is "**A**bsolute." This is the highest positive affinity rating.

An E signifies that the proximity requirement is "**E**xceptional." I shows "**I**mportant" and O is "**O**rdinary." U ratings indicate no flow and hence neutral affinity.

Nonflow factors are seldom quantifiable. A consensus of knowledgeable persons usually rates them. Since nonflow affinities can also be negative, they have an X rating, meaning "not near" and an XX rating, "far apart." The rating and merging of nonflow affinities is explained in Chap. 4.

Transport work.   Total material-handling cost is roughly proportional to the product of flow intensity and distance. In physics, force multiplied by distance defines work. For layout planning, material flow intensity multiplied by distance defines *transport work.*

In an ideal layout, all SPUs with affinities would be adjacent. Since an SPU occupies finite space, proximity is possible, but not necessarily adjacency. Placing two particular SPUs together forces other SPUs further away. The theoretical optimum relative locations occur with equal transport work on all routes where total transport work is at the theoretical minimum. Transport work, then, is a metric for evaluating the layout.

For evaluation, transport work is calculated along every path on a layout and the summation made. Layout options may be evaluated by comparing their total transport work.

The concept of transport work is useful in another way. The graph in Fig. 5.13 plots distance on the vertical axis and flow intensity on the horizontal. Each route on the layout plots as a point. As mentioned above, the ideal layout would have *constant (or iso-) transport work,* such a curve being a hyperbola. Routes with low intensity have long distances; those with high intensity, short distance. The product of distance and intensity for either is then equal.

A "good" layout, from strictly a material flow perspective, is one which has most or all points close to the same hyperbolic iso-transport-work curve. Routes which are significantly distant from the hyperbola indicate an anomaly in the layout.

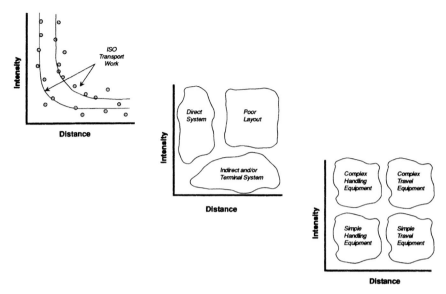

Figure 5.13   The distance-intensity plot.

Step 7—Graphical representation

Several types of charts, plots, and diagrams present material flow information visually. The visual representation assists the layout designer during the creative portion of the layout project. It helps to evaluate layout options and design the material-handling system.

Material handling and facility layout are inextricably intertwined. Layout determines the distances materials must be moved. Handling methods may affect the total handling and cost on any particular route.

Material flow diagrams and *iso-transport-work diagrams* are used to visualize material flow graphically. They show *sequence, distance, intensity,* or a combination thereof. They assist with evaluation and handling-system design. There are at least eight common types of diagrams:

- Schematic
- Quantified schematic
- Locational
- River
- String
- Three-dimensional
- D-I plot
- Animated

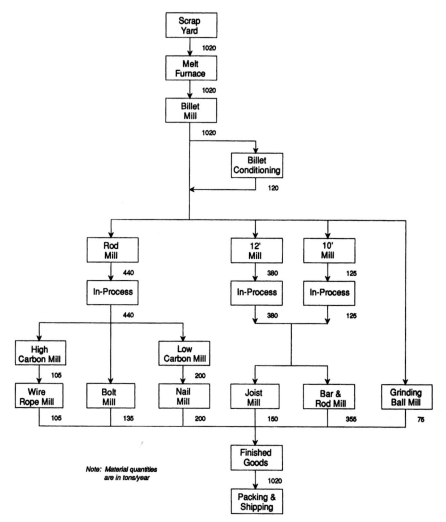

Figure 5.14   Material flow diagram (schematic).

Figure 5.14 is a *schematic diagram*. The blocks represent locations on the layout and the arrows are material move routes. In this example, a single arrow represents all materials. But different line styles or colors might show different materials, or separate diagrams might represent different material classes. Schematic diagrams are most useful in the early stages of a project where they help the analyst document, visualize, and understand the material flows.

Figure 5.15 is a *quantified schematic diagram*. In addition to routes, it illustrates flow intensity. The quantified schematic may derive from the schematic as the project progresses and data become known.

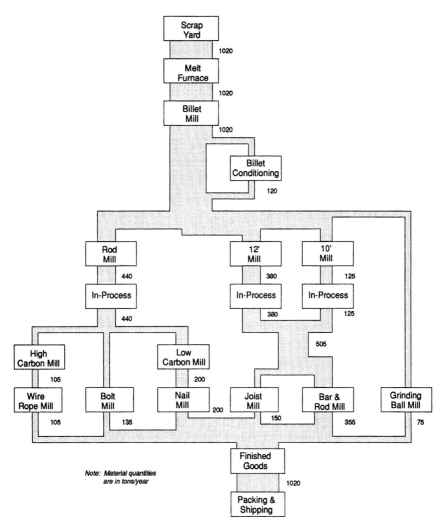

Figure 5.15 Material flow diagram (quantified schematic).

The *locational diagrams* of Figs. 5.16 and 5.17 superimpose material flows on a layout, showing the additional variable of distance. The layout may be existing or proposed. Locational diagrams illustrate the effect that layout has on flow complexity and transport work. The width of the lines is proportional to flow intensity. Colors or patterns can indicate either intensity or material classes. Alternatively, multiple lines may represent various intensities. These examples show no sequence information nor do they indicate the direction of flow. They are appropriate for complex situations.

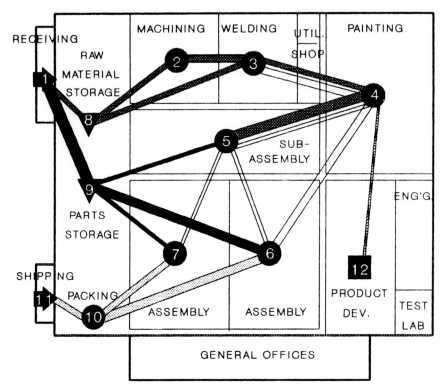

Figure 5.16 Material flow diagram (with shaded lines).

The *river diagram* of Fig. 5.18 presents sequence, intensity, and distance. It is suitable for simple flow situations with few routes and a fixed sequence.

The *string diagram,* Fig. 5.19, traces the path of individual products or items through the plant. A separate line for each item illustrates complexity and total distance. Where flow paths coincide, the lines are thicker. This diagram shows intensity and possibly sequence.

Figure 5.20 is a locational diagram *in three dimensions.* The river and string diagrams can also have three dimensions when vertical movement is important.

Computer simulation and animation software presents flow dynamically. Cinema, SLAM, GPSS, and Witness are popular packages for microcomputers. Simulation is an important tool in demonstrating and selling layouts. It can also assist in designing certain types of handling systems such as automatic guided vehicle (AGV) systems.

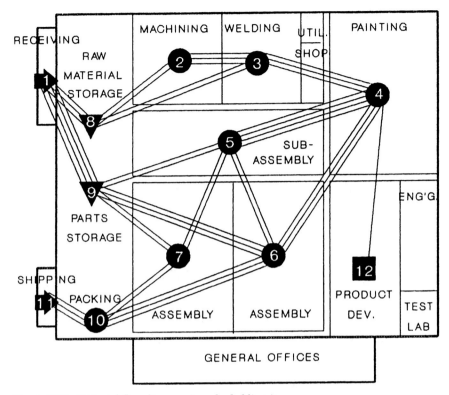

Figure 5.17 Material flow diagram (nonshaded lines).

Macro-level flow patterns

The facility layout affects sequence and characteristics of material flow. Indeed, the flow pattern dictates the shape or arrangement within a facility. Figure 5.21 shows the basic flow patterns. These are:

- Straight-through
- L-shaped
- U-shaped or circular
- Hybrids

With *straight-through flow*, or *linear flow*, material enters and exits at opposite ends of the site or building. Flow deviates little from the shortest straight-line path. Material movement is progressive. Receiving and shipping areas (entrances and exits) are physically separate.

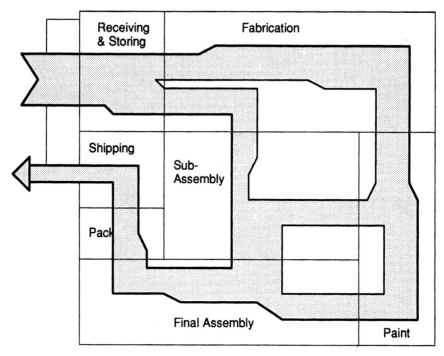

Figure 5.18   Material flow diagram (river).

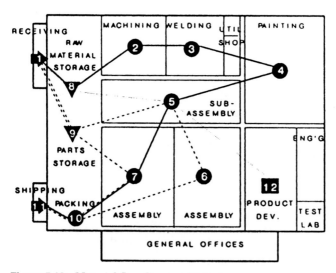

Figure 5.19   Material flow diagram (string).

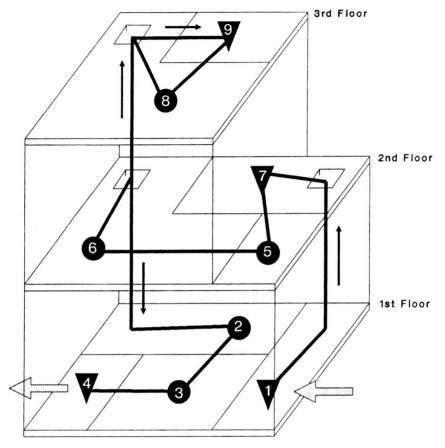

Figure 5.20 Material flow diagram (3D).

Linear flow is simple and encourages high material velocity. Operations are typically sequential. Linear flow has been a hallmark of mass production. With this type of flow, material tracking and handling are relatively simple. In fact, Henry Ford and Charles Sorensen invented the assembly line to solve a material flow problem. Straight-through flow can also be vertical movement in a high or multistory building.

L-shaped flow has a 90° directional change. This results from multiple material entry points along the flow path and a need for direct access.

U-shaped or circular flow is an extension of the L-shaped flow. The loop may be open or closed. Materials return to their starting vicinity. These patterns combine receiving and shipping docks with shared personnel and handling equipment. Conversely, one set of truck docks in a

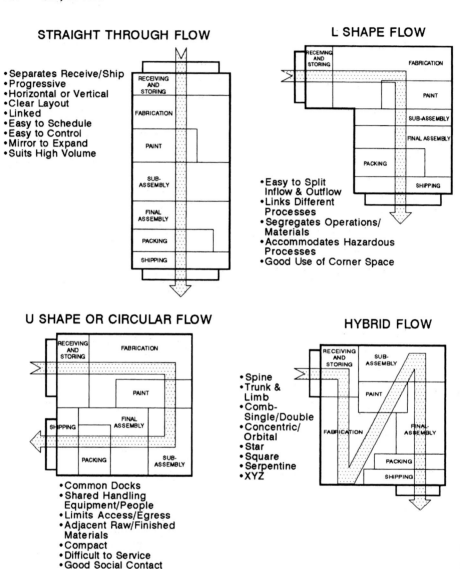

STRAIGHT THROUGH FLOW

- Separates Receive/Ship
- Progressive
- Horizontal or Vertical
- Clear Layout
- Linked
- Easy to Schedule
- Easy to Control
- Mirror to Expand
- Suits High Volume

L SHAPE FLOW

- Easy to Split Inflow & Outflow
- Links Different Processes
- Segregates Operations/ Materials
- Accommodates Hazardous Processes
- Good Use of Corner Space

U SHAPE OR CIRCULAR FLOW

- Common Docks
- Shared Handling Equipment/People
- Limits Access/Egress
- Adjacent Raw/Finished Materials
- Compact
- Difficult to Service
- Good Social Contact
- Less Easy to Expand

HYBRID FLOW

- Spine
- Trunk & Limb
- Comb- Single/Double
- Concentric/ Orbital
- Star
- Square
- Serpentine
- XYZ

Figure 5.21   Macrolayout flow patterns.

building can create a U-shaped or circular flow. This often results in morning receiving and afternoon shipping patterns.

The use of common receiving and shipping personnel is not conducive to good security. In pharmaceutical manufacturing, regulations may require strict separation of receiving and shipping facilities. Incoming material-handling, storage, and material physical characteristic differences may require different personnel skills from those required at shipping.

Hybrids, such as X, Y, Z, or star, are combinations or variations of the basic flow patterns.

Flow complexity

Simple material flow patterns have fewer routes, fewer intersections, and shorter distances. River and locational diagrams show flow complexity. We can use these to evaluate the relative complexity inherent in various layouts.

Figure 5.22 shows locational flow diagrams for two postal facility layouts. A visual comparison indicates that Fig. 5.22*a* has a more complex flow pattern than Fig. 5.22*b*. The diagrams illustrate that the flow from SPU 27 to SPU 2 is the most intense and yet has the longest distance. This is verified by a comparison of total: Transport work = 5.323 million EFUs/day for 5.22*b* versus 8.097 million EFUs/day for 5.22*a*. The option shown in 5.22*b* is superior *from a material flow perspective.*

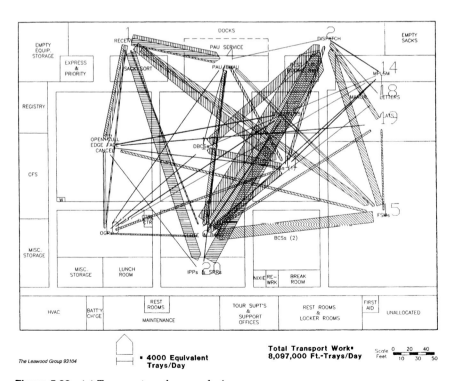

Figure 5.22   (*a*) Transport work example A.

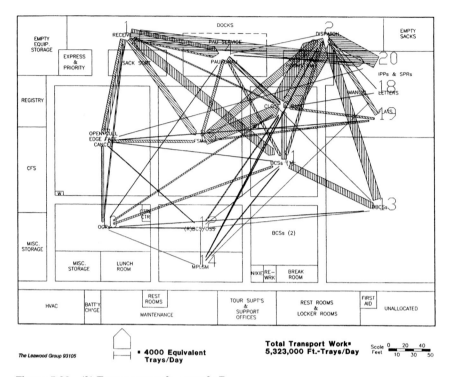

Figure 5.22 *(b)* Transport work example B.

Material-Handling System Design

Optimum material handling requires a *macro, or plantwide, system design.* The system approach examines all materials and all routes. It fits and supports the firm's manufacturing strategy and considers many options.

In the absence of a comprehensive approach, factories usually default to a simple-direct system using forklift trucks. The system designs itself and is quite convenient in that respect. However, the convenience for the designer results in a high-cost system that encourages large lots and high inventories. It seldom supports just-in-time (JIT) and world class manufacturing (WCM) strategies.

The macro-level handling plan specifies the route, container, and equipment for each move. It then accumulates the total moves by equipment type and calculates equipment requirements. To prepare a handling plan:

1. Assemble flow analysis output
   a. Routes and intensities
   b. Material flow diagrams
   c. Layout(s)

2. For each route and material class select

    a. Container
    b. Route structure
    c. Equipment

3. Calculate equipment requirements
4. Evaluate and select equipment

Containers

Materials in industrial and commercial facilities move in three basic forms: *singles, bulk,* and *contained. Singles* are individual items handled and tracked as such. *Bulk materials,* liquids, and gases assume the form or shape of their container. Fine solids such as flowable powders are also bulk.

In *containerized handling,* one or more items are in or on a box, pallet, board, tank, bottle, or other contrivance. The container restrains the items within, and, for handling purposes, the container then dominates.

Some materials take several forms. Rivets and screws can move on belt conveyors almost like powders. Later in the process, handling may be individually or in containers.

Containers offer several advantages. They:

■ Protect the contents

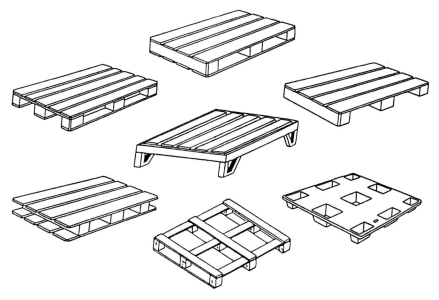

Figure 5.23 Pallet types.

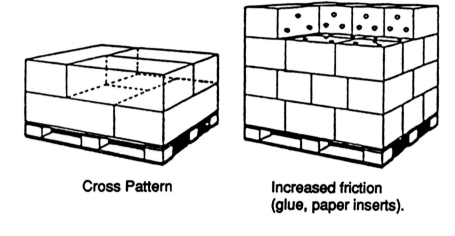

Cross Pattern   Increased friction
                              (glue, paper inserts).

Load strapped to pallet   Stretch or shrink wrapping
(or rubber band around
top-tier).

Figure 5.24 Stabilizing pallet loads.

- Improve handling attributes
- Standardize unit loads
- Assist inventory control
- Assist security

Pallet and palletlike containers have in the past been the most wide-ly used. In many industries they still are. Figure 5.23 shows some of

Figure 5.25   Pallet-size containers.

the pallet types available and their key features. Figure 5.24 shows how to stabilize a load. Figure 5.25 exhibits some palletlike containers.

"Tote pans" and "shop boxes" have evolved into sophisticated container families. They are versatile for internal and external distribution and are an important feature of Kanban systems. Because of

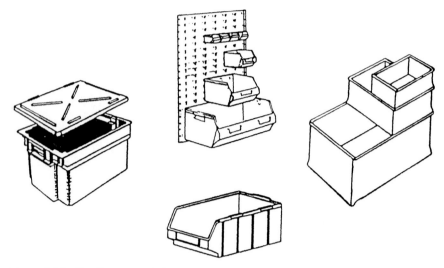

Figure 5.26 Small parts containers.

their wide use they should be standardized and selected with great care. Figure 5.26 illustrates some of the types and features.

JIT, cellular manufacturing, and time-based competition strategies require small-lot-size moves to point of use. This means smaller containers are suitable.

For broad use in large plants, a family of intermodular units is important. The International Standards Organization (ISO) and the American National Standards Institute (ANSI) have set size standards. The most popular families use 48-inch × 40-inch and 48-inch × 32-inch pallet sizes. Figure 5.27 shows one such system.

Larger-than-pallet containers are primarily for international trade. Figure 5.28 shows one type of ISO standardized unit. There is, of course, a large variety of nonstandard loads and containers.

The key to container selection is *integration*. Container, route structure, and equipment are intimately connected. They should fit with and complement each other. Other issues such as process equipment and lot size also influence selection. Unfortunately, most container selections occur by default. Certain containers preexist and new products or items get thrown into them. Existing route structures and equipment may also dictate container selection.

Manufacturing strategy should influence container selection. Conventional cost-based strategies indicate large containers corresponding to large lot sizes; contemporary strategies emphasize variety and response time. Look for the smallest feasible container corresponding to small lot sizes both for process and movement.

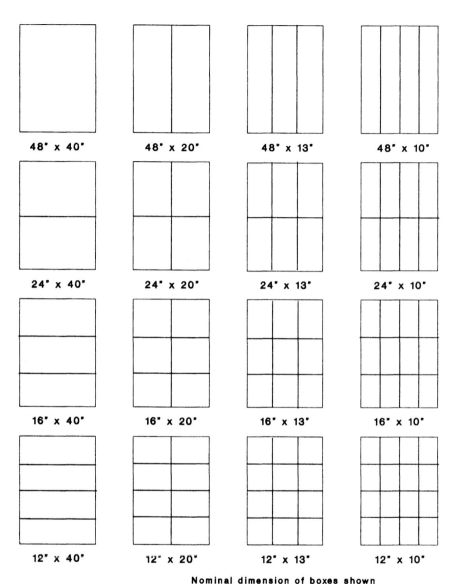

Nominal dimension of boxes shown
Allow 1/2" - 1" clearance between boxes

Figure 5.27 Boxes which fit 48-inch × 40-inch pallets.

Route structure

Route structure influences container and equipment selection. It impacts costs, timing, and other design issues.

Figure 5.29 shows the three basic route structures: direct, channel, and terminal. In a *direct* system, materials move separately and

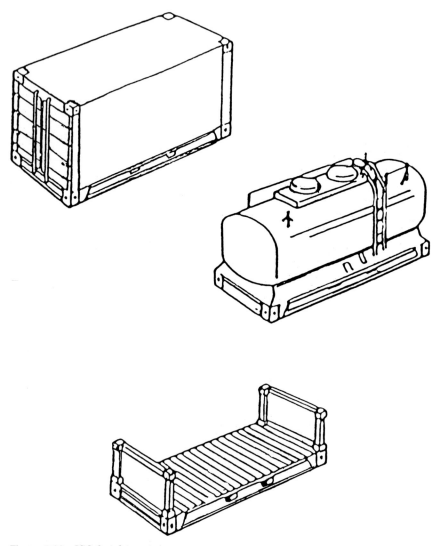

Figure 5.28 ISO freight containers.

directly from origin to destination. The *channel* system has a preestablished route and loads move along it, often commingled with other loads. *Terminal* systems have endpoints where the flow is broken. Materials may be sorted, consolidated, inspected, or transferred at these terminals. Many hybrids and variations occur in practice, as Fig. 5.30 shows.

Direct systems.   Direct systems using fork trucks are common. In operation, a pallet of material needs moving to another department;

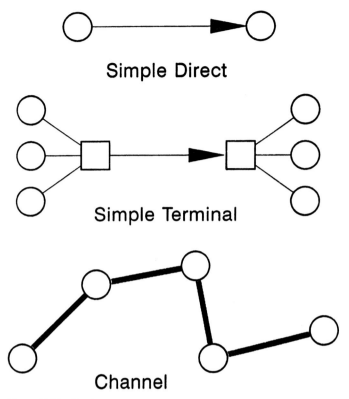

Figure 5.29   Basic route structures.

the supervisor hails a fork-truck driver who moves it to the next department. An analogy for a direct system is how taxis operate, taking their customers directly from one location to another without fixed routes or schedules.

Direct systems are appropriate for high flow intensities and full loads. They also have the least transit time and apply when time is a key factor, provided there is no queuing for transit requests.

Channel systems.   Channel systems use a predetermined path and schedule. In manufacturing, some AGV systems work this way. Trailer trains and industrial trucks with human operators can also fit channel systems. They follow a fixed route, often circular. At designated points they stop to load and unload whatever is destinating or originating at that point. City bus systems and subway systems use the channel structure.

Channel systems are compatible with JIT and WCM strategies. Many JIT plants need to make frequent moves of small quantities in tote boxes. They may use a channel system with electric industrial

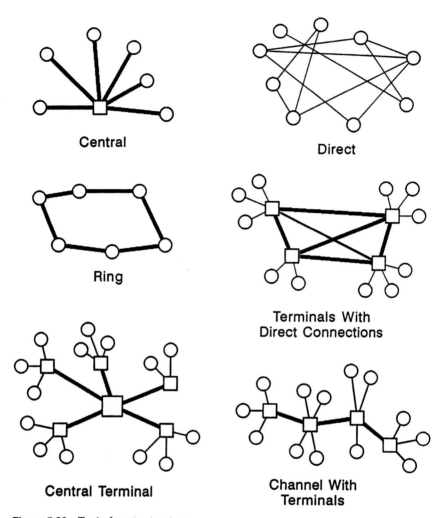

Figure 5.30 Typical route structure.

trucks or golf carts. These carts operate on a fixed route, picking up material and dropping off loads as required. Externally, over-the-road trucks make several stops at different suppliers to accumulate a full load for delivery. Simultaneously, they return Kanban signals and empty containers for refill.

Lower flow intensities, less-than-full loads, and long distances with load consolidation benefit from channel systems. Standardized loads also indicate the use of a channel system.

Terminal systems.   In terminal systems loads move from origin to ultimate destination through one or more terminals. At the terminal material is transferred, consolidated, inspected, stored, or sorted. The U.S. Postal Service, Federal Express, and United Parcel Service all use terminal systems.

A single central terminal can control material well. Multiple terminals work well with long distances, low intensities, and many less-than-full loads. Airlines use terminal systems for these reasons.

Equipment

There are hundreds of equipment types, each with different capacities, features, options, and brands. The designer chooses a type which fits the route, route structure, containers, flow intensity, and distance. These design choices should be concurrent to assure a mutual fit.

Any material move has two associated costs—terminal and travel. *Terminal cost* is the cost of loading and unloading and does not vary with distance. *Transport cost* varies with distance, usually in a direct relationship as Fig. 5.31 illustrates. Equipment is suitable for handling or transporting but seldom both.

Using the D-I plot for selection. The *distance-intensity plot* (D-I plot) is useful for equipment selection. Figure 5.32 is a representative D-I plot with iso-transport-work curves. Each route on a layout plots as a point on the chart.

Routes which fall in the lower-left area have low intensity and short distances. Typically these routes would use elementary, low-cost handling equipment such as hand dollies or manual methods.

Routes in the upper left have short distances but high intensity. These require equipment with handling and manipulating capabilities such as robots, or short-distance travel equipment such as conveyors.

Routes on the lower right have long distances and low intensities. Equipment with transport capabilities like a tractor-trailer train is needed here.

Plots in the middle area indicate combination equipment such as the forklift truck.

In the upper-right area, long distances and high intensities indicate a layout review. These situations require expensive and sophisticated equipment.

Figure 5.33 is from a recent study on handling costs. In this study, representative costs for handling pallets with several common devices were calculated. These costs included direct, indirect, and capital.

Shaded areas on the diagrams show regions where each type of equipment dominates as the lowest-cost solution. This chart is generic and may not apply to a particular situation. Nor does the chart account for intangibles such as flexibility, safety, or strategic issues.

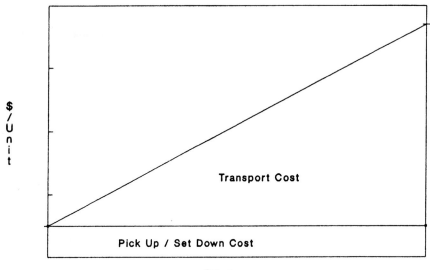

Fork Truck

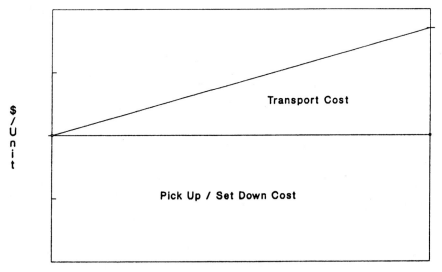

Tractor Trailer Train

Figure 5.31   Terminal/transport cost review.

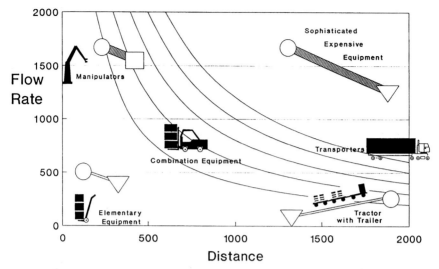

Figure 5.32   Material flow analysis: D-I plot showing iso-transport-work curves.

Using material flow diagrams for equipment selection.   Locational, river, and string diagrams also help with equipment selection. Here, the flow lines indicate distance, flow intensity, and fixed and variable paths. Figure 5.34 shows how to interpret this information.

Equipment selection guide.   The range of equipment choice is broad. There is no substitute for equally broad experience when making selections. Nevertheless, Fig. 5.35 can assist the novice to some extent.

The chart uses a modified Bolz classification system for handling equipment with a three-digit hierarchical code. The first digit represents a primary classification based on design features:

100—Conveyors. Fixed-path equipment which carries or tows loads in primarily horizontal directions.

200—Lifting equipment. Cranes, elevators, hoists, and similar equipment designed to move or position material in a primarily vertical direction.

300—Positioning / weighing / controlling. Handling equipment used for local positioning, transferring, weighing, and controlling of material movement. Included are manipulators, robots, positioning platforms, and transfers. Also included are scales and weighing equipment, float controls, bin indicators, counters, and other control devices.

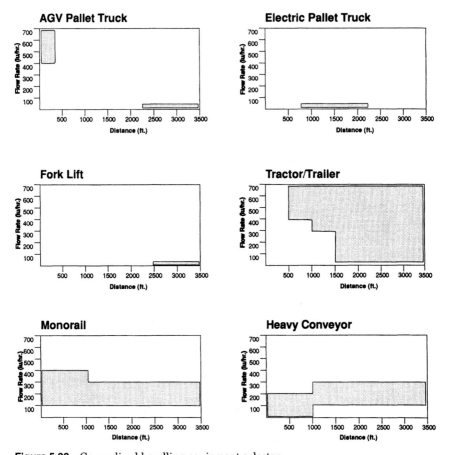

Figure 5.33 Generalized handling-equipment selector.

400—Industrial vehicles. This class includes all types of vehicles commonly used in and around industrial and commercial facilities. Excluded are "motor vehicles" intended for over-the-road use. Examples are forklift trucks, tow tractors, trailers, and excavators.

500—Motor vehicles. Highway passenger and cargo vehicles customarily used on public roads.

600—Railroad cars. All rolling stock suitable for use on national railroads. Excludes narrow-gage cars and locomotives for in-plant use.

700—Marine carriers. All waterborne vessels used on canals, rivers, oceans, and lakes.

800—Aircraft. All types of aircraft used to transport, lift, or position materials.

Short Thin Lines

Simple Handling Equipment
Analysis of Control or Admin.

Short Thick Lines

Complex Handling Equipment
Direct System of Moves

Long Thin Lines

Simple Travel Equipment
Indirect System of Moves

Long Thick Lines

Complex Travel Equipment
Analysis of Layout or Transport Unit

Many Material Classes

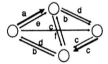

Multi-Purpose Equipment or
Single-Purpose for Combined Classes

Few Material Classes

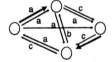

Single-Purpose Equipment or
Single-Purpose Equipment per Class

Defined Routes

Fixed-Path Equipment for
One or Few High Intensity Classes

Undefined Routes

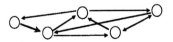

Variable-Path Equipment for
Mixed or Low Intensity Classes

Figure 5.34   Equipment selection using flow diagrams.

900—Containers / supports. Containers, platforms, pallets, coil supports, securement, bulkheads, dunnage, and related items.

The second- and third-digit positions represent finer classifications. For example, the 100 series indicates conveyors. The 110 series indicates belt conveyors. The 111 code indicates a bulk-material belt conveyor. Similarly, 422 indicates a platform hand truck.

The list gives codes that would normally be useful in a commercial or industrial facility. It updates the original Bolz system to include new types of equipment and exclude older, rarely used items.

To use Fig. 5.35, identify the load characteristics in each category across the top of the charts. These characteristics are material frequency, distance, and path. Moving down the columns, note the equipment types which match the required characteristics. The columns for equipment characteristics show additional information.

While the load characteristics shown are important, other factors impact the decision. Moreover, many types of equipment may meet the load requirements. Here are some additional criteria to narrow the selection:

- Equipment reliability
- Capital investment
- Operating expense
- Safety
- Flexibility
- Training requirements
- Maintainability
- Standardization

The final task in system design is to calculate equipment requirements. For fixed-path equipment this is straightforward. For variable-path equipment such as fork trucks, it requires estimating the time and distance on each route and utilization factors. For sophisticated systems such as AGV systems and automatic storage and retrieval systems (AS/RSs) use a computer simulation to test the feasibility.

Industrial forklift trucks (FLTs). *Forklift trucks* (FLTs) are one category of equipment which is so common, versatile, and useful that it warrants further discussion. Figure 5.36 illustrates a few varieties.

The *counterbalanced FLT* is the most universal. These trucks come with many options, and among the more important are:

Figure 5.35 Material-handling selection guide.

Group	No.	Item	Material	Frequency	Distance	Path	Power	Operator	Mobility	Selfload	Tier/Stack
100 Conveyors											
Belt	111	Bulk Material	B	CH	ELM	F	E	N	AN	YN	N
	112	Package	PT	H	LMS	F	E	N	N	N	N
	112	Metallic Belt	PT	H	LMS	F	E	N	N	N	N
	119	Other	·	·	·	·	·	·	·	·	·
Elev	121	Bucket Elevators	B	C	MS	FV	E	N	AN	Y	N
	129	Other	·	·	·	·	·	·	·	·	·
Carrier	131	Apron	PT	H	MS	F	E	N	N	N	N
	132	Slat	PT	H	MS	F	E	N	N	N	N
	133	Crossbar	PT	H	LM	F	E	N	N	N	N
	134	Carrier Chain	P	H	MS	F	E	N	N	N	N
	135	Pallet	P	H	MS	F	E	N	N	N	N
	136	Car	P	HM	LM	F	E	N	N	N	N
	137	Trolley	P	HM	LM	F	E	N	N	N	N
	138	Aerial Tramways	P	HM	EL	F	E	N	N	N	N
Haul	141	Drag	B	C	MS	FV	E	N	AN	N	N
	142	Flight	B	C	MS	F	E	N	N	N	N
	143	Tow	P	CH	LM	F	E	N	N	N	N
	144	Cable Tramways	P	HM	EL	F	E	YN	N	N	N
	145	Car Hauls	P	HM	LM	F	E	YN	N	N	N
	149	Other	·	·	·	·	·	·	·	·	·
Roller	151	Straight Roll	PT	H	MS	FV	ME	N	AN	N	N
	152	Concave Roll	PT	H	MS	FV	ME	N	AN	N	N
	153	Herringbone Roll	PT	H	MS	FV	ME	N	AN	N	N
	154	Skewed Roll	PT	H	MS	FV	ME	N	AN	N	N
	155	Troughed Roll	PT	H	MS	FV	ME	N	AN	N	N
	156	Wheel	PT	H	MS	FV	ME	N	AN	N	N
	159	Other	·	·	·	·	·	·	·	·	·
Screw	161	Horizontal	B	C	MS	FV	E	N	AN	YN	N
	162	Vertical/Inclined	B	C	MS	FV	E	N	AN	YN	N
	163	Feeders	B	CHM	MS	FV	E	N	AN	YN	N
	164	Mixers	B	CHM	MS	FV	E	N	AN	YN	N
	169	Other	·	·	·	·	·	·	·	·	·
Pipeline	171	Hydraulic	B	C	ELM	F	O	N	N	NY	N
	172	Gas	B	C	ELM	F	O	N	N	Y	N
	173	Pneumatic Bulk	B	C	ELM	F	O	N	N	Y	N
	174	Pneumatic Package	I	HM	ELM	F	O	N	N	N	N
	175	Air-Jet Hydraulic	I	HM	·	·	O	N	N	·	·
	179	Other	·	·	·	·	·	·	·	·	·
Vibrate	181	Electric	BIT	C	MS	FV	E	N	AN	YN	N
	182	Mechanical	BIT	C	MS	FV	E	N	AN	YN	N
	183	Oscillating	BIT	C	MS	FV	E	N	AN	YN	N
	184	Reciprocating	BIT	C	MS	FV	E	N	AN	YN	N
	185	Other	·	·	·	·	·	·	·	·	·
Special	191	Feeders	·	·	·	EM	N	·	YN	N	
	192	Screens	·	·	·	EM	N	·	YN	N	
	193	Hopper-Car Shakeouts	·	·	·	EM	N	·	YN	N	
	194	Chutes	·	·	·	M	N	·	YN	N	
	195	Hoppers/Troughs/Spouts	·	·	·	M	N	·	YN	N	
	196	Flumes/Sluices	·	·	·	M	N	·	YN	N	
	197	Air Cushion Transfers	PT	HML	MS	V	EO	YN	UAN	YN	N
	199	Other	·	·	·	·	·	·	·	·	·
200 Cranes/Elevators/Hoists											
Fixed	211	Pedestal	PT	ML	MS	·	EM	Y	N	N	·
	212	Pillar Jib	PT	ML	MS	·	EM	Y	N	N	·
	213	Supported Jib	PT	ML	MS	·	EM	Y	N	N	·
	214	Revolving Jib	PT	ML	MS	·	EM	Y	N	N	·
	215	Derricks	PT	ML	LMS	·	EM	Y	N	N	·
	219	Other	·	·	·	·	·	·	·	·	·
Traveling	221	Overhead Bridge	P	ML	LMS	·	EI	Y	A	N	Y
	222	Gantry Portal	P	ML	LMS	·	EI	Y	A	N	Y
	223	Gantry Storage	P	ML	LMS	·	EI	Y	A	N	Y
	224	Overhead Monorail	PT	HML	LMS	·	EI	Y	N	N	Y
	229	Other	·	·	·	·	·	·	·	·	·
Portable	231	Railroad	PT	ML	EL	F	I	Y	A	N	Y
	232	Crawler	P	ML	MS	V	I	Y	UA	N	Y
	233	Wheeled	PT	ML	LMS	V	I	Y	UA	N	Y
	234	Floor	P	ML	LMS	V	E	Y	UA	N	Y
	235	Floating	P	ML	LMS	V	I	Y	UA	N	Y
	236	Other	·	·	·	·	·	·	·	·	·
Elevators	241	Personnel	·	·	·	F	E	YN	N	N	N
	242	Self-Supported	·	·	·	F	E	YN	N	N	N
	243	Dumbwaiters	PT	HM	·	F	EM	YN	N	N	N
	244	Construction	PT	ML	·	FV	E	YN	AN	N	N
	245	Freight	P	HML	·	F	EO	YN	N	N	N
	246	Skip Hoists	·	·	·	F	E	YN	AN	·	·
	249	Other	·	·	·	·	·	·	·	·	·
Cable, etc.	251	Rope-Trolley Tautline	PT	HM	LM	FV	EI	Y	UAN	N	N
	252	Man-Trolley Tautline	PT	HM	LM	FV	EI	Y	UAN	N	N
	253	Slackline	P	HM	LM	FV	EI	Y	UAN	N	N
	254	Drag Scrapers	B	C	MS	V	EI	Y	A	N	N
	259	Other	·	·	·	·	·	·	·	·	·
200 Continued											
Hoists	261	Cylinder	IT	HM	MS	F	E	Y	N	N	N
	262	Chain	IPT	HML	MS	FV	EM	Y	A	N	NY
	263	Cable	IPT	HML	MS	FV	EM	Y	A	N	NY
	264	Rope	IPT	HML	MS	FV	EM	Y	A	N	NY
	269	Other	·	·	·	·	·	·	·	·	·
Winches	271	Windlasses	WIP	ML	MS	FV	EI	Y	AN	N	N
	272	Capstans	WIP	ML	MS	FV	EI	Y	AN	N	N
	273	Single-Drum	WIP	ML	MS	FV	EI	Y	AN	N	N
	274	Multi-Drum	WIP	ML	LMS	FV	EI	Y	AN	N	N
	279	Other	·	·	·	·	·	·	·	·	·
Aux.Lift	281	Sling	IPT	HML	MS	V	·	Y	·	·	·
	282	Grabs	BI	HML	MS	V	·	N	·	·	·
	283	Buckets	BI	HML	MS	V	·	N	·	·	·
	284	Magnets	BI	HML	MS	V	E	N	·	·	·
	285	Hook Auxiliary	·	ML	·	·	·	Y	·	·	·
	286	Boom Attachments	·	ML	·	·	·	N	·	·	·
	289	Other	·	·	·	·	·	·	·	·	·
300 Position/Weigh/Control Equipment											
Manipul	311	Manipulators	IT	HM	S	FV	EO	N	·	Y	N
	312	Electric Robots	IT	HM	S	FV	E	N	·	Y	NY
	313	Hydraulic Robots	IT	M	S	FV	O	N	·	Y	NY
	319	Other	·	·	·	·	·	·	·	·	·
Dumpers	321	Upenders/Turnovers	IPT	HML	S	FV	EM	YN	AN	YN	N
	322	Car Dumpers	W	ML	S	FV	E	Y	AN	N	N
	323	Truck Dumpers	W	ML	S	FV	E	Y	AN	N	N
	324	Box/Barrel/Bag Dump	PT	HML	S	FV	EM	YN	AN	YN	N
	329	Other	·	·	·	·	·	·	·	·	·
Table	331	Tables	IT	·	·	·	EM	·	·	·	·
	332	Platforms	IPT	·	·	·	EM	·	·	·	·
	339	Other	·	·	·	·	·	·	·	·	·
Transfers	341	Transfer Cars	P	ML	LMS	F	EMI	Y	AN	N	N
	342	Turntables	IPT	HML	S	F	EM	YN	N	YN	N
	343	Ball/Caster	P	ML	MS	FV	M	YN	AN	YN	N
	344	Walking Beam	IPT	CH	S	F	E	N	N	YN	N
	349	Other	·	·	·	·	·	·	·	·	·
Scales	371	Yard	·	·	·	·	EM	Y	N	N	·
	372	Platform	·	·	·	·	EM	NY	N	N	·
	373	Portable	·	·	·	·	M	Y	A	N	·
	374	Counter	·	·	·	·	EM	NY	·	YN	·
	375	Batch	·	·	·	·	EM	NY	N	YN	·
	376	Conveyor	·	·	·	·	E	NY	N	Y	·
	377	Crane	·	·	·	·	M	Y	·	N	·
	378	Spring	·	·	·	·	M	Y	·	N	·
	379	Other	·	·	·	·	·	·	·	·	·
400 Industrial Vehicles											
Power	410	Powered Trucks	IPT	·	LMS	V	EI	Y	U	N	Y
	411		IPT	·	LMS	V	EI	Y	U	N	N
	420	Hand Trucks	IPT	·	MS	V	M	Y	U	N	N
Hand Trucks	421	Two-Wheel	IT	·	MS	V	M	·	U	N	N
	422	Platform	·	·	MS	V	M	Y	U	N	N
	423	Wheelbarrows	·	·	MS	V	M	Y	U	N	N
	424	Carts	IT	·	MS	V	M	Y	U	N	N
	425	Dollies	IT	·	MS	V	M	Y	U	N	N
	426	Bicycles/Tricycles	IT	·	EL	V	M	Y	U	N	N
	429	Other	·	·	·	·	·	·	·	·	·
	430	Indust.Trailers	IPT	·	ELM	V	M	·	U	N	N
500 Motor Vehicles											
600 Railroad Cars/Locomotives											
700 Marine Carriers											
800 Aircraft											
900 Containers/Supports											

Material:
B Bulk
I Single Item
P Pallet/Container
T Tote/Box/Tray
L Liquid
G Gas
W Wagon/Car/Truck

Frequency:
C Continuous
H >100 Loads/hour
M >10 Loads/hours
L >1 Load/hour

Path:
F Fixed
V Variable
A Area

Distance:
S Short < 10'
M Med. < 100'
L Long < 1,000'
E Extra Long < 10,000'
O Over the Road

Power:
I Internal Combustion
S Steam
E Electrical
M Manual or Unpowered
O Other

Mobility:
N None
A Area
U Unlimited
Others are Yes / No

Three- or four-wheel

Battery-driven or internal combustion engine

Rider, stand-up, or walkie

Duplex, triplex, or quad mast

Pneumatic, cushioned, or solid tires

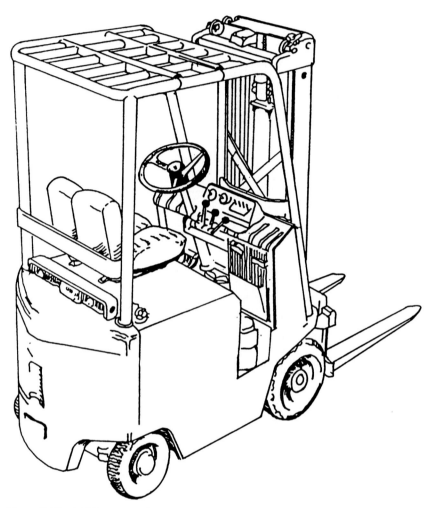

Figure 5.36 (*a*) The counterbalanced FLT.

The counterbalanced design puts a large force on the front wheels and can cause floor-loading problems. Lifting capacity diminishes with height and there is some overbalancing danger. Carton clamps, side shifters, and coil handlers are some of the available attachments.

Reach trucks have small wheels near the forward edge and on each side of the load, and thus require less counterbalancing. In operation, the forks or the entire mast extends to pick up or set down a load. The truck does not travel in this extended position. Some characteristics of reach trucks as compared with counterbalanced trucks are:

Figure 5.36   (Continued)   (b) The reach truck.

- 5 to 15 percent slower
- Nontilting forks
- Require better floors
- Smaller batteries
- Poorer ergonomics
- Lighter weight
- Work in narrower aisles

Pallet trucks are small, inexpensive machines which pick up pallets resting on the floor or on low stacks. There are manual and battery-powered models. Pallet trucks cannot handle double-faced pallets and require almost perfect floor conditions.

Stackers are small manual or electric machines similar to pallet trucks. Unlike pallet trucks, they can elevate and thus stack their loads. Outriggers or legs support the weight of the load. Outriggers

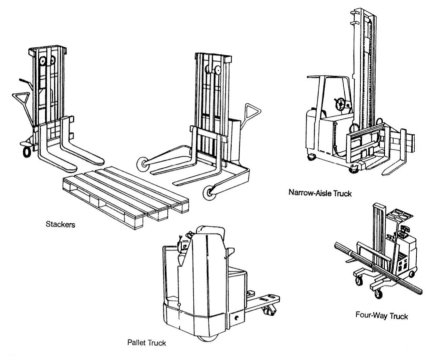

Stackers

Narrow-Aisle Truck

Pallet Truck

Four-Way Truck

Figure 5.36   (*Continued*) (c) Other FLTs.

straddle the load, while legs are underneath the forks. They are often used in maintenance or tool changing.

Four-way trucks are useful for carrying long items lengthwise through relatively narrow aisles. They are variations of the reach truck having rotatable wheels that allow them to travel in four directions.

Turret trucks have a mast that rotates on a track without extending beyond the width of the machine. They can operate in an aisle only 6 inches wider than the truck and can access pallets on both sides. They are used for high-rise storage operations.

Side-loader trucks pick up and carry the load on the side. The forks are at right angles to the travel direction. This is useful for long, narrow loads such as pipe or lumber. The side loader carries such loads lengthwise down the aisle.

Conveyors. *Conveyors* are fixed devices which move material continuously on a preestablished route. Conveyor systems range from short, simple lengths of unpowered conveyor to vast networks of conveyor sections with sophisticated controls.

Belt conveyors have a flexible belt which rides on rollers or a flat bed. The belt may be cloth, rubber, plastic, wire mesh, or other material. Most articles can ride a belt conveyor at up to a 30° inclination.

With *roller* and *skate-wheel conveyors,* objects ride on rollers or wheels. Any objects on the conveyor should span at least three sets of rollers. Movement can come from powered rollers, gravity, or operators.

Chain conveyors carry or push objects with a chain. Many varieties are available.

Overhead conveyors use an I-beam or other shape as a monorail. Carriers roll along the monorail with loads suspended underneath. A chain connects the carriers and pulls them along. In a *power-and-free system,* the chain and carriers are independent. A disconnection mechanism stops the carrier. Power-and-free systems offer more flexibility than standard monorails but at a much higher cost. Recent designs of power-and-free conveyors are inverted and floor-mounted.

Automatic guided vehicle systems. *Automatic guided vehicle systems* (AGV systems) use driverless vehicles to transport materials within an operation. AGVs can vary from small, light-duty vehicles that carry interoffice mail, to heavy-duty systems that transport automobiles during assembly. Several types of guidance are available with a range of sophistication in logic and intelligence.

Most AGVs move along a predetermined track system not unlike a model railroad. Optical tracking systems use reflective tape or paint on the floor to define the track. A photosensitive device on the vehicle detects drift from the track and actuates the steering mechanism for correction. Optical systems are inexpensive and flexible. They are sensitive to dirt, however, and many users consider them unsatisfactory.

Electromagnetic guidance systems follow a magnetic field generated by conductors laid in the floor. The frequency of this field can vary in each track section and thus identify its location. Sensors on the vehicle detect the field, its location, and perhaps the frequency. The guidance system corrects the vehicle's track accordingly. Electromagnetic systems are somewhat expensive to install or relocate. AGV owners generally prefer electromagnetic guidance systems for their reliability.

A newer type of guidance system optically reads "targets" placed high on walls and columns. The system then computes vehicle position with triangulation. In the future, guidance systems may use satellite navigation.

Figure 5.37 illustrates some of the vehicles available for AGV systems. Tractor-trailer systems use a driverless tractor to tow one or more trailers. Such systems may use manual or automatic coupling.

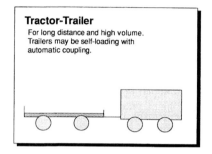

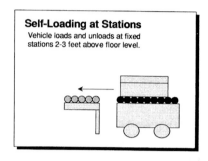

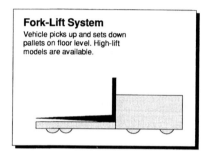

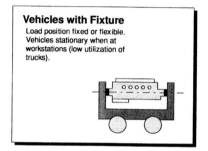

Figure 5.37   Automatic guided vehicles (AGVs).

Such systems are best for larger loads and long distances. Some vehicles serve as assembly stations in addition to moving their loads.

Self-loading vehicles stop at fixed stations and load or unload containers. These are normally pallet-size loads.

Forklift systems use vehicles similar to pallet trucks. They can pick up a pallet, carry it to a new location, and lower it automatically. All or part of the cycle may be automatic.

Special systems may use fixtures to carry engines, automobiles, or other large products through a production process.

At the lowest level of control, vehicles follow a single path in a single direction. They stop at predetermined stations, at obstructions, or when encountering a preceding vehicle. "Intelligent" trucks have preprogrammed destinations. They locate their position by sensing the magnetic frequencies. These vehicles can use multiple paths to navigate and stop at their assigned destination. Centralized control systems use a computer to track vehicles and control movement. Vehicles broadcast their current location and the computer sends control signals back to the vehicle controlling both movement and route.

Towveyors were the precursors to AGVs. They use a cable or chain for power. This cable or chain moves continuously in a floor groove. A pin or grip mechanism connects and disconnects the vehicle.

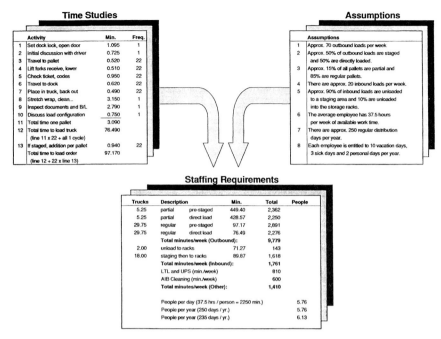

Time Studies

	Activity	Min.	Freq.
1	Set dock lock, open door	1.095	1
2	Initial discussion with driver	0.725	1
3	Travel to pallet	0.520	22
4	Lift forks receive, lower	0.510	22
5	Check ticket, codes	0.950	22
6	Travel to dock	0.620	22
7	Place in truck, back out	0.490	22
8	Stretch wrap, clean...	3.150	1
9	Inspect documents and B/L	2.790	1
10	Discuss load configuration	0.750	1
11	Total time one pallet	3.090	
12	Total time to load truck	76.490	
	(line 11 x 22 + all 1 cycle)		
13	If staged, addition per pallet	0.940	22
	Total time to load order	97.170	
	(line 12 + 22 x line 13)		

Assumptions

	Assumptions
1	Approx. 70 outbound loads per week
2	Approx. 50% of outbound loads are staged and 50% are directly loaded.
3	Approx. 15% of all pallets are partial and 85% are regular pallets.
4	There are approx. 20 inbound loads per week.
5	Approx. 90% of inbound loads are unloaded to a staging area and 10% are unloaded into the storage racks.
6	The average employee has 37.5 hours per week of available work time.
7	There are approx. 250 regular distribution days per year.
8	Each employee is entitled to 10 vacation days, 3 sick days and 2 personal days per year.

Staffing Requirements

Trucks	Description		Min.	Total	People
5.25	partial	pre-staged	449.40	2,362	
5.25	partial	direct load	428.57	2,250	
29.75	regular	pre-staged	97.17	2,891	
29.75	regular	direct load	76.49	2,276	
	Total minutes/week (Outbound):			9,779	
2.00	unload to racks		71.27	143	
18.00	staging then to racks		89.87	1,618	
	Total minutes/week (Inbound):			1,761	
	LTL and UPS (min./week)			810	
	AIB Cleaning (min./week)			600	
	Total minutes/week (Other):			1,410	
	People per day (37.5 hrs / person = 2250 min.)				5.76
	People per year (250 days / yr.)				5.76
	People per year (235 days / yr.)				6.13

Figure 5.38   Calculating requirements.

System design and documentation

When the flow analysis is complete and a layout selected, it is time to prepare the macro-level material-handling plan. This is Task 2.23 on the generic project plan (GPP) described in Chap. 4.

When the handling for each route has been identified, estimate equipment requirements. In the case of fixed-path equipment such as roller conveyors, this estimation is simple and straightforward. Where variable-path equipment is used on multiple routes, estimate the total time required for each route and material class as well as the effective equipment utilization.

In Fig. 5.38 an example estimate is shown for a bakery ingredients warehouse.

Warehousing and Storage

The most successful manufacturers and distributors now recognize that inventory usually camouflages some form of waste. The causes of inventory are in the structure of their systems. They strive to restructure and eliminate all storage of products.

Restructuring for minimum inventory is usually more fruitful than pursuing better storage methods. In real situations compromises must be made and a requirement for some storage usually exists. This section explains how to design optimum storage systems for the inventory which remains after a suitable restructuring effort.

Stores activities

Storage operations have two main functions: *holding* and *handling*. *Holding* refers to the stationing of materials in defined storage positions. *Handling* is the movement to and from the storage position. Ancillary activities such as inspection, order picking, or receiving are also part of handling.

Average turnover is the ratio of annual throughput to average inventory over the same period. Throughput and inventory may be in dollars, production units, or storage units ($, pieces, pallets, cartons).

$$\text{Turnover} = \frac{\text{annual throughput}}{\text{average inventory}}$$

The relative importance of holding and handling in a particular situation guides the analysis. With high turnover, handling dominates. With low turnover, holding dominates operations.

Handling-dominated warehouses call for detailed analysis of procedures and material handling. They use more sophisticated handling devices such as AS/RSs and automated conveyors.

Holding-dominated warehouses call for simple, inexpensive, and flexible handling equipment. They often require high-density storage methods such as drive-through racking.

Storage equipment

The types of storage equipment available are almost as diverse as the types of containers and handling equipment. The selection of storage equipment and containers is interrelated.

Analysis and design of storage systems

The design of storage systems should coordinate with the layout design of the total facility. Layout planning has four phases: site orientation, macro layout, detail layout, and implementation.

Orientation. Storage planning during this phase is at a high level. In this phase the planners are oriented to the scope of the project. This includes building size estimates, planning assumptions, project staffing, policies, and strategies.

Macro layout. This is the main planning phase where the major storage area SPUs are determined. As well as storage, these SPUs can include pick-and-pack, docks, and receiving areas although some of them may be in a separate location. The designer refines estimates of storage space and coordinates them with other design and strategic decisions.

SPU detail layouts. This is the phase where the location of each piece of equipment within each SPU is determined. Detail planning of designated storage occurs during this phase. This may or may not include the identification of storage locations for each item number. In situations where handling and picking dominate, the specific storage location may be very important for effective operation of the overall system. In other situations, assignment of storage locations is considered an operating decision.

Implementation. In the final phase, equipment is purchased and installed, and operations initiated.

The detailed storage plan should have the following:

- Detail layout
- Material-handling plan
- Equipment requirements summary
- Information systems plan
- Staffing plan

Detail layouts show the location of all racks, aisles, doors, offices, and other features. The layout should have sufficient information to prepare architectural and installation drawings.

The material-handling plan for the storage operations is similar to that made for the macro layout of any facility. It shows all originating and destinating points for materials. It shows flow rates, equipment, and containers used on each route. For many warehousing operations, the material-handling plan is simple and can be overlaid on a layout plan.

The *equipment requirements summary* lists the types and numbers of storage and handling equipment. It should also include a summary specification for each type.

The *information systems plan* specifies the type of information which is to be available, equipment required, and other data necessary to purchase equipment and set up systems. It should include manual as well as computer-supported systems.

To prepare a complete storage plan:

1. Acquire data/information.
2. Classify storage materials.
3. Calculate material and order flows.
4. Calculate storage requirements.
5. Select equipment.
6. Plan the layout.
7. Specify information procedures and systems.

Step 1—Acquire data/information.   Information required for the storage analysis covers the following: products, volumes, inventory, orders, and current and past operations.

Products and volumes.   Information on products includes general orientation material on the types of products to be stored and any special characteristics. Include a detailed list or database with every item number. Identify products by size, brand, or other classification.

Volume information should include historical sales (or throughput) volumes for each line item or product group as well as total sales. This is often the same PV information used for facility planning and material-handling analysis. A *product profile* showing items or groups and their volumes on a ranked bar chart is useful. Forecasts by product group should be obtained or prepared.

Inventory.   Historical inventory information may be available when there is a similar existing operation. It should include average and peak inventory for each item or product group over a meaningful period. When historical information does not apply, policies or judgment must suffice. A decision to "keep two months on hand" or "maintain an average 10 turns" can help establish inventory requirements.

Orders.   *Order* refers to any withdrawal request. It may be a sales order, production order, or verbal request for incidental supplies. An *order profile* shows the average line items and line item quantity per order. It may also include weekly or seasonal order patterns. It should include forecast trends and changes. It may be necessary to identify urgency or delivery requirements for various orders.

Current and past operations.   This information includes staffing, space usage, procedures, operation sequence, equipment, policies, and any other pertinent information not included above.

Unit		Description	Factor
Large Unit	Extra Large Unit	Footprint 60-70 sq. ft. Plates, extrusions, etc. normally supported on stringers.	6
	Long Unit	Footprint 30-40 sq. ft. Plates, extrusions, weldments, etc.	3
Pallet Load	Oversized Palletload	Pallet and pallet boxes greater than normal pallet. 20-25 sq. ft.	1-1/2
	Normal Palletload	Palletloads and pallet boxes 40" x 44"	1
Small Unit	Lewis Box	Tote bins (22" x 12" x 7") used for small parts.	1/12
Equivalent Storage Unit		Normal palletload 40" x 44" x 40" 40 cubic feet	1

Source: Haganaes KE 111

Figure 5.39 Example: Classification of material.

Step 2—Classify materials. The classification of materials is similar to that used for an MFA. There may be slight differences, however, since the primary concern here is storage characteristics. Figure 5.39 shows one classification scheme. Other categories might be function, destination, work in process, finished goods, high-turnover items, and slow movers.

Step 3—Calculate material flows and order flows. Material flows for a storage operation are calculated in the same way as for any other layout. Orders are an additional parameter. Order flows in a storage operation affect the timing of an order and picking patterns.

Step 4—Calculate storage requirements. For each storage class, the storage space requirement must now be calculated. This may be done using turnover rates, existing data, or computer simulation.

Step 5—Select equipment. In a warehouse operation, handling and storage equipment are interrelated and should be selected together.

Storage Method	Required Ceiling Ht.	Space per Pallet Position	Utiliz. Factor	Accessib. Factor
Floor Stacking 5 pallets high, 4 pallets deep Long side of pallet facing the aisle Counterbalanced fork-lift truck	21' 6.3m	3.8 s.f. .40 s.m.	.60 - .70	0.05
Pallet Racks 4 pallets high Short side of pallet facing the aisle Reach truck	21' 6.3m	7.2 s.f. .65 s.m.	.85 - .90	1.00
Flow-Through Racks 4 pallets high, 10 pallets deep Short side of pallet facing the aisle Counterbalanced fork-lift truck	23' 6.8m	5.0 s.f. .45 s.m.	.65 - .75	0.10
Drive-In-Racks 4 pallets high, 5 pallets deep Long side of pallets facing the aisle Reach truck	21' 6.3m	5.5 s.f. .50 s.m.	.60 - .70	0.05
Hi-Bay Turret Truck 7 pallets high, pallet racks Short side of pallet facing the aisle Narrow aisle turret truck	37' 11.0m	3.5 s.f. .30 s.m.	.85 - .95	1.00
Hi-Bay Stacker Crane 12 pallets high, pallet racks Short side of pallet facing the aisle Stacker crane (autom. or manual)	67' 20.0m	2.3 s.f. .20 s.m.	.90 - .95	1.00

Calculations based on pallet size of 48" x 32" (1.2 x .8m) total height 48" (1.2m)
Clearance under roof 12" (.3m)

Source: Haganaes KE 114

Figure 5.40   Storing unit loads.

Handling equipment types were discussed previously. Storage equipment types will be discussed in a later section of this chapter.

Step 6—Plan the layout.   Planning a storage or warehouse layout follows the same procedure as for factory layouts. Figure 5.40 is helpful for estimating space for storage.

Step 7—Specify management/operating systems.   Figure 5.41 illustrates the external flows of material and information from and to the warehouse. Figure 5.42 traces the high-level internal flows within a typical warehousing operation. In designing the storage system, these flows should be refined and specified. Operation process charts and information flowcharts are useful for this documentation.

Information systems require specification. For simple, manual systems, a notation on the flowchart may suffice. For computer-based systems a more detailed specification will be required. Figure 5.43 illustrates the overall operation of a computerized warehouse information system.

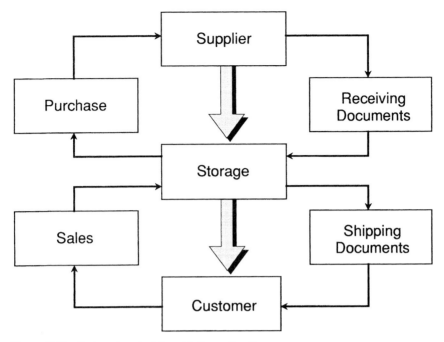

Figure 5.41   External material and information flows.

A staffing study should also be performed. This may range from an informal estimate to detailed work-measurement study. Figure 5.44 is a time standard for unloading a typical truck. This was generated using the MTM-based EASE* software system. From such time studies and throughput information, an estimate of staffing is generated.

Pallet storage

Floor stacking is the simplest way to store pallet loads. It utilizes floor space and building volume effectively and can tolerate pallets with overhang. To achieve these advantages requires stacking three to five pallets deep and hence a high storage volume per item. In the right location, with access from both sides, floor stacking can operate on a first-in–first-out (FIFO) basis. Floor stacking also requires strong, stable, stackable, and unbroken loads as illustrated in Fig. 5.45.

Typical *pallet racks* are shown in Fig. 5.46. These ordinary pallet racks are used when loads are unstackable or storage volume is too small for deep floor stacking. *Double-deep racks* are also available

*EASE is a trademark of EASE Inc.

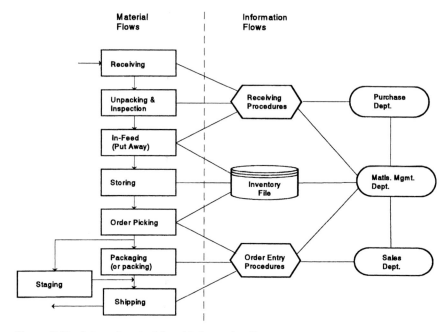

Figure 5.42 Internal material and information flows.

and achieve higher-density storage but require a reach truck, causing problems of access to the rear pallet.

Flow-through racks are used when FIFO is important and turnover is high. In these racks, the pallets or cartons ride on rollers or wheels and flow by gravity from the load to the unload end. They require high-quality pallets and the racks are relatively expensive.

Use *drive-in racks* for large quantities of unstackable pallet loads. The rack openings must be wide enough for the truck, and special narrow trucks may be necessary. Since pallets are supported only on their edges, pallet quality must be high. A limited FIFO is possible if there is access to both sides.

Small-parts storage

Small-parts storage systems are either static or dynamic. Static systems include shelving and drawers in various configurations. Dynamic systems are vertical carousels, horizontal carousels, minitrieves and movable-aisle systems.

Shelving is a basic and inexpensive storage method that is flexible. It often does not use space effectively and is costly for intensive picking operations.

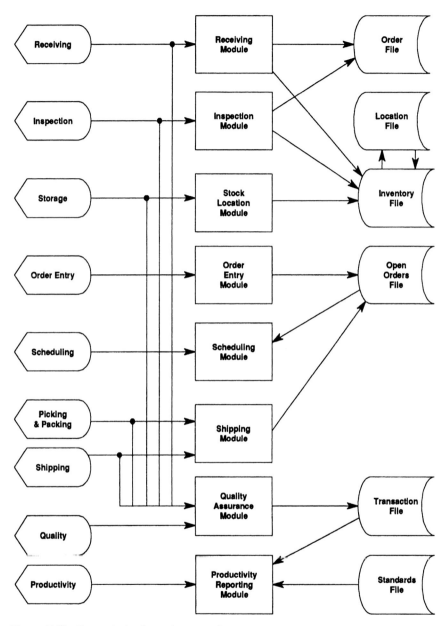

Figure 5.43 Computerized warehouse system.

Activity: Unloading a Truck

No.	Element	Freq.	Min.
1	Prepare for truck arrival	1	0.79920
2	Get pallet jack	1	0.05520
3	Lift pallets with jack (avg. load = 30 pallets)	30	5.52600
4	Walk with pallet truck (distance = 35 ft.)	30	4.41000
5	Lower pallet	30	3.94200
6	Return pallet truck unloaded	30	0.12600
	Total:		14.85840

Figure 5.44   Time standard generation.

Practical Rules

- No more than 5 pallet-spots in a row

- At least 2 rows for each article

- Rows of varying capacity

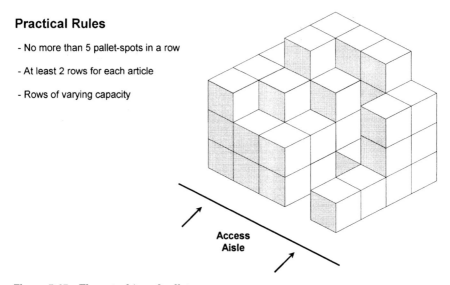

Access Aisle

Figure 5.45   Floor stacking of pallets.

Modular drawer systems offer denser storage than shelving. They are more expensive than shelves and more difficult for picking.

Automatic storage and retrieval systems

Automatic storage and retrieval systems (AS/RSs) store materials in a high-density configuration. They have a stacker-crane or other mechanical device to carry each load to its location and place it in storage. The same crane retrieves loads as required and delivers

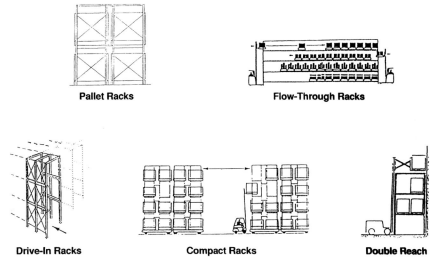

Pallet Racks   **Flow-Through Racks**

Drive-In Racks   **Compact Racks**   **Double Reach**

Figure 5.46 Pallet racks.

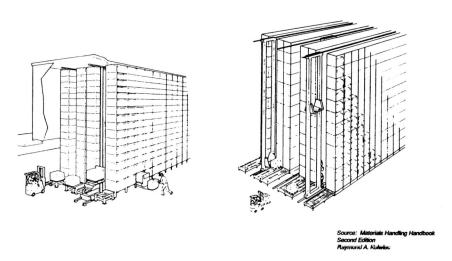

Source: *Materials Handling Handbook*
Second Edition
Raymond A. Kulwiec

Figure 5.47 Automated storage and retrieval systems.

them to an output station. A computer system controls movements
and tracks location. The AS/RS computer often is in communication
with a production control system such as MRP. Figure 5.47 shows an
example of an AS/RS installation.

Such systems usually work with pallet-size loads. Mini-trieve systems are similar in concept but use smaller loads such as tote boxes.

Conclusion

Materials movement is a key consideration in facility planning. The analysis of this material flow, MFA, is necessary for proper facility design and is a prerequisite for the design of material-handling systems and storage areas. It is also an important means of evaluating design options.

Following the step-by-step procedures outlined in this chapter will support the operating strategy by reducing costs, time, and material damage. This is basic to achieving "world class" and being a time-based competitor.

Further Reading

Allegri, T. H., Sr.: *Materials Handling Principles and Practice,* Van Nostrand Reinhold, New York, 1984.

Apple, J. M.: *Materials Handling Systems Design,* The Ronald Press, New York, 1972.

Bolz, Harold A., and George E. Hagemann: *Materials Handling Handbook,* The Ronald Press, New York, 1958.

Dock Planning Standards, Kelly Company, 1988.

Hannon, N. L.: "Layout Needs: An Integrated Approach," *Modern Materials Handling,* April 1986.

Haynes, D. O.: *Materials Handling Applications,* Chilton Company, Philadelphia, 1958.

Haynes, D. O.: *Materials Handling Equipment,* Chilton Company, Philadelphia, 1957.

Hodson, W. K.: *Maynard's Industrial Engineering Handbook,* 4th ed., McGraw-Hill, New York, 1992.

Hulett, M.: *Unit Load Handling,* Gower Press, London, 1970.

Krippendorff, H.: *Wirtschaftlich Lagern,* Wolfgang Dummer & Company, München, 1969.

Krippendorff, H.: *Integrierter Materialfluß,* Wolfgang Dummer & Company, München, 1965.

Kulwiec, R. A.: *Materials Handling Handbook,* 2d ed., Wiley-Interscience, New York, 1985.

Lee, Quarterman: "Layout for the Just-in-Time Plant," presented at *National Plant Engineering & Maintenance Conference,* March 16, 1987, Chicago IL.

Lee, Q., and M. Niedenthal: "Layout for the Just-in-Time Plant," *National Plant Engineering and Maintenance Conference Proceedings,* March 16, 1987, Chicago, IL.

Lee, Quarterman: "Computer-Aided Plant Layout with Real-Time Material Flow Evaluation," *Journal of Engineering Computing and Applications CAD/CAM Management Strategies,* vol. 1, 1988, p. 559.

Lee, Quarterman: "Quality & Productivity Trends," presented at *Latin American Quality Control Congress,* 1986, Imecca, Mexico, DF.

Lee, Quarterman: "Integrated Plant Layout for JIT & World Class Manufacturing," *IIE International Engineering Management Conference Proceedings,* Orlando, FL, October 1991.

Lee, Quarterman: "Reconfiguring a Warehouse for JIT Operations," presented at Society of Manufacturing Engineers, Schaumburg, IL, November 19–20, 1991.

Material Handling Engineering 1990–1991 Handbook and Directory, vol. 44, no. 13, Penton Publishing, Cleveland, 1990.

Morris, W. T.: *Analysis for Materials Handling Management,* Richard D. Irwin, Homewood, IL, 1962.

Moura, Reinaldo A.: *Manual de Movimentacao de Materiais,* catalogacao-na-publicacao, Brasil, 1951.

Müller, DR-ING, Thomas: *Automated Guided Vehicles,* IFS Publications, New York, 1983.

Muther, Richard: *Systematic Layout Planning,* 2d ed., CBI Publishing, Boston, 1973.

Muther, Richard, and Knut Haganas: *Systematic Handling Analysis,* Management and Industrial Research Publications, Kansas City, 1969.

Salvendy, Gavriel: *Handbook of Industrial Engineering,* 2d ed., Wiley-Interscience, New York, 1992.

Tompkins, J. A., and J. D. Smith: *The Warehouse Management Handbook,* McGraw-Hill, New York, 1988.

White, J. A.: *Production Handbook,* 4th ed., Wiley, New York, 1987.

Wrennall, W.: "The Warehouse Layout Design Process," presented at Society of Manufacturing Engineers, Schaumburg, IL, November 19–20, 1991.

Wrennall, W., and M. McCormick: "A Step Beyond Computer Aided Layout," *Industrial Engineering,* vol. 17, no. 5, May 1987, pp. 40–50.

Wrennall, W., and Q. Lee: "Achieving Requisite Manufacturing Simplicity," *Manufacturing Technology International,* Sterling Publications, London, 1989.

6

Workcell Design for World Class Manufacturing

Quarterman Lee
Vice President, The Leawood Group, Ltd.
Leawood, Kansas

Many manufacturers attempt *cellular manufacturing* as part of their integrated manufacturing system. It is an important element of just-in-time (JIT) and world class manufacturing (WCM) strategies. Cellular operations can yield large improvements in productivity, quality, and response time when compared to conventional manufacturing practices.

However, they have had mixed success. Workcells appear simple, but they are sophisticated socio-bio-technical systems. Proper functioning depends on subtle interactions of people and equipment. Each component must fit with the others in a smoothly functioning, self-regulating, and self-improving operation.

A *workcell* is a work unit larger than an individual machine or workstation but smaller than the usual department. Typically, it consists of a primary work group with several machines or operations grouped in a compact location.

Workcells often involve teams as a fundamental part of their design. Such design relies on the principles of socio-technical systems developed by Eric Trist in the 1960s. Many of the benefits of workcells derive from the "socio" side of this design.

The *cell synthesis algorithm* (CSA) (Fig. 6.1) is a step-by-step approach. It leads the cell designer through the logic required for optimum workcell design.

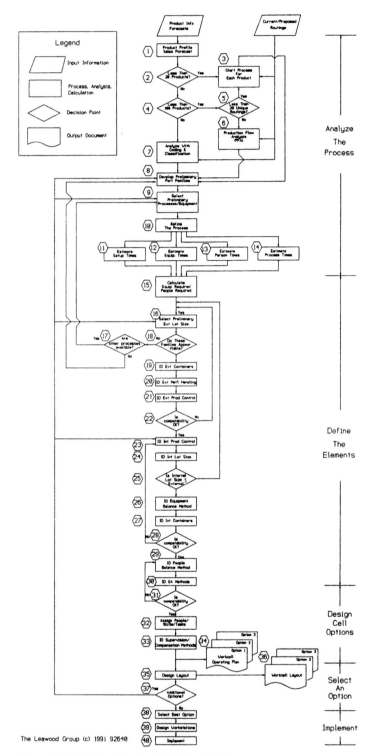

Figure 6.1 Cell synthesis algorithm (CSA).

Benefits of Work Cells

The benefits of workcells are many and varied. Properly designed workcells:

- Improve productivity
- Improve quality
- May eliminate or reduce indirect labor
- Simplify material flows
- Require less supervision
- Simplify costing systems
- Enhance members' self-esteem
- Minimize cumulative trauma disorders
- Facilitate self-determining work teams
- Provide an integratable module

Manufacturing Workcells

Key characteristics

The key characteristics of a workcell are:

- Operated by 2 to 12 persons
- Consists of 2 to 10 machines or processes
- People and processes are co-located
- Compact in size

The number of people in a workcell usually ranges from 2 to 12. This corresponds to the size of a primary group. One-person work-cells lose the benefits of mutual motivation and problem solving. Larger groups cannot coordinate themselves well and are cumbersome for discussion and problem solving.

The number of machines or processes relates to the number of people. The usual workcell has more workstations than people. This achieves maximum people utilization at the expense of equipment utilization.

Workcells should have people and machines co-located in a contiguous, compact area. Distance is important for people to communicate properly in a team environment. Electronics engineers have yet to provide a good substitute for eye contact.

Workcell types

Workcells come in many varieties. To classify and consider this wide range, we should first look at the focus or organization of the workcell. Several other chapters in this book address manufacturing focus in considerable depth.

Functional. Functional workcells have a single process or function. A heat-treat cell might have several furnaces and heat-treat several products. Operations on several products often occur in parallel. The cell may be part of a larger functionally organized facility. It might also serve several product-focused cells with a centralized skill or process. Functional cells are appropriate for large-scale processes that do not fit well into product cells.

Many products pass through the functional cell. Since pure functional cells have only a single function or operation, teamwork is often unnecessary. They are compatible with functional organization styles.

Functional workcells concentrate expertise or allow the use of large-scale equipment which serves multiple products. They provide, theoretically, better equipment utilization.

Functional workcells have serious disadvantages which often overshadow the advantages. They increase coordination difficulties and require far more inventory than product-focused cells. They do not fit well with teams and employee involvement (EI) approaches. They often inhibit total quality (TQ) approaches.

Group technology. Group technology (GT) cells process a family of similar products. They usually have multiple processes or machines arranged in a series. GT cells, like product cells, both promote and require high teamwork. They need consultative/participative organization styles. They can use either single- or multiskilled workers, depending on their design.

In practice, GT cells are nearly as advantageous as pure product cells. They have the additional advantage of flexibility for new products and product mix changes. This flexibility of GT cells duplicates some of the advantages of functional cells.

Dedicated. Dedicated workcells are highly product focused. In pure form, they manufacture one product with little or no variation. They should have all processes required to produce the product arranged in a series. Such cells demand and promote high teamwork. Such teamwork requires a participative/consultative organization style. Depending on the cell design, workers may require single or multiple skills.

Project. A project cell builds one product at a time. Each product is often unique in design and may use only a few of the processes available in the cell. A home workshop is an example of a project cell. A small tool and die shop is often a project cell.

With multiple operators, these cells require a high level of teamwork. Management within the cell requires a participative/consultative style. However, such cells often exist in a larger environment that is highly bureaucratic.

Elements of a Workcell

An optimum workcell requires engineering. As a science, workcell engineering demands a knowledge of principles. As an art, workcell engineering needs an eye for unique combinations and applications of basic principles.

Every workcell has primary elements of software and hardware. The designer should explicitly specify each element. This helps to ensure that all elements are appropriate to their purpose and consistent with each other. All too often, we see workcells where some elements, such as layout, derive from careful design and other elements come about by happenstance. We suggest the following as fundamental to the design of every workcell:

Hardware

- Product grouping/selection
- Process selection
- Containers
- Material-handling methods
- Layout

Software

- People selection/training
- Lot sizing
- Production control
- Equipment balance
- People balance
- Task assignment
- QA

The Cell Synthesis Algorithm

The Leawood Group's cell synthesis algorithm (CSA) takes a step-by-step approach to workcell design. It structures the design process and addresses all the normally important issues, including:

- Product grouping/selection
- Process selection
- Containers
- Material handling
- Layout
- People selection
- Lot sizing
- Production control
- Equipment balance
- People balance
- Task assignment
- QA

The CSA has the following elements:

- Logic flowchart
- Compatibility chart
- Forms
- Checklists
- Guides

The *logic flowchart* directs the user through a series of analyses and decisions. Users define and document the cell design, step by step.

The logic flowchart can help aspiring workcell designers structure their thought processes. Many of those faced with a cell design task have little experience with high-performance cells. They are often unfamiliar with techniques such as GT and *production flow analysis* (PFA). Those with an appropriate technical background rarely have knowledge of socio-technical systems. Yet they need socio-technical principles to structure a team's supervisory methods and compensation.

The flowchart also structures the design process. Interlinking the many workcell elements makes the cell design process complex. The selection of a particular element may constrain or dictate the choice of other elements. [For example, selection of a medium external lot size (EXLS) disallows large external containers.] This complexity, coupled with the usual lack of experience and inadequate reference material, makes a structured design process essential.

The forms and procedures in the CSA document the design as well as assisting with the design process. At completion, the designer has a layout, process charts, standard times, and an operations summary. This is often sufficient for implementation.

There are several options within each design element of a workcell. But these options are not completely independent. Selection of a large batch size, for example, limits the options for containers, material handling, and task assignment. The CSA compatibility chart, Fig. 6.2, advises the designer when elements are discordant.

At the left of the compatibility chart are the cell design elements. The chart also shows the available options for each element. For a

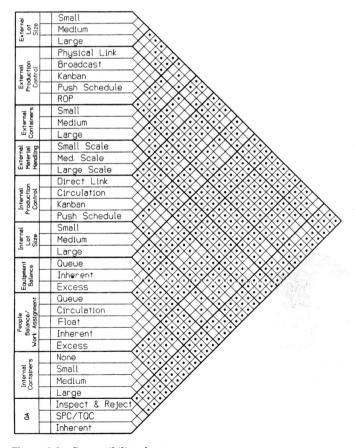

Figure 6.2 Compatibility chart.

given element, bullets on the diagonal lines indicate compatibility with other elements. The compatibility chart is valid most of the time for common situations. It is not infallible or a substitute for experience. Rather it is a guide for the novice who would use it provisionally and with caution.

As an example, a selection of Kanban for the external production control is one of five choices. Bullets to the upper right indicate that only medium and small EXLSs have compatibility with Kanban systems. The bullets to the lower right indicate that only medium and small containers have compatibility with Kanban systems. Continuing down the lower-right diagonal tells the designer which other design elements are compatible with the selection of Kanban for external production control.

In addition to the logic flowchart and compatibility chart, the algorithm includes a series of forms, guides, and checklists. These present the designer with the widest range of choices, and thus help ensure unbiased consideration of all possible solutions.

The CSA measures lot size and container size with the convention of small, medium, and large. It defines *small* as less than 0.5 hours of normal production. *Medium* represents 4.0 hours or less of normal production. *Large* is a quantity of product which requires more than 4.0 hours of production.

The output of the CSA is an *equipment list,* a *layout,* and a cell operations plan. The *workcell operations plan,* Fig. 6.3, identifies each cell element, task assignments, and people assignments. It defines how the workcell functions. The equipment list identifies the type and number of pieces of equipment. The layout shows the cell's physical arrangement. Figure 6.4 is an example.

Use of the CSA helps to assure a complete and integrated design in minimal time. With the CSA, inexperienced designers produce viable workcell designs. Their designs are, in fact, often superior to those of highly experienced manufacturing and plant layout people.

A workcell design should start with the logic flowchart of Fig. 6.1. The legend in the upper left shows symbols for information input, processes, decisions, and outputs. Each symbol on the chart represents a step in the design process. It has a number (1 to 40). Labels on the far right group the steps into broad, largely sequential tasks:

- Analyze the process.
- Define the elements.
- Design cell options.
- Select an option.
- Implement.

Description:		Proj. No.
Prototype Mfg. Cell - Flexible Cable		*1125*
Company:		Prep By:
Easy Glide Control Company		*HT*

Cell ID:	Location:	Page: of:	Date:
Brake Cable	*Main Plant*	*1 1*	*12 Jul.*

External Lot Size
- ☐ Single Piece
- ☐ Small <.5 hours
- ☐ Medium < 4 hrs.
- ▨ Large

Prod.Units: *2100 units/ship week*

External Containers
- ☐ Small
- ▨ Medium (70 each)
- ☐ Large

Type: *2100 / 70 = 30 boxes*

External Material Handling

Scale:
- ☐ Small
- ▨ Medium
- ☐ Large

Route Structure:
- ▨ Direct
- ☐ Channel
- ☐ Terminal

Equipment:
- ▨ Hand Carry
- ☐ Other

Kanban items into cell
Forklift boxes on pallets
away from cell

External Prod. Control Method
- ☐ Physical Link
- ▨ Broadcast
- ▨ Kanban
- ☐ Push Schedule
- ☐ Re-Order Point

Internal Prod. Control Method
- ▨ Direct Link
- ☐ Circulation
- ▨ Kanban
- ▨ Push Schedule

Internal Lot Size
- ▨ Single Piece
- ▨ Small
- ☐ Medium
- ☐ Large

Equipment Balance Method
- ☐ Queue
- ☐ Inherent Balance
- ▨ Excess Capacity

Internal Containers
- ▨ Small
- ☐ Medium
- ☐ Large

Type: *carts*

People Balance Method
- ☐ Queue
- ☐ Circulation
- ☐ Float
- ▨ Inherent Balance
- ☐ Excess People

Quality Assurance
- ☐ Inspect & Reject
- ☐ SPC/TQC
- ▨ Inherent

Supervision
- ▨ Self-Determination
- ☐ Kanban Signal
- ☐ Command & Control

Compensation
- ▨ Hourly / Salary
- ☐ Individual Incentive
- ☐ Group Incentive

Operator Assignments & Skill Matrix

	Skills				Operations																
Operators	Setup	QA Test	Die Cast	Assemble		Obtain Matl.	Deburr	Press & Adj.	String Cable	Press Tube	Position Adp.	Off Cast	Form Tube	Sprg & Cage	Die Cast	Clip & Assbl.	Insp.& Pack		Setup	QA Test	Runner	Make Boxes
Alan	X	X	X	X		X	X	X	X	X	X	X	X	X	X	X	X		X	X	X	X
Betty		X	X	X		X	X	X	X	X	X	X	X	X	X	X	X			X	X	X
Bonnie	X	X	X	X		X	X	X	X	X	X	X	X	X	X	X	X		X	X	X	X
Doug		X	X	X		X	X	X	X	X	X	X	X	X	X	X	X			X	X	X

The Leawood Group 92615 / 93133

Figure 6.3 Workcell operations plan.

Analyze the process

Steps 1 to 14 start with primary input data. They then examine the products and processes and make preliminary design decisions on product range, processes, and resource requirements.

Primary input data include information on all products which are initially open for possible inclusion in the cell or cells. This category

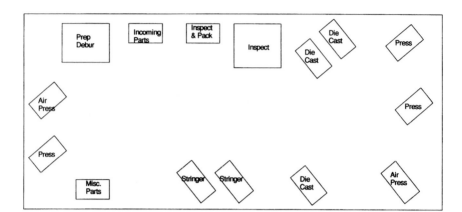

Figure 6.4   Cable cell configuration.

also includes current and prospective sales volume for each product. If the products (or similar products) are currently in production, primary input data include routings and information on current production methods.

Minimum information includes a product list with current sales, current processes, and process times. Additional information is often helpful. This may include product drawings, process charts, current lot sizes, cost information, supplier lists, plant layouts, photographs, product samples, GT codes, current container information, and alternate process information.

Experience is the best guide for determining information requirements. Too much information may bring confusion and wasted effort. Too little may result in a suboptimal design.

Selecting a product or compatible set of products for the workcell is primary. Workcells perform best when they have low variety. It follows that more workcells with narrower product ranges are better. This is generally true until decreasing utilization of equipment and space overcomes the advantages of lower variety.

Variety may or may not have significance for manufacturing. Variety in marketing terms may involve minor changes in color or options which have no impact on manufacturing. For workcell design, consider only the variety that has impact on workcell operations.

The science of GT is helpful when selecting products for workcells. PFA, coding and classification, manual classification, or instinctual classification are all effective grouping methods.

A project example

Commercial Fixtures Corporation manufactures commercial and architectural lighting fixtures. The firm wants to install one or more manufacturing workcells in its fabrication shop. We will follow the Commercial Fixtures cell design through the algorithm. The reader then sees a complete cell design.

Figure 6.5 shows the output from Step 1 for the Commercial Fixtures project. It includes a product profile and forecast. One or more cells should produce these 30 part numbers. Step 2 is a decision point. Our project has more than 20 products so we proceed to Step 4 and then to Steps 5 and 6. Figure 6.6 illustrates the PFA of Step 6.

PFA uses existing or proposed routings to group parts in similar families. Each family is then the nucleus for a workcell design.

Figure 6.6a lists the part names and numbers down the left side. Across the top are possible processes. A bullet or "X" indicates that a part visits a process. The processes across the top are not always in sequence. In Fig. 6.6b, a rearrangement of rows and columns groups processes with part families. Each machine group is a potential workcell. The third cell in Fig. 6.6b has welding as a dominant and unique process. Thus the name *weld cell* is given to this cell in Fig. 6.7.

After grouping the products and processes, the PFA analyst prepares a process chart as shown in Fig. 6.7. A single chart represents all parts in the family. A family usually has a dominant process sequence. Minor variations in sequence frequently do not matter.

Another method for developing part families is classification and coding. Figure 6.8 shows a simple coding system along with four parts

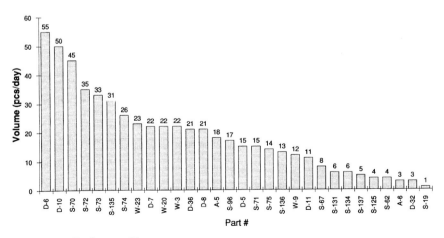

Figure 6.5 Product profile.

Description: Plant Layout	Proj. No. 123
Company: Commercial Fixtures Corp.	Prep By: MJN
Location: U.S.	Page: 1 Date: 15-Jan.

Process Area:
Small Parts Machine & Fab

Name	Matl. Part #	Spin	Drill	Shear	Weld	Saw	Assemble	Trim	Miter	Brake	Tap	Punch	Polish	Countersink	Roll	Rgh.Grind
Canopy	A-5				X		X						X			X
Shield (10")	D-10	X	X	X			X									
Base Plate	S-70	X		X			X									
Mounting Adapter	D-6		X			X		X								
Recessed Rnd. (10")	S-72	X	X	X			X									
Recessed Rnd. (8")	S-73	X		X			X				X					
Extruded Housing	S-19					X							X			X
Shallow Recess	A-6				X	X							X			X
Reflector	D-11	X	X	X			X						X			
Die Case Housing	D-32		X		X					X						
Housing Trim	W-23				X								X			
Housing Trim	W-20				X								X			
Attachment Clip	S-131			X					X			X				
T-Bar Attachment	S-134			X						X		X				
Shield (6")	S-67	X	X	X												
Mounting Box	D-5		X		X	X							X			
Recessed Sq. (10")	W-9		X		X						X		X			
Mounting Adpater	D-7		X		X	X							X			X
Face Plate	S-77	X	X	X			X									
Fixture Hanger	S-96			X									X			
Pendant Mount	S-137			X							X	X				
Yoke Mount	W-3		X		X								X			
Louver	S-136			X									X		X	X
Face Plate	S-75	X		X			X									
Suspension Hanger	S-125			X							X	X				
Recessed Sq. (10")	D-36		X		X	X					X	X				
Wall Bracket	S-135		X	X							X	X				
Surface Plate	S-74	X		X			X				X			X		
Ceiling Bracket	S-62		X	X							X	X				
Mounting Adapter	D-8		X			X	X					X				X

Figure 6.6 (*a*) Production flow analysis.

and their code numbers. In this system, each character position represents a part characteristic. The first position contains information on shape; the second position is for holes; the third and fourth positions are for overall dimensions. Practical systems have more than four positions and contain far more information than this simple illustration.

Coding systems use a database which contains every part number and its associated code. The database may also have information on processes, volumes, lot sizes, containers, and costs. Sorting and

Description:	Proj. No.
Plant Layout	123
Company:	Prep By:
Commercial Fixtures Corp.	MJN
Location:	Page:	Date:
New	1	15-Jan.

NOTE:
Grouping by common operations.

Name	Matl.	Part #	Trim	Spin	Drill/Tap/Csink	Shear	Punch	Brake	Miter Saw	Drill/Tap/Cs #2	Rgh.Grind	Polish	Drill/Tap/Cs #3	Weld	Assemble	Saw	Roll
Base Plate		S-70	X	X		X											
Face Plate (10")		S-75	X	X		X											
Surface Plate		S-74	X	X	X	X											
Face plate (6")		S-77	X	X	X	X											
Shield (10")		D-10	X	X	X	X											
Recessed Rnd. (10")		S-72	X	X	X	X											
Shield (6")		S-67		X	X	X											
Reflector		D-11	X	X	X	X					X						
Recess Rnd. (8")		S-73	X	X		X	X										
Pendant Mount		S-137							X	X	X						
Attachment Clip		S-131							X	X	X						
Suspension Hanger		S-125							X	X	X						
T-Bar Attachment		S-134							X	X		X					
Fixture Hanger		S-96							X	X							
Wall Bracket		S-135							X	X			X				
Ceiling Bracket		S-62							X	X			X				
Louver		S-136							X	X	X						X
Recessed Sq.(10")		W-9									X	X	X				
Mounting Box		D-5									X	X	X	X			
Mounting Adapter		D-7								X	X	X	X	X			
Mounting Adapter T		D-8								X		X	X	X			
Canopy		A-5	X							X	X		X				
Extruded Housing		S-19								X	X					X	
Shallow Recess		A-6								X	X		X	X			
Housing Trim		W-23								X	X						
Housing Trim		W-20								X	X						
Yoke Mount		W-3								X	X	X					
Recessed Sq. (12")		D-36								X	X	X	X				
Diecast Housing		D-32								X	X						
Mounting Adapter		D-6								X	X	X					

Figure 6.6 (b) Production flow analysis (grouped).

retrieving on various character positions is the most common method to develop part families. PC-based database managers are suitable for this. Alternatively, sophisticated software is available for large, complex situations. Step 7 in the logic flowchart represents this analysis.

For simple situations, "instinct" grouping can be by inspection. With this method, the analyst uses process charts for each part, and manually groups parts with similar processes and sequences. Prints or part samples can also serve as the basis for grouping. This is Step 3.

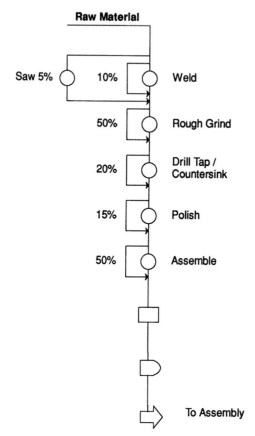

Figure 6.7   Operation process chart weld cell.

In Step 8, the analyst selects a preliminary part family. The analyst then decides on a process methodology for this part family in Step 9. This selection determines cell focus: dedicated, product, process, or project.

Process selection relates closely to product selection. Wide product ranges typically require more flexible, complex, and expensive processes than narrow product ranges. Large-scale processes require larger batches and higher volume than small-scale processes. Conversions from functional to cellular layouts are difficult. The large-scale processes used in functional layouts do not lend themselves to smaller-scale cellular operations.

After the initial process and equipment selection, the designers should review and refine the process in Step 10. To refine the process, use *work simplification* to question every aspect of the

	Digit			
Value	First	Second	Third	Fourth
1	Rotational Part	No Holes	Largest Dimen. LD < 1"	Second Largest Dimen. SLD < 1"
2	Flat Part	1 Hole at Center	Largest Dimen. 1" < LD < 5"	Second Largest Dimen. 1" < SLD < 5"
3	Box-shaped Part	1 Hole Off-center	Largest Dimen. 5" < LD < 10"	Second Largest Dimen. 5" < SLD < 10"
4	Other Main Shape	2 or More Holes	Largest Dimen. LD > 10"	Second Largest Dimen. SLD > 10"

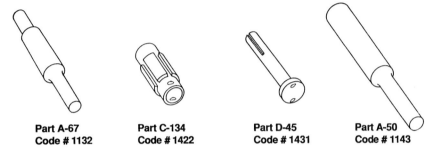

Part A-67	Part C-134	Part D-45	Part A-50
Code # 1132	Code # 1422	Code # 1431	Code # 1143

Figure 6.8 Elementary coding system.

process. For each element on the process chart ask: What? Why? Where? When? Who? and How?

What?

What is done?

Why?

What is the purpose?

Is the purpose accomplished?

Why is it necessary?

What if it were eliminated?

What would make it unnecessary?

Where?

Where is it performed?

What other locations would work?

When?

What other sequences would work?

Can it be combined with another element?

Who?

Who now performs this element?

Who else could perform it?

How?

What other methods would work?

How can we simplify this element?

Use these techniques for analyzing and improving setup times. Figure 6.9 (*a, b,* and *c*) is a tabular process chart for a setup that was reduced from 7.25 minutes to 7.1 seconds. Thirty-two process elements (*a* and *b*) were reduced to five (*c*).

Steps 11 through 15 in Fig. 6.6*a* calculate important time elements. The time required to accomplish the various manufacturing tasks, along with the expected volume, determines:

- Number of machines required
- Number of people required
- Lot sizes
- Production mode

Estimate, synthesize, or measure these times. High accuracy is not important for several reasons:

- In a workcell, several mechanisms accommodate deviations.
- A good team will constantly improve their productivity and quickly render established time standards obsolete.
- People have wide variations in ability and speed.

The CSA method uses four times for each product and each operation in the workcell:

1. *Setup time:* The basic expected time to prepare the machines, equipment, fixtures, and people to produce a particular product.

2. *Operator time:* The time required by an operator to complete an operation on one unit of product. This should include only time that the operator is working. Time spent by an operator waiting for a machine to complete its cycle is excluded.

TABULAR PROCESS CHART	Job: _National Exhaust_	By: _MJN_	Date: _2 - Feb._		Page: _1_	Of: _2_
	Product/Person: _Operator_	Dept: _Fabrication_			Start: (AM) _9:35_	End: :

Description:
Seam Rolling Machine Setup (PRESENT)

No	Description	Notes	Time (min.)	Distance	Quantity	Operation	Transport	Inspection	Delay	Storage
1	Lift machine hood		.02			●	⊃	□	D	▽
2	Select tool		.07			○	■	□	D	▽
3	Loosen adjustment rod clamp		.07			●	⊃	□	D	▽
4	Replace tool		.02			○	■	□	D	▽
5	Loosen screw		.07			●	⊃	□	D	▽
6	Turn adjustment wheel		.15			●	⊃	□	D	▽
7	Select tool		.07			○	■	□	D	▽
8	Position upper flange roller		.20			●	⊃	□	D	▽
9	Loosen set screws		.32			●	⊃	□	D	▽
10	Position lower flange roller	Difficult working position	.27			●	⊃	□	D	▽
11	Loosen set screws	Difficult working position	.42			●	⊃	□	D	▽
12	Replace tool		.03			○	■	□	D	▽
13	Turn adjustment wheel		.33			●	⊃	□	D	▽
14	Insert blank material / measure	Visual inspection[1]	.07			○	■	■	D	▽
15	Turn adjustment wheel	Adjustment plate jams	.10			●	⊃	□	D	▽
16	Replace material		.03			○	●	□	D	▽
17	Retrieve tool	Tool not located at work station	.73			○	■	□	D	▽
18	Free-up adjustment plate	Temporary fix	.33			●	⊃	□	D	▽
		Totals								

Notes:
(1) _Blank material used as gauge for determining width adjustment._

The Leawood Group (c) 1988 92044/92065-1

Figure 6.9 (*a*) Tabular Process chart showing steps 1–18.

| TABULAR PROCESS CHART | Job: National Exhaust | By: MJN | Date: 2 - Feb. | Page: 2 | Of: 2 |
| | Product (Person:) Operator | Dept: Fabrication | | Start: : | End: (AM) 9:43 |

Description:
Seam Rolling Machine Setup (PRESENT)

No	Description	Notes	Time (min.)	Distance	Quantity	Operation	Transport	Inspection	Delay	Storage
19	Insert blank material / measure	Visual inspection[1]	.16			○	●	■	D	▽
20	Turn adjustment wheel		.30			●	▷	□	D	▽
21	Select tool		.07			○	●	□	D	▽
22	Tighten adjustment rod clamp		.20			●	▷	□	D	▽
23	Replace material		.08			○	●	□	D	▽
24	Position upper flange roller		.42			●	▷	□	D	▽
25	Tighten set screws		.33			●	▷	□	D	▽
26	Position lower flange roller	Difficult working position	.62			●	▷	□	D	▽
27	Tighten set screws	Difficult working position	1.12			●	▷	□	D	▽
28	Replace tool		.02			○	●	□	D	▽
29	Lower machine hood		.07			●	▷	□	D	▽
30	Turn on machine		.08			●	▷	□	D	▽
31	Insert material blank		.18			○	●	□	D	▽
32	Inspect finished material	Visual inspection	.14			○	▷	■	D	▽
						○	▷	□	D	▽
						○	▷	□	D	▽
						○	▷	□	D	▽
						○	▷	□	D	▽
	Totals		7.09			19	12	3	5	0

Notes:

Figure 6.9 (b) Tabular Process chart showing steps 19–32.

3. *Machine time:* The time required by a piece of equipment to complete an operation on a single unit. This includes load and unload time while the equipment is not free to produce other items.

4. *Process time:* The time required for a single unit of product to complete an operation. This may include portions of load and unload

TABULAR PROCESS CHART	Job: _National Exhaust_	By: _M1N_	Date: _2 - Feb._	Page: _1_	Of: _1_
	Product/Person: _Operator_	Dept: _Fabrication_		Start: (PM) _2:13_	End: (PM) _2:13_

Description:
Seam Rolling Machine Setup (PROPOSED)

No	Description	Notes	Time (Min.)	Distance	Quantity	Operation	Transport	Inspection	Delay	Storage
1	Lift machine hood		.02			●	⟳	☐	D	▽
2	Release cam lock		.03			●	⟳	☐	D	▽
3	Turn adjustment handle		.04			●	⟳	☐	D	▽
4	Tighten cam lock		.03			●	⟳	☐	D	▽
5	Lower machine hood		.03			●	⟳	☐	D	▽
						○	⟳	☐	D	▽
						○	⟳	☐	D	▽
			Totals	0.15			5 ∅ ∅ ∅ ∅			

Notes:

The Leawood Group (c) 1988 92044/92065-3

Figure 6.9 (c) Tabular Process chart showing reduction of 32 steps to 5.

time as well as machine time. This is often called *cycle time*.

In situations where the operator must tend a machine constantly, operator time, machine time, and process time are equal. You may obtain these times from any of several sources and methods:

- Stopwatch study
- Predetermined-motion-time systems (e.g., MTM)
- Computer-based time standard systems (EASE)
- Videotapes
- NC tapes
- Experienced opinion
- Historical data
- Equipment specifications

Figure 6.10 illustrates these times.

Once calculated, insert the times into a spreadsheet similar to Fig. 6.11. This figure contains the times for the Commercial Fixtures weld cell. At the left is each part name and number determined from the PFA. The next columns show the average production in pieces per day, an EXLS, and the number of setups per day. Note that the lot size, in every case, is one day's production. This is for trial and may not be the final lot size.

Across the top are the processes: grind, polish, drill, weld, assemble, and saw. The process time per piece is below each operation. On the right are times required for setup. Note that only the drill, weld, and saw operations require setup time. On all operations, machine time, operator time, and process time are identical.

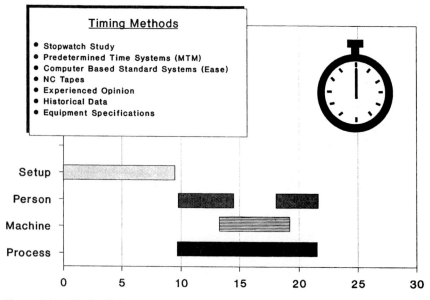

Figure 6.10   Workcell times.

(ALL TIME IN MINUTES)

NAME	PART NMBR	PCS/ DAY	EXT LOT	SU/ DAY	PROCESS TIME/PC							TIME PER SETUP			
					GRND	POL	DRLL	WELD	ASSM	SAW	TOTAL	DRLL	WELD	SAW	TOTAL
RECESSED SQ	W-9	12	12	1.00	0.00	1.30	0.98	3.75			6.03	0.50	0.36		0.86
MOUNT BOX	D-5	15	15	1.00	1.90	1.60	0.36	2.40	0.40		6.66	0.00	0.36		0.36
MOUNT ADPTR	D-7	22	22	1.00	0.36	0.36	0.42	1.75	0.40		3.29	0.00	0.36		0.36
MOUNT ADPTR(T)	D-8	21	21	1.00	0.36	0.00	0.42	1.75	0.40		2.93	0.00	0.36		0.36
CANOPY	A-5	18	18	1.00	3.75	4.56		12.80			21.11		2.00		2.00
EXTRUDED HSG	S-19	1	1	1.00	2.06	1.50				1.25	4.81		0.00	0.25	0.25
SHALLW RECESS	A-6	2	2	1.00	1.06	2.16		4.00	0.76		7.98		0.36		0.36
HSG TRIM	W-23	23	23	1.00		0.66		1.50			2.16		0.75		0.75
HSG TRIM	W-20	22	22	1.00		0.66		1.50			2.16		0.75		0.75
YOKE MOUNT	W-3	22	22	1.00		0.98	0.98	2.12			4.08	0.50	0.36		0.86
RECESS SQ(12)	D-36	21	21	1.00		1.56	0.67	2.12	0.66		5.01	0.75	0.36		1.11
DIECAST HSG	D-32	2	2	1.00			0.98	1.50			2.48	0.75	0.36		1.11
MOUNT ADPTR	D-6	55	55	1.00			0.42	1.50	0.40		2.32	0.00	0.66		0.66

					TOTAL DAILY PROCESS TIME (MIN)										
NAME	PART NMBR	PCS/ DAY	EXT LOT	SU/ DAY	DAILY PROCESS TIME/PC							DAILY SETUP TIME			
					GRND	POL	DRLL	WELD	ASSM	SAW	TOTAL	DRLL	WELD	SAW	TOTAL
RECESSED SQ	W-9	12	1	12.0	0.0	15.6	11.8	45.0	0.0	0.0	72.36	6.0	4.3	0.0	10.32
MOUNT BOX	D-5	15	1	15.0	28.5	24.0	5.4	36.0	6.0	0.0	99.90	0.0	5.4	0.0	5.40
MOUNT ADPTR	D-7	22	1	22.0	7.9	7.9	9.2	38.5	8.8	0.0	72.38	0.0	7.9	0.0	7.92
MOUNT ADPTR(T)	D-8	21	1	21.0	7.6	0.0	8.8	36.8	8.4	0.0	61.53	0.0	7.6	0.0	7.56
CANOPY	A-5	18	1	18.0	67.5	82.1	0.0	230.4	0.0	0.0	379.98	0.0	36.0	0.0	36.00
EXTRUDED HSG	S-19	1	1	1.0	2.1	1.5	0.0	0.0	0.0	1.3	4.81	0.0	0.0	0.3	0.25
SHALLW RECESS	A-6	2	1	2.0	2.1	4.3	0.0	8.0	1.5	0.0	15.96	0.0	0.7	0.0	0.72
HSG TRIM	W-23	23	1	23.0	0.0	15.2	0.0	34.5	0.0	0.0	49.68	0.0	17.3	0.0	17.25
HSG TRIM	W-20	22	1	22.0	0.0	14.5	0.0	33.0	0.0	0.0	47.52	0.0	16.5	0.0	16.50
YOKE MOUNT	W-3	22	1	22.0	0.0	21.6	21.6	46.6	0.0	0.0	89.76	11.0	7.9	0.0	18.92
RECESS SQ(12)	D-36	21	1	21.0	0.0	32.8	14.1	44.5	13.9	0.0	105.21	15.8	7.6	0.0	23.31
DIECAST HSG	D-32	2	1	2.0	0.0	0.0	2.0	3.0	0.0	0.0	4.96	1.5	0.7	0.0	2.22
MOUNT ADPTR	D-6	55	1	55.0	0.0	0.0	23.1	82.5	22.0	0.0	127.60	0.0	36.3	0.0	36.30
DAILY TIMES:		236			116	219	96	639	61	1	1132	34	148	0.3	183

WORK TIME: 1132
SETUP TIME: 183
TOTAL TIME: 1314

Figure 6.11   Weld cell production data.

Next multiply the process time per piece by pieces per day. The result shows the total minutes per day required for each part number. On the far right are the setup times for each part per day assuming a single-piece lot size. This is the most extreme case and is done as a trial.

Define the elements

Step 15 calculates the number of people and machines required for the cell. In the lower-left corner of Fig. 6.11 are total times for work and setup as well as the grand total. At Commercial Fixtures, the total time available for each worker is 420 minutes per day. The worst-case setup time will therefore require 3.1 persons. If no setups were necessary, 2.6 persons would suffice. The designers selected 3.0 persons to run the cell.

Looking at daily machine times for each operation shows that except for "weld" the other operations are well below the 420 minutes of a seven-hour shift, or day. The designers decided to provide two welding machines. Other solutions, such as overtime, were working possibilities.

In Step 16, we select a preliminary EXLS. *External lot size* is the amount of a particular product manufactured in a single run or after a single setup or changeover. Figure 6.12 shows a form which can assist. It contains data for the Commercial Fixtures D-7 adapter.

Small lots require less work in process (WIP). They smooth production, enhance quality, and improve stability of the manufacturing system. The ideal EXLS is one unit or the customer order quantity.

Several factors may prevent the use of an ideal lot size. Among these are:

- High setup cost
- Large containers
- Large-scale handling equipment
- Large-scale/high-volume processes
- Lot control
- Compensation plans

Perform the following steps for lot-size analysis using Form 92613:

1. Fill in the project header information.
2. For a workcell which produces many items, select a small number representing the range of products.
3. Calculate the *piece cost*. This is the direct cost associated with each piece or production unit. It includes direct labor, material, overhead, and other costs which do not vary with lot size. Do not include any setup or warehouse costs.
4. Calculate the *setup cost*. This includes all costs associated with a single setup. Labor, setup scrap, and setup materials are examples. Where equipment has high utilization and setup time reduces output, you may include lost profit from the lost production.
5. Calculate *storage cost*. This will include all costs associated with warehousing finished product or WIP for one year. Examples are insurance, capital charges for the building, lease payments, and labor to track and maintain the warehouse inventory. Also include the working capital charge for the product. This is the current interest rate multiplied by the piece cost. These costs are often difficult to track. You may have to estimate them.
6. After calculating the total annual warehouse cost, determine what portion of it applies to a single product unit. For example, a warehouse may contain 10,000 units. The allocated cost per unit is therefore 0.01 percent of the total annual warehouse cost. Without better information, use 30 to 40 percent of the piece cost as an estimate for total annual warehouse cost per piece.

Lot Size Analysis		Description:	Proj:
		Weld Cell	1086
Part Name:		Company:	By:
Mount Adapter		Commercial Fixtures	QL
Part No:		Location:	Date
D-7		Spring Green	21 Jan.

PIECE COST			SETUP COST	
Item	**$ Amount**		**Item**	**$ Amount**
Labor (hours x rate)	$0.58		Labor (hours x rate)	$3.81
Material / Parts	$1.38		Setup Scrap	$2.75
Overhead	$0.00		Setup Materials	$0.00
			Lost Production Profit	$0.00
Total Cost / Piece	**$1.96**		**Total Cost / Setup**	**$6.56**

ECONOMIC LOT SIZE

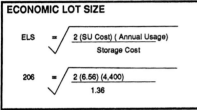

$$ELS = \sqrt{\frac{2\ (SU\ Cost)\ (Annual\ Usage)}{Storage\ Cost}}$$

$$206 = \sqrt{\frac{2\ (6.56)\ (4,400)}{1.36}}$$

STORAGE COST	
Annual Warehouse Cost	**$ Amount**
Lease / Capital Charge	$58,500
Labor (hours x rate)	$42,280
Insurance	$3,280
Maintenance	$500
Taxes	$300
Utilities	$800
Obsolescence	
Subtotal	**$105,660**
% or Whse / Piece	1.16
Piece Cost x Int. Rate	0.20
Annual Whse Cost / Piece	**$1.36**

TOTAL PIECE COST

Lot Size	Piece	Setup	Store	Total
100	1.96	0.066	0.03	2.06
206	1.96	0.032	0.06	2.05
300	1.96	0.022	0.09	2.07
400	1.96	0.016	0.12	2.10
500	1.96	0.013	0.15	2.12

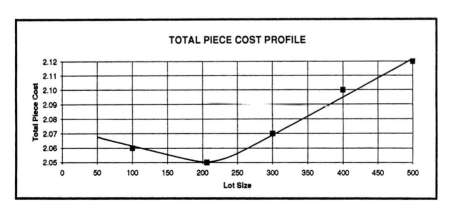

Figure 6.12 Lot size analysis.

7. From the piece cost, setup cost, and storage cost, calculate the ideal *economic lot size* (ELS). Use the formula shown in Fig 6.12.

8. Calculate the *total cost per piece* for various lot sizes. These lot sizes should range from one to about two times the ELS. Use the formula:

$$\text{Setup cost per piece} = \frac{\text{total setup cost}}{\text{lot size}}$$

$$\text{Storage cost per piece} = \frac{(\text{annual warehouse cost per piece})\,(\text{lot size})}{2\,(\text{annual usage})}$$

Total cost per piece = piece cost + setup cost per piece + storage cost per piece

9. Plot the total cost per piece. Highlight the range of viable lot sizes. Normally a total cost increase of ±5 percent represents a good range of lot sizes. External factors such as those listed above may also limit the viable range.

10. Select, as the preliminary EXLS, the smallest lot size within the viable range.

This completes your preliminary selection of EXLS. Container sizes, material-handling methods, and other factors may revise this number.

With adequate experience, you may find this detailed procedure unnecessary.

For the Commercial Fixtures weld cell, it was decided to use a lot size of one day's production—whatever the daily demand might be. In this way, downstream operations receive replenishment at least once per day.

Steps 17 and 18 are decision points which check design progress and verify decisions to this point. Step 18 checks the current design against the design criteria:

Workcell design check criteria:

- 1 to 7 persons (12 max)
- 2 to 12 workstations
- Co-located people/processes
- 70 to 90 percent utilization on high-cost equipment
- 10 to 60 percent utilization on low-cost equipment
- Moderate skill range

■ Compatible processes

If the design in progress does not meet check criteria, the designer must review process or product selection. If the cell has too many people and duplicate machines, the product selection is too wide. If the cell has low equipment utilization, the product range is too narrow for the selected process. First review alternate processes; then, if necessary, expand the product range.

Once the design criteria are satisfactory, proceed to Step 19 and select the external containers. External containers carry product to the workcell and away from it. External containers may or may not serve as containers for movement within the cell. The external container size should be equal to or smaller than the EXLS.

Most manufactured items move in containers of various types. Containers serve the following purposes:

■ Protect the contents

■ Standardize loads for automated handling

■ Improve the handling characteristics of irregular or other difficult materials

■ Serve as Kanban signals

■ Combine many small items into a larger unit load

Containers vary greatly in both size and design. Selecting no container is also an option. For cell design, think of containers as small, medium, or large. *Small* containers hold less than 0.5 hours of production. *Medium* containers hold less than 4.0 hours. *Large* containers hold 4.0 hours or more.

Containers external to the cell should be equal to or smaller than the EXLS. Internal containers may differ from external ones, particularly when external and internal lot sizes differ.

Step 20 selects the external material handling. This selection includes the route structure and equipment. Chapter 5, "Material Management and Storage," explains in detail how to make this selection and integrate it with other parts of the plantwide handling system.

In Step 21 we select the method for external production control. Production control addresses the issue of coordinating multistep processes, often with multiple products. Some regard production control issues as a question of which computerized scheduling software is best. In a broader sense, production control is the question of how to coordinate multistep processes. Computerized scheduling is only one method of doing so.

Physical linking of processes as in an automotive transfer line is one means of coordination. Here, each part in the process moves in sync. WIP inventory is nonexistent. Physical linkage requires that all processes have the same lot size. It also requires co-location of processes.

If processes are not co-located but can nevertheless operate with the same lot size, a broadcast system may work well. In broadcast, all concerned operations receive a "broadcast" of the final assembly schedule. Final assembly then builds directly to schedule. Subassembly and supply operations build in "line-set order." They use a small time offset to allow for delivery. Figure 6.13 illustrates this.

Kanban systems are useful when lot sizes differ between process steps, with unbalanced processes. They also work well when distance introduces too much time lag and/or variability. In a Kanban system, a small stock of material resides between the processes. The downstream process withdraws from this stock as needed. With each withdrawal, a signal informs the producer. The signal is authorization for the producer to replace withdrawn stock. This stock and signal system introduces a slight decoupling between the processes, allowing them to operate with some independence. Figure 6.14 shows how Kanban works throughout the process stream.

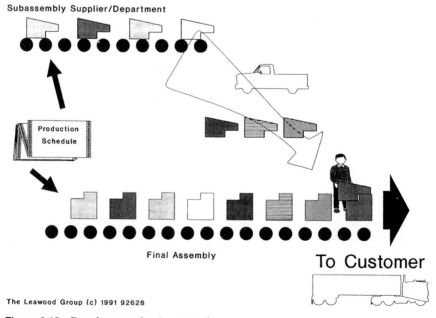

Subassembly Supplier/Department

Production Schedule

Final Assembly

To Customer

The Leawood Group (c) 1991 92628

Figure 6.13   Broadcast production control.

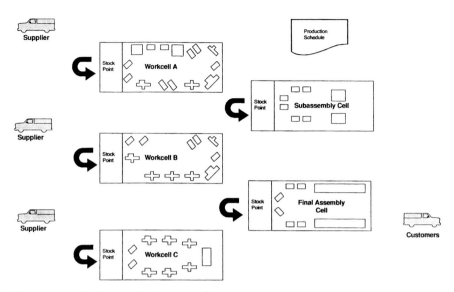

Figure 6.14   Kanban production control.

Materials requirements planning systems (MRP systems) work from bills of material (BOM), routings, and forecasts. They plan each stage of the production process for each product, subassembly, and item. These systems then accumulate production needs for each work center and each period. MRP allows planning to proceed independent of upstream or downstream operations. Figure 6.15 shows how an MRP system accumulates lead times for several production stages. MRP systems allow effective scheduling even with unconnected work centers, markets, and suppliers.

The ability to connect a disjointed production system comes with a price. It allows mediocre and amateurish manufacturing engineering. The administrative costs are high. Throughput times are long and inventory turns low. More than 50 percent of installed MRP systems do not meet their user's expectations.

Figure 6.16 illustrates *reorder point* (ROP) control. ROP systems store each required item and issue to a downstream work center on demand. This is somewhat like a Kanban system. ROP, however, signals a resupply only when the inventory is at a critically low stage. Typically these systems also use large reorder quantities.

ROP systems are simple. Under special conditions, they can be effective. They require a very steady and predictable withdrawal rate and a predictable replenishment time. Unfortunately, these conditions are rare and ROP systems operate with very high levels of inventory and frequent stockouts.

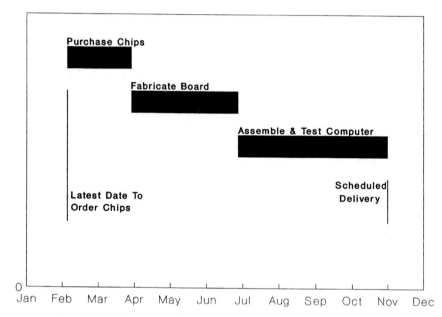

Figure 6.15 Material requirements planning.

In Fig. 6.17, flow line, broadcast, Kanban, MRP, and ROP methods of external production control form a hierarchy of desirability. When selecting the external production control method, make every effort to use flow line, broadcast, or Kanban. MRP and ROP systems are least desirable and are generally incompatible with WCM strategies.

Step 22 checks the compatibility of selections. In Step 23 the designer selects a method for internal production control. Here we decide how to coordinate operations within the cell. This internal production control method controls the sequence and timing of processes. It also controls the timing and assignment of people.

Internal production control has a different set of options. Broadcast, ROP, and MRP are not practical within a cell; Kanban and direct linkage are. Circulation becomes viable within a cell.

The circulation method of internal production control uses mobile operators. These operators physically carry or follow a workpiece or batch of product through a series of processes. If several operators are necessary, they follow each other around the cell. Circulation is more than an internal production control method. It also balances people. Since each operator performs all jobs, the work always balances.

Step 24 selects internal lot size. *Internal lot size* is the amount of product which moves as a lot between operations. *Transfer batch size*

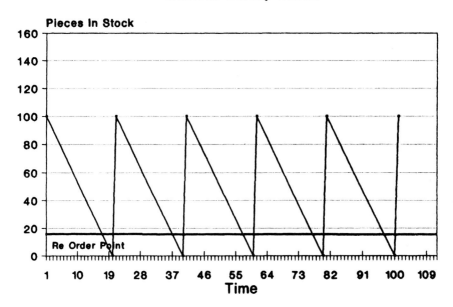

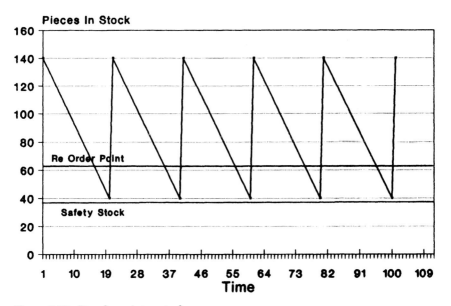

Figure 6.16   Reorder point control.

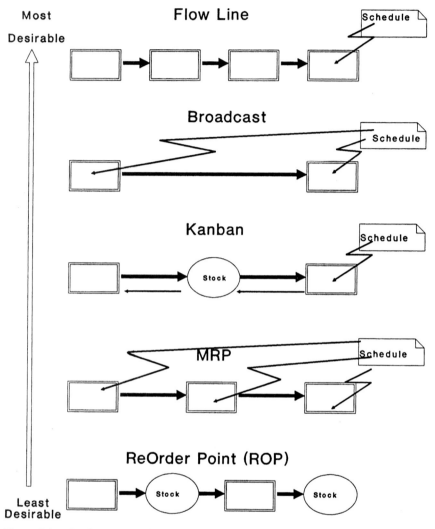

Figure 6.17 Production coordination.

is another term for internal lot size.

As with the EXLS, we select the smallest practical amount. The constraining factors here are move time, cycle time, and equipment. When the time required to move between operations nears the operation time, it indicates a larger internal lot size. Very fast processing times or large move times can lead to this.

Equipment may have a batch size and process more than one piece. In this situation, the internal lot size should be equal to or larger than the equipment lot size.

Step 25 verifies that the internal lot size is smaller than the EXLS. If this is not the case, review the external lot sizing or process selection.

In Step 26 we identify the method for balancing equipment requirements. Equipment balance distributes work to the various machines and ensures that some machines are not idle while others carry too much load. This selection closely relates to the choice of internal production control and also to internal material handling.

One approach negates the issue by providing excess equipment and workstations. While financial controllers cringe at the concept, it is often most sensible. Where equipment is inexpensive, excess capacity has advantages over elaborate scheduling and balancing methods. Often a cell has one or two items of high-cost equipment and many auxiliary items which are inexpensive. In such situations, "key machines" determine the cell capacity and secondary operations have excess capacity. This approach is also good where fast response to uncertain orders dominates.

Queuing is one equipment balance method. Small queues of product accumulate between operations. This compensates for short-term variation in process time. It cannot compensate for a chronic capacity deficit in one or more machines. This method is good for short process times and fixed task assignments. The method is suitable for mechanized handling systems such as roller conveyor.

Operations with identical process and setup times and no significant variation have inherent balance. This allows process synchronization as well.

Another compatibility check occurs in Step 29. Step 30 decides how to balance the total workload among people within a cell. This relates closely to the method for assigning work tasks.

Inherent balance uses a fixed assignment of tasks or equipment. It attempts to give each person the same amount of work. This is the traditional method using time study.

The *queuing method* maintains balance by allowing products to accumulate within the cell at various points. When a queue is full, reassigned tasks work the overloaded queues down. Alternatively, workers may float between primary and secondary tasks to maintain the queue at an acceptable level.

With the *circulation method,* people follow a product through the process, performing multiple operations. Several persons may operate in the cell simultaneously. This method accomplishes both people balance and task assignment. It requires either close inherent balance between operations or significant surplus capacity on some operations.

Quality assurance (QA) considers three parts: methods, responsibility, and attitudes. Under methods, the fundamental choices are process control versus inspection-rejection. In addition, the designer should specify specific instruments and techniques.

Operators, supervisors, or a separate department may have primary responsibility. Attitudes reflect the predominant approach and may or may not conform to official policy.

Step 30 decides these QA issues. A compatibility check follows.

The issue of task assignment closely relates to people balance, equipment balance, and production control.

Task assignment decides the question of who does what and when. A supervisor may assign tasks. A self-directed work team assigns tasks within itself. A signaling method such as internal Kanban may assist with task assignment.

Step 32 makes assignments of people to one or more tasks. The lower section of the workcell operations plan (Fig. 6.3) documents this. First, list all required skills in the cell. List the operations in the appropriate space. List operators at the left. Operators may be listed by name, number, or code. Use a solid bullet to associate particular operators with primary skills and with specific operations. Use an open bullet to identify secondary skills required for operators.

Supervision is an issue. A good way to supervise a workcell is *not to*, at least in the conventional sense. The team decides and takes responsibility. Supervisors and managers then take a different role. Rather than telling and ordering, they advise, consult, and assist. Cybernetic, self-correcting work teams require fewer supervisors. When selecting a supervisor, look for an older, mature person. Avoid young, brash individuals who need to prove themselves.

Step 33 identifies the supervisory method. "Command and control" can work with fixed task assignments and long cycle times. An internal Kanban system can fill some supervisory tasks as well as control production. Self-determining (cybernetic) work teams can be very successful when properly implemented. Self-determining teams, however, are sensitive to management styles and are quite delicate. Do not use them without experience and training.

Step 34 documents the design decisions on the workcell operations plan.

Design cell options

Layout at the workcell level requires detail and a good fit with the design of individual workstations. For simple cells, an experienced designer can proceed from the process chart to the layout, placing equipment and workstations in sequence. An abbreviated version of the layout procedure detailed in Chap. 4, "Factory Layout and Design," is a more rigorous approach.

Features

- Multiple Material Entry Points

- Good Material Flow

- Short Distances

- Difficult People Balance

- Poor Flexibility

- Poor Communication

- Difficult Macro Layout

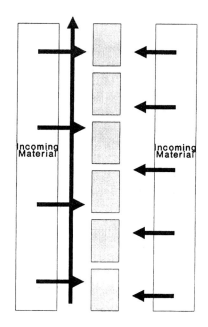

The Leawood Group (c) 1991 92638

Figure 6.18   Straight line cells.

Some authors and designers place undue emphasis on the U-shape of many workcells. In fact, effective cells can have several shapes. While shape is a factor in cell functioning, it is only one of several important factors. Shape should reflect and integrate with other elements of the cell design. Figures 6.18 through 6.21 show the most common shapes and their key characteristics.

The U-shaped workcell is perhaps the most versatile. It accommodates a wide range of production volumes and, as volume changes, allows people to float easily between primary and secondary operations. It can use the circulation method of people/task balance and switch to fixed assignments if production increases. Communication between people is good, as is material flow. The traditional U-shape may offer limited storage when many items enter the product stream.

The inverse U-shape has many advantages of the traditional U-shape. It is more difficult for people to move and rebalance themselves. However, it allows for better communication, since the people face each other. It also allows for parts and material storage around the entire periphery. It lends itself to complex assembly operations.

Straight-line cells have good material flow and provide space for many incoming parts or materials. Henry Ford's first assembly line at Highland Park was, in fact, an attempt to resolve the material flow problem. Straight-line cells (and assembly lines) are difficult to

Features

- High People Flexibility
- Easy Balance/Rebalance
- Good Communication
- Good Material Flow
- Short Distances
- Single Material Entry Point

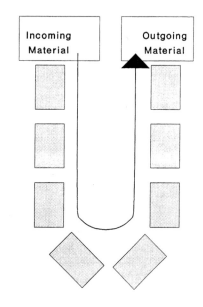

The Leawood Group (c) 1991 92636

Figure 6.19 U-shaped cells.

Features

- Good Communication
- Good Material Flow
- Short Distances
- Multiple Material Entry Point

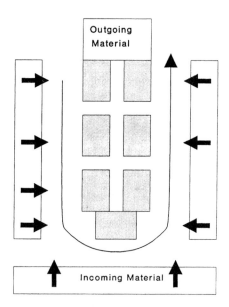

The Leawood Group (c) 1991 92637

Figure 6.20 Inverse U-shaped cells.

Features

- Several Entries
- Good Material Flow
- Short Distances Within Zones
- Easy People
- Good Flexibility
- Fair Communication
- Easy Macro Layout
- Accommodates Many Processes

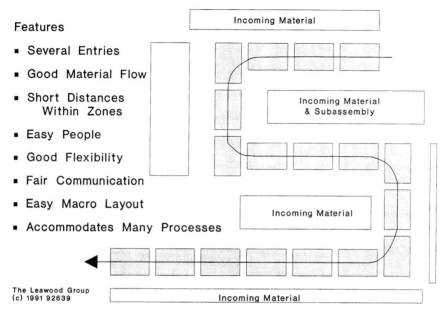

The Leawood Group
(c) 1991 92639

Incoming Material

Incoming Material & Subassembly

Incoming Material

Incoming Material

Figure 6.21 Serpentine cells.

balance, lack flexibility, and offer poor communication. They are sometimes difficult to integrate into a macro layout if the resulting shape is long and thin.

Serpentine cells take the traditional assembly line and intertwine it. Many complex arrangements are possible with a long line. Such cells overcome some of the disadvantages of the traditional assembly line.

At Step 37, the logic flowchart may take another iteration. These iterations should provide two to five design options from which to choose.

Chapter 11, "Evaluating and Selecting Facility Design Options," discusses the advantages of multiple options. It also explains the tools and techniques for selection. This is Step 38.

Workstation design is the lowest level of facilities design. Much of the best literature on workstation design is from the 1930s and 1940s. A summary of Barnes' principles of motion economy is shown in Fig. 6.22. Figure 6.23 contains hints on workplace design. Figure 6.24 shows the range of normal hand functions along with selected ergonomic dimensions.

Implementation

We noted earlier that workcells are socio-bio-technical systems. People, therefore, are an integral part of workcell design.

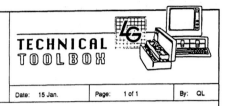

The Principles of Motion Economy

TECHNICAL TOOLBOX

Date: 15 Jan.	Page: 1 of 1	By: QL

A number of "principles" concerning the economy of movements have been developed as a result of experience and they form a good basis for the development of improved methods at the workplace. They may be considered in three sections:

- Use of the human body
- Arrangement of the workplace
- Design of tools and equipment

These "principles" can be made the basis of a summary questionnaire which will help, when laying out a workplace, to ensure that nothing is overlooked.

Use of the Human Body

When possible:

1. Begin and complete movements requiring two hands at the same time.

2. Keep at least one hand in motion at all times except during periods of rest.

3. Keep arm motions simultaneous, symmetrical and in opposite directions.

4. Design motions at the lowest classification possible to do the work satisfactorily.

5. Maintain movement momentum.

6. Plan continuous and curved movements. Avoid straight-line motions involving sudden and sharp changes in direction.

7. Use "Ballistic" (ie. free-swinging) movements. They are faster, easier and more accurate than restricted or controlled movements.

8. Maintain a natural rhythm to smooth performance of an operation.

9. Relieve the hands of work that can be done by other parts of the body.

Arrangement of the Workplace

1. Provide definite and fixed locations for tools and materials to permit habit formation.

2. Pre-position tools and materials to reduce searching.

3. Deliver materials close to the point of use.

4. Locate tools, materials and control as near to the worker as possible.

5. Arrange materials and tools to permit the best sequence of motions. Provide "drop deliveries" of ejectors wherever possible.

6. Provide adequate lighting and environmental controls.

7. Provide flexible workstations and chairs to reduce fatigue.

8. Contrast color of workplace and the work to reduce eye fatigue.

Design of Tools and Equipment

1. Relieve hands of "holding" the workpiece where this can be done by a jig, fixture or foot-operated device.

2. Combine tools wherever possible.

3. Distribute workload by inherent capacities of the fingers when hands are used.

4. Design tool handles to permit as much of the surface of the hand as possible to come into contact with the handle. This is especially necessary when considerable force has to be used.

5. Place levers, crossbars and handwheels so the operator can use them with the least change in body position and the greatest "mechanical advantage".

The Leawood Group 93128

Figure 6.22 The principles of motion economy.

Inadequate integration of people into the design is a common cause when workcells fail to reach their potential. Improper supervision is another.

The key criterion for selecting people for the cell is their ability to work in a team environment. Many employees who are outstanding

Classification of Movements

TECHNICAL TOOLBOX

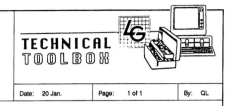

Date:	20 Jan.	Page:	1 of 1	By:	QL

Motion economy in the use of the human body calls for movements to be of the lowest classification possible. This classification is built up on the pivots around which the body members must move.

Classification of Movements

Class	Pivot	Body Member(s) Moved
1	Knuckle	Finger
2	Wrist	Hand, fingers
3	Elbow	Forearm, hand, fingers
4	Shoulder	Upper arm, forearm, hand, fingers

Each movement above Class 1, will involve movements of all classes below it. Placing everything required within easy reach will minimize the class of movement which the work demands from the operator.

Notes on Workplace Layout

1. If similar work is done by each hand, there should be a supply of the same materials or parts for each hand.

2. If the eyes are used to select material, the latter should be kept as far as possible in the area where the eyes can locate them without turning the head.

3. The nature and the shape of the material influence its position in the layout.

4. Hand tools should be picked up with the least possible disturbance to the rhythm and symmetry of movements. The operator should be able to pick up or put down a tool as the hand moves from one part of the work to the next. Natural movements are curved, not straight. Tools should be placed on the arc of movements, but clear of the path of movement of any material which has to be slid along the surface of the bench.

5. Tools should be easy to pick up and replace. They should have an automatic return, or the location of the next piece of material to be moved should allow the tool to be returned as the hand travels to pick it up.

6. Finished work should be:

   ■ dropped down a hole or chute;

   ■ dropped through a chute when the hand is starting the first motion of the next cycle;

   ■ put in a container placed so hand movements are kept to a minimum;

   ■ if the operation is an intermediate one, the finished work should be placed in a container in such a way that the next operator can pick it up easily.

7. Always look into the possibility of using pedals for operating locking or indexing devices on fixtures or devices for disposing of finished work.

Figure 6.23 Classification of movements.

performers in a conventional environment are counterproductive in a workcell. Employees who seem mediocre in a conventional environment can be the key to cellular performance.

Training is essential for cell performance, but again conventional wisdom fails. Training in the specific direct tasks is least important. The workcell employees will usually do that themselves. The impor-

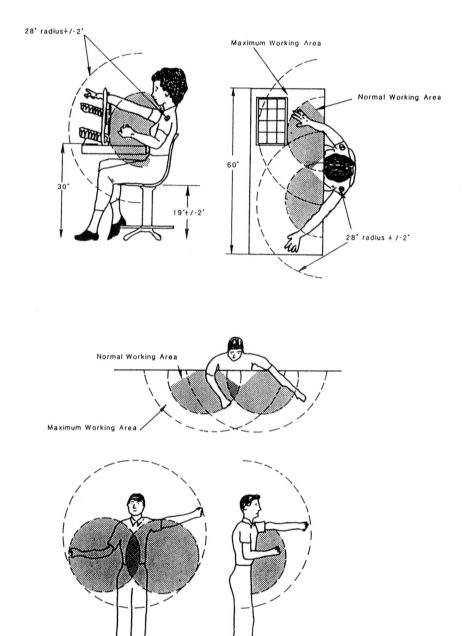

Figure 6.24   Working areas for hands.

tant training is in teamwork, problem solving, and statistical process control.

Summary

Manufacturing workcells offer many advantages for a broad range of products, volumes, and processes. They are an important part of JIT and WCM strategies. The social impact of these coupled concepts will have economic, social, and even geopolitical consequences.

Workcells and cybernetic teams are the most significant change in the workplace since Henry Ford and Charles Sorensen invented the moving assembly line. They promise to make even mass production factory jobs rewarding and satisfying. With them, work can fill personal and social as well as the financial needs of many workers. Simultaneously, these concepts can be an important part of making manufacturing competitive.

Further Reading

Burbridge, J. L.: *The Introduction of Group Technology,* Wiley, New York, 1975.

Burbridge, John L.: *Production Flow Analysis,* Institution of Production Engineers Journal, London, vol. 42, no. 12, 1963, pp. 139–152.

Dumolien, William J., and William P. Santen: *Cellular Manufacturing Becomes Major Management Philosophy at One Plant,* Institute of Industrial Engineers, Norcross, GA, November 1983.

Gallagher, C. C., and W. A. Knight: *Group Technology,* Butterworth, London, 1973.

Gombinski, J.: *Fundamental Aspects of Component Classification,* Annals of the C.I.R.P., vol. XVII, 1969, p. 367.

Ham, I.: *Introduction to Group Technology,* Pennsylvania State University, University Park, 1975.

Harmon, Roy L., and Roy L. Peterson: *Reinventing the Factory,* The Free Press, New York, 1990.

Lee, Quarterman: "Computer-Aided Plant Layout with Real-Time Material Flow Evaluation," *The MTM Journal,* vol. XIV, 1987, pp. 33–38.

Lee, Quarterman: "Layout for the Just-in-Time Plant," National Plant Engineering and Maintenance Conference Proceedings, March 25, 1986, Chicago, IL.

Lee, Quarterman: "Integrated Plant Layout for JIT & World Class Manufacturing," presented at the nineteenth annual CSI Conference, October 6–8, 1991, *IIE International Engineering Management Conference Proceedings,* Orlando, FL, October 1991.

Lee, Q., and M. Niedenthal: "Layout for the Just-in-Time Plant," presented at the National Plant Engineering and Maintenance Conference, March 16, 1987, Chicago, IL.

Lovelace, Giles: "The Use of Classification and Coding in Component Standardization," *Standard Engineering,* October–November 1967, p. 5.

Marion, David, J. Rubinovich, and Inyong Ham: *Developing a Group Technology Coding & Classification Scheme,* Systems Integration Series, Part 3, Pennsylvania State University, University Park, July 1986.

Nyman, Lee R. (ed.): *Making Manufacturing Cells Work,* 1st ed., Society of Manufacturing Engineers, Dearborn, MI, 1992.

Opitz, H.: *A Classification System to Describe Workpieces—Part I,* Pergamon Press, 1970.

Ranson, G. M.: *Group Technology,* McGraw-Hill, New York, 1972.
Schonberger, Richard J.: *World Class Manufacturing,* The Free Press, New York, 1986.
Wrennall, W., and M. McCormick: "A Step Beyond Computer Aided Layout," *Industrial Engineering,* vol. 17, no. 5, May 1987, pp. 40–50.
Wrennall, W.: "Facility Layout—The Key to Workflow, Cashflow and Profit," *Manufacturing Technology International,* Sterling Publications, London, 1988.
Wrennall, W.: "Manufacturing Work Cell Design: Getting It Right—Step By Step," presented to Institute of Industrial Engineers, Chicago, IL, May 19, 1992.

Office Space Planning

William Wrennall
President, The Leawood Group, Ltd.
Leawood, Kansas

Offices are those facilities or parts of a facility where clerical, administrative, and commercial transactions are performed. They house design, creative, and professional activities. The work ranges from the mental (information and knowledge processing) to the paper factory.

Office space planning is the physical arrangement of space for people, furniture, hardware, and their supporting services. This chapter describes:

- Generic project plans for office layouts
- Procedures for the space-planning process
- How to develop the office design to support the overall objectives of the company

The Benefits of a Sound Space Plan

An office space-planning project is a productivity improvement plan at the macro level. It will:

- Simplify work flow
- Provide flexibility
- Optimize equipment requirements
- Make good use of human skills
- Minimize work-in-process (WIP)
- Reduce space requirements
- Compress throughput time

- Promote quality
- Promote employee health and welfare
- Implement appropriate technology
- Support operations strategy
- Fit organization structure
- Derive logically from a clear and simple approach
- Enjoy broad support

Who Should Design Office Space Plans?

Office space plans are designed by a range of people. The degree of importance given to the process varies widely. Some view office layout projects as exercises in interior design. Others recognize a strategic opportunity to improve customer satisfaction. The following resources have been used:

- Existing staff
- Facility planners
- Architects
- Interior designers
- Facility-planning consultants
- Office equipment suppliers

The best results come from the participation of those whose lives are affected by the arrangement of their work environment.

Corporate mission statements of world class companies emphasize responsiveness to customers, flat organizations, and employee empowerment. Office space planning is a suitable project for an empowered team. Specialist support in the form of training in, or application of, proven techniques can be enlisted by the team. To be effective, the planner should understand the current office processes and procedures before proceeding with a revised space plan.

Frequent office rearrangements often follow dissatisfaction with an existing arrangement. Office relocation and relayout is expensive. The benefits and consequences of a sound, strategic, and flexible layout are too great to ignore.

The Generic Office Layout Project Plan

Office space plans serve:

- A temporary or short-term purpose
- A long-range facility plan
- A new business
- A new building
- Additional operations
- New occupants
- Relocation of major operations
- A new site with no existing buildings.

Whatever the reason for the project, the design should meet current conditions as well as future needs. Each project is unique, but there are common tasks for all office layout designs.

In Fig. 7.1 a *generic office layout project plan* structures a project into 21 tasks. Every project will have its own task breakdown to suit the specific purpose.

The following is a sample list of tasks and deliverables that modify and elaborate the generic project plan. They are a sample list to choose from.

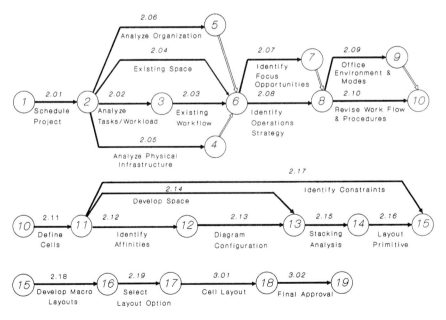

Figure 7.1 Office layout generic project plan.

1. Define, scope, and schedule project. List objectives, assumptions, and limitations.
2. Collect input data including:

   - Organization structure, present and future
   - Current and planned personnel by category
   - Human relations policies and practices
   - Furniture and equipment needs

3. Review business strategy and site missions.
4. Identify focus opportunities.
5. Establish and approve:

   - Capacity levels
   - Operating hours
   - Staff job grades
   - Information systems policy

6. Develop office space-planning unit (OSPU) list:

   - Major operations
   - Support services
   - Special features
   - Utilities

7. Determine affinities between OSPUs.
8. Generate configuration diagram.
9. Establish office types:

   - Open plan
   - Conventional
   - Modular
   - Panel-hung
   - Combination

10. Develop space classes and planned-for amounts of space by class.
11. Allocate OSPUs to sites, buildings, and floors.
12. Generate layout primitive(s).
13. Identify constraints:

- Sites and buildings
- Financial policies
- Technical/technological
- Ecological
- Environmental
- Regulatory
- Security
- Location preferences
- Fixed features

14. Develop macro or block layouts.

15. Refine and review alternative office layout plans.

16. Present alternative layouts for evaluation and selection of preferred option.

17. Confirm personnel establishment and policies as still valid.

18. Flush out latent agenda.

19. Reconfirm planning basis of:

- Office configuration
- Furniture systems
- Information technology
- Support systems
- Utilities
- Space availability
- Long-range business plan

20. Modify and detail selected macro layout with options.

21. Present, reevaluate, and confirm preferred space plan. Obtain authority to proceed.

22. Prepare implementation plan.

23. Update facility database.

24. Generate layout, construction, and utilities drawings.

25. Produce equipment, furniture, utilities, and personnel register.

26. Prepare project report, and obtain financial approvals.

27. Select contractor.

28. Develop and implement plan.

Data Acquisition

Tasks 2.02 and 2.03 of the generic project plan determine the quantity and type of office work. In these tasks review the work to be done in-house and what to contract out. For example, will annual reports, sales literature, and office newsletters be internally printed or purchased? Will office cleaning, catering, mail room operations, messenger, and copying services be done in-house or subcontracted?

From the above decisions determine how many people, by grade, will be employed. The *staff survey form,* Fig. 7.2, is suitable for recording this information.

In Task 2.05 examine the physical infrastructure requirements. The physical infrastructure consists of hardware space users that support the work flow and processes. Examples are elevators, telephone exchanges, and utilities. A more complete list is given in the checklist shown in Fig. 7.3.

In Task 2.06 analyze the present and planned-for organization structure. This becomes one basis for quantifying space needs. An example organizational chart is given in Fig. 7.4.

Description: Office Layout		Proj.No:	1162
Company: XYZ Corporation		Date:	1/15
Location: Central Office	Pg: 1	By:	KB

Area	Title	Current Staff				Chg.	Planned Staff			
		Type 1	Type 2	Type 3	Type 4		Type 1	Type 2	Type 3	Type 4
Order Entry	O.E. Manager	1	4	0	0	3	1	6	1	0
Marketing	Marketing Mgr.	1	5	0	0	3	1	8	0	0
Accounting	Accounting Mgr.	1	3	0	0	0	1	3	0	0
Purchasing	Purchasing Mgr.	1	2	0	0	0	1	2	0	0
Totals:		4	14	0	0	6	4	19	1	0

>>>>>>

Comments:

Figure 7.2   Staff survey (example).

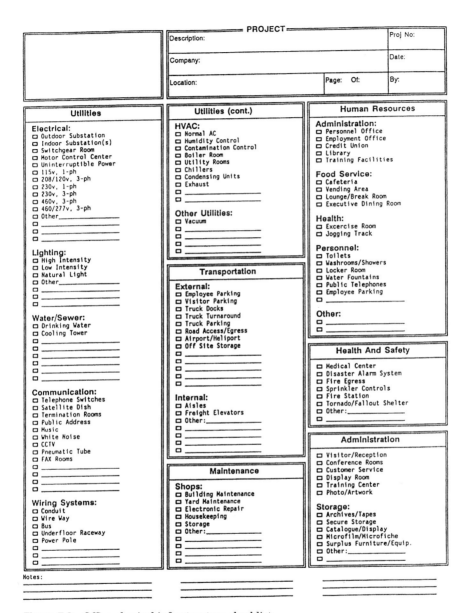

Figure 7.3 Office physical infrastructure checklist.

Existing Space

Few layouts start with totally new processes and procedures. To introduce new procedures and systems at the same time is risky. With an existing office, record and summarize the current space.

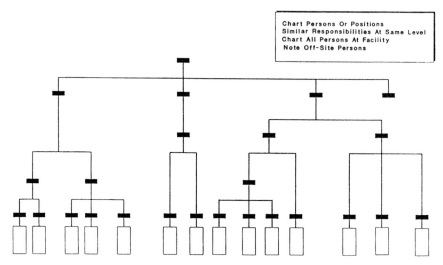

Figure 7.4   Organization analysis—Preparing the chart.

This is Task 2.04. Space determination is easier if data are on a facility database. Checks for needed space, by department and in total, against space available provide a firm planning base.

For space allocations, comparisons, and projections, analyze existing space by department and class. Typical classifications of office space are:

- Executive
- Clerical operations
- Professional
- Design
- Support services
- Circulation
- Storage
- Conference and meeting
- Utilities

A space analysis of a similar existing layout is useful to:

- Identify OSPUs for the new layout
- Confirm space allocations and proportions
- Identify improvement opportunities
- Avoid known problems or difficulties

Basic Planning Symbols & Colors

Process Chart Symbols & Action*	Symbols Extended to Identify Activities/Cells & Areas	Color Ident. **	Black & White **
◯ Operation	◯ Open Office Areas	Green	
	◯ Enclosed (walled) Office Areas	Red	
⟱ Transportation Circulation	⟱ Entrances, Exits, Docks	Orange Yellow	
▽ Storage	▽ Storage Activities/Areas	Orange Yellow	
◷ Delay	◷ Set-down or Hold Areas	Orange Yellow	
☐ Inspection	☐ Inspect, Test/ Check Areas	Blue	
Adapted from *A.S.M.E. Stand. **I.M.M.S. Stand.	◠ Service & Support Activities/Areas	Blue	
	⌂ Building Features	Brown (Gray)	

For Product Focus use separate colors on an uncolored layout to identify all activities/cells related to specific product lines or families.

Ratings

Rating			Proximity	Color Code **	Evaluating Results	Rating
A	4	////	Absolute	Red	Almost Perfect (Excellent)	A/4
E	3	///	Exceptional	Orange Yellow	Exceptionally Good (Very Good)	E/3
I	2	//	Important	Green	Important (Good)	I/2
O	1	/	Ordinary	Blue	Ordinary (Fair)	O/1
U	0		Unimportant	Uncolored	Unimportant (Poor)	U/0
X		⋏	Not Near	Brown	Not Acceptable (Unsatisfactory)	X/?
XX		⋏⋏	Apart	Black		

Figure 7.5 Office space-planning conventions.

A space class analysis superimposes color or patterns on the layout. This illustrates the amount, proportion, and distribution of space use. Symbols and codes identify the functions and the product groups in each space. The conventions for office space planning are provided in Fig. 7.5.

The procedure is: First, list each current activity or feature and its present amount of space under the appropriate classification. Then determine total space by class. This is a convenient time to note new activities. Since they do not yet exist, show them on the list but with zero space.

Prepare a space profile

A *space profile* is a bar chart showing the current proportions of space by class. The proportions (particularly aisles and circulation) vary widely from organization to organization and will be affected by company practice. The space profile is useful for comparison with the new layout.

Infrastructure Analysis

Infrastructure may be physical or nonphysical. It consists of operating system elements which support processes and tasks.

Physical infrastructure

Physical infrastructure includes activities, equipment, and locations which are outside the principal activities. In offices their space requirements are often, but not always, small compared to other areas, and it is easy for the layout designer to overlook them. Examples of physical infrastructure are buildings and land that house utility systems supplying electricity, water, and other widely used essentials. Other examples are roads, parking lots, and docks supplying the offices.

The *physical infrastructure checklist,* Fig. 7.3, is a tool which helps the designer include the appropriate items in the layout. It also documents many design requirements for infrastructure elements. The task of identifying physical infrastructure is completed by filling out the checklist, noting the items required in the new layout. Later, this checklist will help guide OSPU definition.

Nonphysical infrastructure

Nonphysical infrastructure includes the organization structure, policies, procedures, and information systems. The construction of the organization chart is shown in Fig. 7.4. This chart verifies site population, and establishes consistency and OSPU requirements. A study of the policies, systems, or procedures will indicate their impact on the development of the new layout.

Office Environments and Modes

An office building design is based on structural, technical, and economic considerations. The design may have a high ego input that has little to do with office work.

Office space planning integrates the building design with the purpose of the office. Because office buildings have a life that is longer than their current use, the layout is often a compromise.

Buildings should be as flexible as possible to maximize their effectiveness and efficiency over time. Special-purpose or custom-built buildings seldom suit other uses, with the result that the office planners then face severe building constraints.

Prior to the Industrial Revolution, offices were rooms of moderate size. The prototype was the library, study, or consulting room of professionals. This is the origin of the *private office*. Adjacent smaller but still private space would be occupied by a secretary or private assistant.

With the trend toward bigger organizations, another kind of office emerged for staff, with people sharing space. These shared offices evolved into large *general offices*. Executives and senior managers were located on the periphery in private or semiprivate rooms. These large offices became common in the United States, and were typical of organizations that employed many clerical workers. This system is known as the *American,* or *"bullpen,"* pattern.

In the 1960s the so-called *open office plan* was developed. This planning approach was first introduced in Germany and was called *"office landscape."* The concept was based on studies by Everhard and Schnelle of the Quickborner consulting group. Their studies concluded that office work was inhibited and confused by illogical office layouts. They found that:

- People who frequently communicated were located too far apart.
- Office workers with infrequent communication were often grouped together.
- Seniority often determined office layouts.
- Practical and logical needs were often ignored.

The *Quickborner offices* were free from walls, fixed partitions, and corridors. Office staff locations were in communication and paper flow sequence. Managers were available, and working groups developed. This generated a sense of cohesiveness and made layout changes easier. Movable screens, potted plants, trees, and background sound provided a degree of privacy.

The introduction of new office furniture systems modified "office landscape" designs. In the Probst *action office,* desks and credenzas are replaced by work surfaces, storage units, and lights hung on interlocking walls or screen panels. This is known as "panel-hung furniture." The panels or screens provide privacy, work area definition, sound absorption, and conduits for power, telephone, and computer cabling.

Modular office furniture systems make use of:

- Technological advances in panel-mounted furniture
- Interlinked panel dividers
- Various types of freestanding furniture and equipment

Many organizations will have hybrid systems. These could be mixtures of:

- Full panel-hung
- Some panel-hung with freestanding furniture
- Private offices with floor-to-ceiling walls
- Freestanding furniture
- Movable screens

A full panel-hung system requires 15 to 20 percent less space than dividers with freestanding furniture. To save space in new offices, use a full panel-hung system. But the conversion of existing conventional offices to panel-hung just to gain a few more workstations is very expensive and seldom worthwhile.

Prewired floor-to-ceiling movable walls for private offices are more flexible than the more permanent stud-and-sheetrock construction. This is particularly true when the movable walls are mounted on top of one-color carpet that covers the entire area.

Financing and taxation can directly affect the type of office furniture system selected. For example, it is common to construct office space without ceiling lights. *Systems furniture* has built-in indirect lighting. The system is depreciated as "furniture and equipment" and not as "building structure." Similarly, floor-to-ceiling movable wall panels can be classed as "furniture and equipment," whereas studs and sheetrock are classified as "building structure."

Open planning lends itself to:

- Rapid layout changes
- Varying levels of privacy
- Technology change

- Easy wiring and lighting changes

Open offices and traditional offices both have supporters and opponents. New patterns are continuously evolving with the availability of many new furniture systems. It is essential that the office space planner get the full support of top management before installing a new and different furniture system.

Reuse of existing furniture and systems

Office furniture systems are expensive, so the office planner will often have to integrate new furniture with the old. Repainting or refinishing of the existing can give it a new lease on life.

Office operating modes

Operation modes are the fundamental groupings of work into layout modules. The most common are:

- Process or functional
- Product
- Cellular
- Linked cellular
- Line
- Project

A *functional* or *process* group is the classic mode. Examples are accounts payable, personnel, computer center, or drawing office. The workers specialize in a few operations. The work is processed in batches and the layout is arranged by function or procedure.

A *product mode* is typical of the strategic business unit (SBU) where all office functions related to the SBU are grouped in one area.

An office using a *cellular mode* has a cluster or natural team working on similar but not necessarily identical transactions. The equipment is selected for a few basic processes. An example is home-loan-processing work that is performed in batches grouped by similarity into larger quantities.

Cells add value progressively. There may be redundant equipment and surplus reliable capacity to create a "pull system." A cell can be straight-line, U, or collapsed U, S, Y, or circular design. Flow is always positive, and operators may be stationary or mobile.

The *linked cellular mode* consists of sequences of cells interlinked to provide a smooth flow of work throughout the office.

A *line mode* dictates a balanced sequence of linked workstations, sometimes with a conveyor belt, to transfer documents from one workstation to another. This mode is the office equivalent of the automotive assembly line.

An example of an office in a *project mode* is a team of designers grouped and organized for the purpose of producing an office layout or office system. So is a team of software programmers designing an accounting package. The team is formed for a specific purpose, and disbanded after the project has been completed.

Office Layout Focus

An operation strategy statement is a prerequisite to developing a layout. It determines how an office layout will be focused. The concept of focused operations was introduced by Skinner about twenty years ago. It is typically applied to manufacturing environments, but the concept also applies to commercial and office operations. Offices that do many things are not as productive as those that focus on fewer products or services. Adequate variety should be grouped into focused *offices-within-offices* (OWOs).

Focus as a concept spans from the whole business to the workplace. Office operations can be focused into the process, product, family, or cellular operating modes. They can also be focused by family (similar to cellular), market, or customer.

A typical office product focus would integrate for one product line the processes and transactions of:

- Order receiving
- Credit approval
- Pricing
- Specification
- Material and production planning
- Shipping and invoicing

It follows that other particular types of focus decided upon will demand a compatible organization structure. Examples are:

- When a family of similar transactions or documents are processed by a small group of pieces of equipment and office workers, this is called *family,* or *cellular, operations.*
- When office work is organized and the layout is matched to process documents—for example, government, catering, or retail—this is known as *market focus.* Work and equipment grouped to carry out transactions or document preparation for West Coast customers would be an example of regional market focus.

- When, for example, insurance work for private automotive policies is focused on a particular customer grouping, this is known as *customer focus.*

- A measure of an OWO is when all the activities that are clustered or grouped together could be moved into a separate building or site and still function. Of course, they do not have to be physically moved; they can be separate, but within a bigger operation.

- The ultimate in *operation focus* is when office-type and non-office-type activities are clustered into an organizational whole—for example, purchasing, receiving, storing, and repacking or assembly, shipping, and invoicing.

The *organization focus algorithm,* Fig. 7.6, illustrates the decision process leading to how the business should be organized. This output is fundamental to the office layout process. Figure 7.7 shows exam-

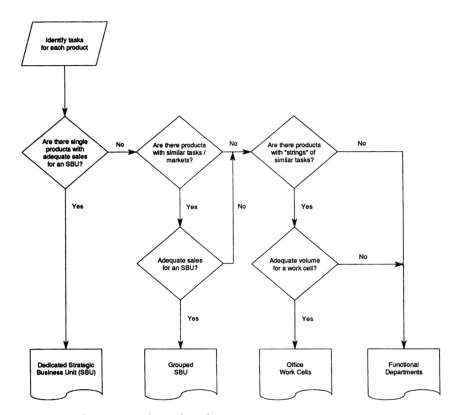

Figure 7.6 Organization focus algorithm.

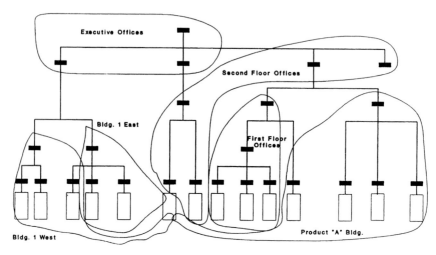

Figure 7.7   Organization analysis: (*a*) poor.

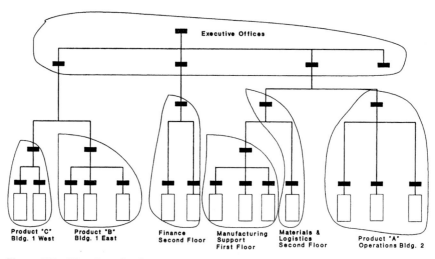

Figure 7.7   (Continued)   Organization analysis: (*b*) good.

ples of poor (*a*) and good (*b*) fit between office layouts and organization structure.

The Office Space-Planning Process

Office space plans are derived from the following elements:

- Purpose and mission

- Tasks and products
- Volumes and variety
- Strategy and policy

This leads to a balanced interactive model. Each element will contribute to the layout design, and changes will call for updates to the layout. There is no one correct office layout because:

- Inputs are subjective
- Inputs are imprecise
- Options are infinite
- Data is dynamic
- Constraints apply and are inconsistent

Generating a few sound options:

- Flushes out the uncertainties
- Develops the preferred option
- Builds in flexibility for updates and upgrades

The data collection phase, Tasks 2.01 through 2.06, and the operation analysis phase, Tasks 2.07 through 2.10 of the generic project plan in Fig. 7.1, are now complete. The office space planner or project team should now make a summary of key points and preliminary findings. This is a milestone for management review and concurrence. While this information forms a sound and approved base for the project, new information will continue to accrue.

The new layout development part of the overall project will also follow the logic of the generic project plan, given in Fig. 7.1.

Hannon suggests a seven-step layout development approach which is illustrated in Fig. 7.8. This encompasses Tasks 2.11 through 3.02 of the generic plan. Each of these seven layout development steps is discussed in the following sections, while the illustrations show steps in a small carry-through project.

Step 1—Identify office space-planning units

There are 10 tools for defining and developing OSPUs. These are:

1. *Existing space summary*—The first way to develop OSPUs is to look at the existing ones. These may be functional, product, support, storage, or feature areas.

2. *Procedure analysis*—This method uses procedure analysis charts as a way to determine what OSPUs are needed.

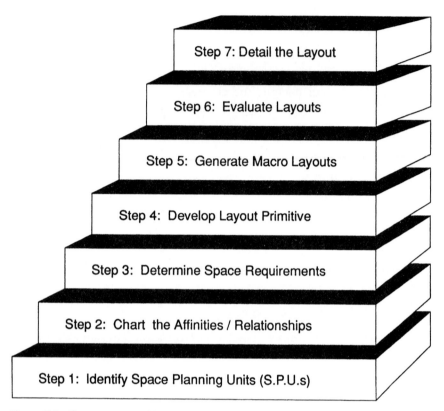

Figure 7.8   Seven steps to office layouts.

3. *Group technology (GT) or family group analysis*—Here the planner studies the product families and procedures of an organization and its offices to develop SPUs.

4. *Organization analysis*—Another method is to look at the organization charts. Analyzing the organization structure tells a planner what functions can be grouped together and what supporting services are needed.

5. *Technology forecasting* will indicate the need to modify existing areas or provide for additional ones to accommodate future technology.

6. *Research and development* intentions will indicate what changes to products, processes, and systems will need to be accommodated.

7. *Infrastructure* is the source of the seventh method. A study of the organizational infrastructure will help determine what level of physical facility infrastructure is required.

8. *Company policy changes* will indicate if they affect SPU definition. For example, are there plans for some office functions to be moved to other locations?

9. *Strategy*—A company's policies may overlap with the ninth method, consideration of the strategic business plans. The planner must know the strategic plans to be able to determine how they will influence SPU definition.

10. *Codes and regulations* may be imposed on the business by government regulatory agencies, existing laws, and/or professional guidelines. These may call for special space provisions in the layout.

The *SPU definition summary,* Fig. 7.9, brings together all the OSPU components necessary for definition.

Step 2—Develop affinities

Affinities represent a requirement for proximity or closeness between a pair of OSPUs in a layout. It can also be important to keep activities physically apart. Affinities may be positive, negative, or neutral.

The most widely used rating system is the "vowel scale." Here, A-E-I-O represent four levels of positive affinity; U represents a neutral

Description:	Office Layout	Proj.#	1162
Company:	XYZ Corp.	Date:	1-15
Location:	Central	By:	KB

page 1 of 2

NOTE: All areas include working aisle space.

No.	Name	Inclusions	Exclusions	Exist	Oper.	Org.	GT	Tech	R&D	Infra	Polic.	Code	Strat.	Proc.	Prod.	Other	None	Note
1	Order Entry	Order Entry workstations File cabinets, reference table, etc.	Sales and marketing catalog area	x		x								x				
2	Credit	Credit workstations File cabinets, reference bookcases, etc.	Order entry references	x		x								x				
3	File Storage	Storage for historical files (centralized)	Active files in various areas	x		x								x				
4	Telemarketing	Telemarketing workstations	Other marketing areas	x		x								x				
5	Design	Design engineers Workstations, files, references, etc.	R&D production and testing labs	x		x									x			
6	Retail Coordinators	Workstations, files, references, etc.	Field sales offices	x		x								x				
7	Senior Retail Coordinator	Desk and workstation, Group seating for 4 etc.	Field sales offices	x		x								x				
8	Manager of Customer Service	Desk and conference table, bookcases, workstation, etc.	Other manager offices	x		x										x		
9	Admin. Asst. - of Customer Service Mgr.	Workstation, file cabinets, bookcases, network printers for area, etc.	Other administration offices and computer equipment areas	x		x										x		

Figure 7.9 SPU definition summary (example).

affinity; X and XX are two levels of negative affinity, which encourage distance between cells.

The vowel scale works well where experience and opinion are primary inputs for rating affinities. The letters easily relate to descriptive terms. Some designers use a numerical scale corresponding to the vowels. The integers 4 through -2 correspond to A-E-I-O-U-X-XX. The numerical scale is easier to use with computer applications although care must be taken when applying the negative figures.

Affinities arise for many reasons. Among these are:

- Interpersonal communication
- Personnel movement
- Paper/material movement
- Supervisory duties

Rating	Pairs of Activity Areas		Reasons for Closeness or Separation
A	President Purchasing Restrooms	Exec. V.P. Lobby Lunchroom	Personal contact Visitors Movement of people, common utilities, convenience
E	Sales VP Central Stores Accounting	General Sales Truck Dock Data Processing	Personal contact Movement of supplies and equipment Personal contact, shared supervision, paperwork flow
I	President President Operations V.P.	Sales V.P. Lobby Plant	Personal contact Convenience Personal contact
O	Accounting Operations V.P. Prod. Planning	Purchasing Lobby Restrooms	Paperwork flow, convenience Convenience Convenience
U	President Central Stores Purchasing	Mail Room Restrooms General Sales	No contact No contact Little contact
X	Conference Room Data Processing Lunchroom	Production Areas Lobby Board Room	Noise, vibration Security Appearance

Benchmark affinities, as shown in these examples, can guide in rating other activity areas.

Figure 7.10 Benchmark affinities.

■ Environmental hazards/conditions

■ Security

The above list is only a sample. The office analyst often identifies other factors during the project. When developing affinities, it helps those involved to have examples. Some "benchmark" examples are shown in Fig. 7.10, but customized examples are the most useful.

In the office environment material flow rarely dominates. Office layout affinity factors are typically:

■ Many and varied

■ Intangible

■ Controversial

■ Political

Information required for determining office affinities is obtained from surveys, interviews, and consensus meetings. These are reviewed as follows:

Survey methods—The analyst sends survey forms to selected managers and workers in each layout OSPU. They are asked to rate their required proximity to other OSPUs.

Interview method—To use interviews as the basis for affinity development, the analyst discusses the layout and its development process with knowledgeable individuals. Among the topics are affinities and their underlying factors. The understanding of the discussions is then translated by the analyst into affinities.

Consensus meeting method—In this method a group of people who represent all users of the facility meet to identify affinities. Through discussion and compromise, they develop affinities between all layout OSPUs. Many veteran layout designers consider the consensus meeting the most productive and satisfactory method.

Whatever method is used, the analyst should check the distribution of affinities before proceeding to a configuration diagram. This should show a small number of A's, with progressively increasing numbers of E's, I's, and O's. X's and XX's will be infrequent affinities. An *affinity chart* is the Step 2 output for the illustrated project shown in Fig. 7.11.

Step 3—Determine space

Space is a fundamental element of every layout. The space on existing layouts is a matter of record, but the space for new layouts must be determined.

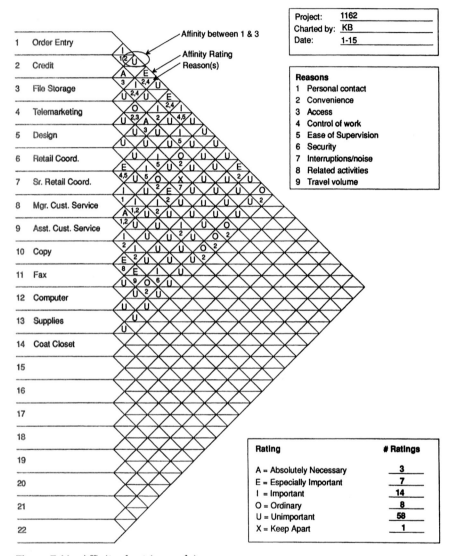

Figure 7.11 Affinity chart (example).

Six methods are available for computing space, each having advantages and disadvantages. These methods are:

- Elemental calculation
- Standard data calculation
- Transformation

- Visual estimating
- Proportioning
- Ratio forecasting

The above methods give the designer a toolbox from which to select the most appropriate method, or combination of methods. The accuracy level required for space calculations varies widely. The level of effort, credibility, and time horizon also vary for each method. The designer takes account of these factors when choosing a method for each cell.

The features of the six methods are discussed briefly below. A full discussion on space determination is given in Chap. 4, "Factory Layout and Design."

The accuracy of *elemental calculation* reflects the accuracy of the forecasted headcount, equipment, and furniture schedules. The computed space can be quite accurate when based on an accurate and stable list. Forecasts for office needs, however, are notoriously inaccurate. They may reflect a manager's opinion of the department's importance rather than true space needs. Head count projections seem to vary from month to month, demanding recalculation of the elemental totals every time there is a change.

Standard data is a close cousin to elemental calculation. With the standard data method, the designer takes known units which have been developed and approved and uses these to build up the appropriate space requirements.

Office space standards depend on many factors, such as:

- Management philosophy
- Furniture style
- Office automation
- Image requirements
- Inconsistent inclusions and exclusions

Some examples of net space, excluding main aisles, mechanical rooms, conference rooms, supply stores, computer rooms, cafeterias, and so forth, are (in square feet):

Senior executive	150 to 600
Managers	100 to 200
Secretaries	65 to 120
Clerical	35 to 75
Professionals	60 to 120

Gross space standards for general offices vary widely. They range from 180 to 320 square feet per person. This includes conference rooms, supply stores, restrooms, copiers, and computer rooms. Mechanical rooms and large cafeterias are excluded. The lower end of the range is spartan, and this will be the projected company image.

Nevertheless, gross space standards can be very useful when projecting future office needs. For example, five persons at 200 square feet per person will require 1000 square feet. At 250 square feet, the same space will only accommodate four persons.

Standard data calculation is the method of choice for large firms with many rearrangements. Once the standard data are complete and approved, space calculations are fast and accurate. Moreover, standard data calculation eliminates many of the status and ego issues that accompany office layout projects.

We recommend the following procedure for developing new space standards:

1. Identify complaints/symptoms of nonstandardization.

2. State goals/objectives.

3. Survey current situation and reconfirm goals/objectives.

4. Develop data using regression analysis and synthesis.

5. Gain approval.

A management may impose new space standards on a nonstandardized office environment. If so, some people find their individual space improved and others find it diminished. This introduces an emotional element and requires patience and maturity from the analyst.

Under item 1 above, the analyst reviews and lists the complaints or symptoms voiced that seem to relate to standardization. Some examples are:

"Every office move becomes a tug-of-war for the best view."

"The assistant sales manager has a large office and conference table and his position doesn't justify it."

"Engineering people are piled on top of each other while everyone in production control has a desk and reference table."

"Every office has a different color, different furniture and a different look."

"I'm tired of listening to arguments from my subordinates over space!"

"Office space for Level 4 Managers varies from 128 square feet to 245 square feet."

"Our space planners spend excessive time calculating space and then managers always argue with the space we give them."

Such a listing is nonthreatening, allows people to relieve frustration, and establishes fact and opinion. It documents complaints and leads to clear objectives.

Under item 2, the analyst clearly states the objectives and obtains preliminary management approval. Some of these objectives may be unrealistic and the analyst should reconfirm them after the initial investigation.

Under item 3, an initial survey determines whether the goals and objectives are feasible. Space standards are rarely practical for all space calculations. This initial survey identifies the areas, activities, equipment, furniture, or positions which the standards will cover.

Development of the standards uses either regression analysis or synthesis. Testing and approval by senior management completes the process.

Transformation is perhaps the most versatile of the six methods of space computation, particularly for macro- and site-level layouts. With an experienced layout designer, the method is reasonably accurate, simple, and does not require complete furniture, equipment, or personnel schedules. It is not recommended for the inexperienced layout designer.

To estimate the space for a particular OSPU, where there is an existing layout, the analyst first lists the existing space. This may be tight, loose, or just right for today's activities, so it should be *normalized* to what it ought to be for current use. This requires experience. The analyst then reviews business forecasts, plans, and task definitions. The normalized space is adjusted to allow for projected changes in activity levels or other conditions which affect the space required. It is not necessary to change the space projections when minor changes in head count occur.

Visual estimating uses layout templates and the experience of the designer to estimate space. The method is fast and has moderate accuracy and high credibility. The designer does not require a complete furniture and equipment schedule. Visual estimating does require a high level of experience on the part of the designer.

The designer starts with a set of scale templates based on a preliminary furniture and equipment schedule. The templates are arranged on a grid sheet or screen, leaving appropriate space for aisles, circulation, and miscellaneous equipment. The gross space occupied is the estimate for that particular area. Note that this estimate does not consider affinities or constraints. It just visually indicates the space required and is not the proposed layout.

Proportioning is the fifth method of space computation. The development of a space profile was discussed earlier in the chapter. The space profile shows the current proportions of space by class. This can be used as the basis for projecting proportions of space by percent for each class. Proportions may be expected to vary with changes in policy, etc. However if the projected needs for one class can be established, then the needs of other classes can be determined with fairly high accuracy by proportional allocation. Without known and tested proportions, only the experienced designer should use this method. Standard data systems often use proportioning as one element.

Ratio forecasting has the lowest near-term accuracy of the six methods but the best long-term projection base. Its primary use is for site planning and long-range planning. In stable industries it can produce adequate accuracy on twenty- to forty-year forecasts.

The analyst must begin with historical data, then classify and measure the space used at several past intervals. These intervals vary, but typically are five to twenty years. The analyst then relates the space in each class to some basis parameter such as sales dollars or site population. The ratios of space class to basis parameter often have clear long-term trends which can be projected. The projected ratio can then be converted to future space required in gross terms.

Selecting a method.   When computing space, the analyst or layout designer may use one method for the whole project or different methods for each OSPU. More than one method may be used as a cross check.

Several factors affect an office space planner's choice of method. Among these are:

- Accuracy
- Computation effort
- Planning horizon
- Data completeness
- Experience of designer
- Credibility to users

Figure 7.12 compares the six methods on each of these dimensions.

A *space forecast sheet* summarizes the space calculation output, and this is shown in Fig. 7.13. A sample space-planning allocation sheet, Step 3, for the Customer Service Department project, is given as Fig. 7.14.

Methods	Accuracy	Effort	Planning Horizon	Input Data Complete	Experience Req'd	Credibility
Elemental	High	Very High	Short	Very	Low	High
Standard Data	High	High	Short	Very	Low	High
Visual Estimate	Mod.	Mod.	Med.	Mod.	High	High
Transformation	Mod.	Low	Med.	Mod.	High	Mod.
Proportioning	Mod.	Low	Long	No	Mod.	Mod.
Ratio Forecast	Low	Mod.	Very Long	No	Mod.	Mod.

Figure 7.12 Computing space—Selecting a method.

Step 4—Develop the layout primitive

This step consists of three parts. First, the configuration diagram is developed. Next comes the stacking, which leads into developing the layout primitive.

Developing the configuration diagram. In this process transform the information gathered and entered on the affinity chart into a *configuration diagram*. This is done using the symbols and the number of "rating lines" shown in Fig. 7.5.

Consider the rating lines as rubber bands, with the A rating giving a strong pull like four bands. OSPU pairs with A ratings should be close together. Those with E (3), I (2), or O (1) ratings would have lesser pulls and can be proportionately further apart. X or negative ratings should be connected by a wiggly line representing a spring pushing them apart.

Figure 7.15 (*a* to *d*), Diagrams 1 to 4, shows the follow-through procedure for a Customer Service Department project. Start with diagramming the A's and E's. Next add the I's, rearranging for better fit if necessary. Then add the O's and X's, again rearranging for best fit. This can be done either manually or on a computer. The configuration diagram is now complete.

Stacking. *Stacking* is the allocation of OSPUs to sites, buildings, or floors. Horizontal stacking allocates OSPUs to sites and buildings. Vertical stacking assigns OSPUs to floors in a multistory building.

Description:	Office Layout		Project: 1162
Company:	XYZ Corp.		Date: 1-15
Location:	Central Office	Page 1	By: KB

SPU#	Space Planning Unit Name	Current Now	Planning Horizon + 2 yrs.			Method Elemental	Transform	Visual	Std. Data	Proportion	Ratio
1	Order Entry	500	650			X	X				
2	Credit	90	150			X	X				
3	File Storage	150	320			X	X				
4	Telemarketing	600	780			X	X				
5	Design	350	450			X	X				
6	Retail Coordinators	800	960			X	X				
7	Sr.Retail Coordinator	110	110				X				
8	Mgr. Customer Service	140	140				X				
9	Asst.Customer Service	90	90				X				
10	Copy	40	40				X				
11	Fax	10	10				X				
12	Computer	100	140			X	X				
13	Supplies	30	30				X				
14	Coat Closet	30	30				X				
	Totals:	3,040	3,900								

Notes:

Figure 7.13   Space forecast (example).

The layout primitive. The third part of this step is to add the space, to scale, on the configuration diagram. A *layout primitive* example is shown in Fig. 7.15e, Diagram 5. Adding the different sizes of the OSPUs to the configuration diagram to produce the layout primitive will modify the configuration.

Description:	Office Layout	Project:	1162
Company:	XYZ Corp.	Date:	15-Jan
Location:	Central	Page: 1	By: KB

SPU#	Space Planning Unit Name	People			Space Type	Area (s.f.)	Aspect Ratio			Special Requirements
		Male	Female	Total			Min.Length	Min.Width	Clear.Height	
1	Order Entry	0	10	10	O	650	*	10	9	
2	Credit	0	2	2	O	150	*	10	9	
3	File Storage	0	0	0	G	320	*	10	9	Max. filing height should be 8 ft.
4	Telemarketing	4	8	12	O/T	780	*	10	9	
5	Design	2	3	5	O/T	450	*	10	9	Space for CAD stations
6	Retail Coordinators	3	9	12	O/T	960	*	10	9	
7	Sr.Retail Coordinator	1	0	1	T	110	11	10	9	
8	Mgr. Customer Service	0	1	1	P	140	14	10	9	
9	Asst.Customer Service	0	1	1	T	90	9	10	9	
10	Copy	0	0	0	S	40	8	5	9	
11	Fax	0	0	0	S	10	2	5	9	
12	Computer	0	0	0	P	140	14	10	9	Enclosed to reduce printer noise
13	Supplies	0	0	0	S	30	6	5	9	
14	Coat Closet	0	0	0	G	30	6	5	9	
	+ Aisles, etc.:					475				
	Totals:	10	34	44		4,375				

Notes:

Space Type: P-Private, O-Open, T-Partitions, S-Service, G-General

* No minimum length stated - length dependent upon final width of designed space

Figure 7.14 Office space-planning sheet.

Step 5—Generate macro layouts

Constraints. OSPUs, affinities, and space lead to the development of one or more space configurations referred to as *layout primitives*. Office space-planning projects have additional information which puts limitations on the ideal or conceptual design of the office layout.

Some examples of these constraints are:

1. Floor loading in a certain area will accommodate desks and furniture but not a heavy piece of equipment. This restriction does not apply to other parts of the building.

2. Directors demand offices with windows, but any window office is acceptable.

3. A small automatic guided vehicle (AGV) is planned to deliver mail. This requires that main aisles be a minimum of 5 feet wide and arranged for routing the AGV.

4. The architect has designed a luxurious area from the office entrance through the center of the building. Management decrees this area be restricted to executive secretaries.

5. Superior office space is provided on the first and second floors. The other floors have not been remodeled. Departments with no outside visitors have to be located on these floors.

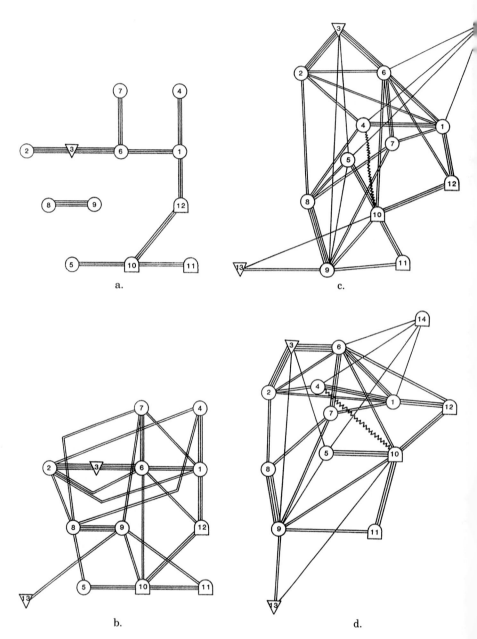

Figure 7.15 (a) Diagram 1: A's and E's; (b) Diagram 2: Add I's; (c) Diagram 3: Rearrange and add O's and X's; (d) Diagram 4: Rearrange for best fit.

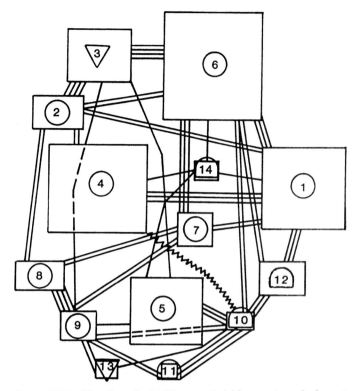

Figure 7.15 (*Continued*) (*e*) Diagram 5: Add space to make layout primitive.

6. Tradition dictates that certain departments should not be on a particular floor.

The *constraint summary,* shown in Fig. 7.16 is an example of how formally to collect the constraint data for use in developing the macro office space plans. When evaluating layout options, refer to the summary to see how well the various options adapt to the constraints.

Developing the macro or block layouts. A number of macro layouts should be developed, by constraining the layout primitive. They should be significantly different space plans, not just offer cosmetic differences. An example of an office macro space plan with options is shown in Fig. 7.17.

Step 6—Evaluate and select preferred option

Several sound space plan options will be generated from the macro or block planning process. These result from:

Notes

1. Customer Service area only
2. _____
3. _____
4. _____

Description: Office Layout   Proj. No. 1162
Company: XYZ Corp.   Prep By: KB
Location: Central   Page: of 1 1   Date: 1-15

SPU	SPU Name	Constraints	Site Conditions	Visitor Access	Building Features	Columns	Utilities	Commun. Lines	Traffic	Personnel	Paper / Printouts	Organization	Appearance	Security	Confidential	High Value	Proprietary
1	Order Entry				X			X							X		
2	Credit				X			X						X			
3	File Storage				X					X				X	X	X	
4	Telemarketing				X			X						X			
5	Design				X									X	X	X	
6	Retail Coordinators				X			X		X				X			
7	Sr. Retail Coordinator		X		X			X		X				X			
8	Mgr. Customer Service		X		X			X		X	X		X	X			
9	Asst. Customer Service		X		X			X		X	X			X			
10	Copy							X									
11	Fax						X	X									
12	Computer				X			X		X	X			X			
13	Supplies							X									
14	Coat Closet							X									

Figure 7.16   Constraint summary.

- Differing constraints
- Rotations
- Combination or splitting of OSPUs
- Horizontal or vertical stacking
- Different buildings
- Financial considerations
- Technology policies
- Process options
- Different OSPU focus

The generation and presentation for evaluation of three to five sound office space plans is desirable. This satisfies a management need to exercise choice and avoids the one-option rejection.

A participative selection process generates an innovative environment. This results in a release of new information, constructive and critical input, and thus further options. Generating new layout option ideas during the selection process wins acceptance. The selling is done and the selectors claim ownership of their choice.

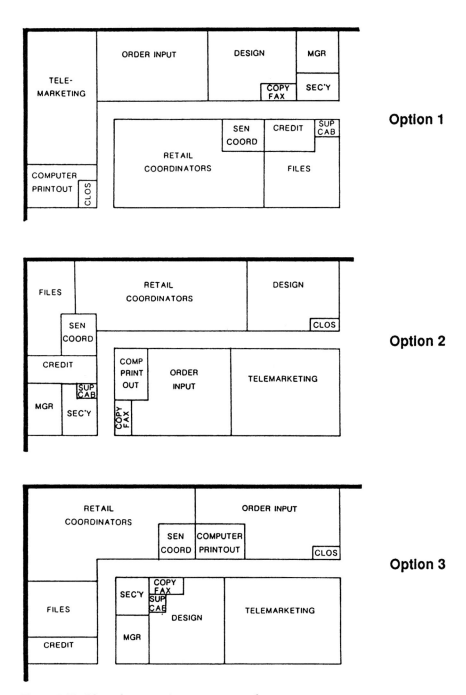

Figure 7.17   Macro-layout options are generated.

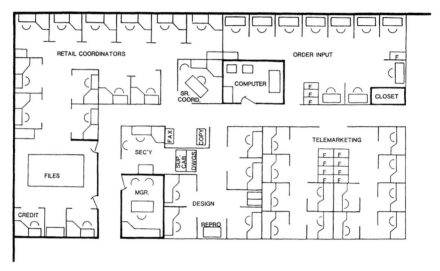

Figure 7.18 Detailing the layout.

The first challenge in a selection and evaluation process is *to qualify*. Certain basic or given requirements are not for evaluation. They are design specifications that must be met. Examples of such criteria are:

- The office will process 600 claims per day.
- The office is capable of processing a mortgage loan application within twelve hours.
- The office rearrangement will not cause loss of capacity.

Quantitative considerations. Every organization has its own methods of making financial comparisons. Methods of financial appraisal are beyond the scope of this chapter. The reader is advised to refer to specialist texts or the company's finance department.

Qualitative considerations. Frequently the quantitative considerations are not significant discriminators. When they are dissimilar, their importance may be outweighed by other factors.

Qualitative evaluation methods include:

- The sensory method
- PNI
- Weighted-factor analysis
- KORDANZ

The above methods of evaluation are covered in Chap. 11.

Step 7—Detail the layout

Detailed layouts are now prepared for the selected and approved macro or block space plan option. In this step the office equipment and furniture are arranged and located for each OSPU or block of space. This can be done by a repeat of the seven-step process for each office area.

The office detail may be designed either by interior designers or furniture systems suppliers. Planning from the macro to the micro level ensures that the plan is based on a logically developed and approved *macro layout*. An example of a detailed layout is shown in Fig. 7.18.

Recently, greater importance has been placed on *detailing the layout*. This is the application of the principles of *cellular manufacturing* to office operations. Cellular modes bring productivity gains to office operations. They are focus opportunities at the lower end of the focus spectrum. A separate chapter (Chap. 6) is devoted to workcell design for world class operations.

Summary

The procedures for office space planning described in this chapter provide a basis for significant productivity improvements in office operations. Other chapters in this book extend those opportunities.

The office layout project plan and the space-planning procedure provide a systematic and strategic approach to office designs. They can support your efforts to be and continue to be world class.

Global competitiveness requires that businesses restructure to provide a customer-driven advantage. Unfortunately, manufacturing advantages are often lost by sluggish office responses. The office should enhance the firm's competitiveness.

Further Reading

Cohen, Elaine, and Aaron Cohen: *Planning the Electronic Office,* McGraw-Hill, New York, 1983.

Fernberg, Patricia M.: "Designing the Workplace for the Knowledge Worker," *Modern Office Technology,* September 1985, pp. 73–82.

Fredrickson, Jack M.: *Designing the Cost Effective Office: A Guide for Facilities Planners and Managers,* Quorum Books, New York, 1989.

Hannon, N. Leslie: "Simplified Space Planning," Unpublished monograph, The Leawood Group, Ltd., Leawood, KS, 1986.

Hannon, N. Leslie: "The Well-Planned Office Is a Matter of Compromise," *The Office,* March 1983, pp. 104, 106.

Hooper, J. E.: "Determining Office Space Needs," *Area Development,* Easton, PA, March 1985, p. 70.

Katz, Stanley: "Outsourcing the Mail Room," *Chief Executive,* June 1990, pp. 32–35.

Knobel, Lance: *Office Furniture; Twentieth Century Design,* Dutton, New York, 1987.

Larson, Richard W., and David J. Zimney: *The White-Collar Shuffle,* Amacom, New York, 1990.

Murrey, J. M.: "Determining Office Space Needs: A Professional Approach," *Area Development,* Easton, PA, October 1985, p. 62.

Patri, Piero, and Jay R. Hendler: "Accommodating Office Automation in Facility Design," *Industrial Development,* March–April 1985, pp. 356–357.

Schafer, G.: *Functional Analysis of Office Requirements: A Multiperspective Approach,* Wiley, New York, 1988.

Stalk, George, Jr., and Thomas M. Hout: *Competing Against Time,* The Free Press, New York, 1990.

Steele, Fritz: *Making and Managing High-Quality Workplaces: An Organizational Ecology,* Teachers College Press, New York, 1986.

Stone, P. J., and Robert Luchetti: "Your Office Is Where You Are," *Harvard Business Review,* vol. 63, no. 2 (1985), pp. 102–117.

Swift, W. B.: "Outfitting Your Office," *US AIR,* vol. 12, no. 4 (1990), p. 40.

Taylor, Dwight: "Customizing Offices for High-Tech Industry," *Area Development,* Easton, PA, November 1989, pp. 25, 108.

Terjesen, James H.: "A Computer Room for Machines & People," *Modern Office Technology,* March 1985, pp. 64–78.

Trude, Joan: "Planning the OA Assault," *Modern Office Technology,* April 1985, pp. 71–76.

Tweedy, Donald B.: *Office Records Systems and Space Management: A Guide for Administrative Services Managers,* Quorum Books, New York, 1986.

Tweedy, Donald B.: *Office Space Planning and Management: A Manager's Guide to Techniques and Standards,* Quorum Books, New York, 1986.

Wineman, Jean D.: *Behavior Issues in Office Design,* Van Nostrand Reinhold, New York, 1986.

Wrennall, William: "Evaluation of Management Design Options," *Manufacturing Technology International: Europe 1989,* Sterling Publications, London, 1989, pp. 50–52.

Wrennall, William: "Kordanz Application," *The Leaword,* The Leawood Newsletter, vol. 3, Fall 1987, pp. 2–3.

Wrennall, William: "Productivity of Capital, Some Benefits from Facilities Planning," *The Journal of Methods-Time Measurement,* vol. XI, no. 3, M.T.M. Association, Fairlawn, NJ, 1986.

Wrennall, William: "Office Layout," *Handbook of Industrial Engineering* (ed. Salvendy, G. H. B., 2d ed.), Wiley-Interscience, New York, 1991.

Facilities Planning for Laboratories

N. Leslie Hannon
Vice Chairman/Director
The Leawood Group, Ltd.
Leawood, Kansas

Introduction

Facility design procedures apply to a range of operations. But laboratories are different from factories, offices, warehouses, and stores. They warrant special consideration. Laboratories are unlike manufacturing, storage, or offices, yet they may include them. Design issues may involve national security or unusual environmental problems. They accommodate people who generate thoughts, ideas, and concepts. Concepts are transformed into practical manufacturing reality.

This chapter discusses how laboratories fit into the corporate strategy and the special factors in facilities design, as well as:

- The phases of a laboratory layout project
- A generic project plan
- Layout features peculiar to laboratory design
- A laboratory layout example

What are laboratories?

Laboratories are described in Webster's *New World Dictionary,* 3d. College Edition (1988), as:

1. A room, building, etc. for scientific experimentation or research

2. A place for preparing chemicals, drugs, etc.

3. A place where theories, techniques, and methods, as in education or social studies, are tested, analyzed, demonstrated, etc.

4. A room, often containing special equipment and materials, in which students work to enhance skills, remedy deficiencies, etc. in a particular subject, as a foreign language

5. A class period during which students perform experiments or work in a laboratory

This chapter focuses on laboratories which:

- Carry out research activities
- Design and develop new materials, processes, and products
- Test materials, processes, and products
- Monitor processes

Such laboratories are in schools, colleges, industrial plants, commercial facilities, and government departments.

Laboratory focus

Focus is the accumulation or clustering of activities based on some criteria. Laboratories are typically focused by:

1. *Divisions of science*—chemistry; electronics; pathology; physics, either experimental or theoretical

2. *Material type*—lumber; plastics

3. *Product group*—automotive; dyestuffs; paint; paper

4. *Industry*—agricultural; chemical; petrochemical

5. Individual or ongoing *repetitive project*—particularly for 2, 3, or 4 above

6. *The "skunk works"*—the nonconforming autonomous innovative organization which is left on its own with its own image or lack thereof

Within each focus group laboratories often specialize in research, development, monitoring, or testing. The focus and specialty influences:

- The type of equipment housed
- Their supporting physical infrastructure
- The workplace design
- The building structure

■ Site features

Range and size

Laboratories range in size from relatively simple, small ones for manufacturing quality assurance (QA) to large, sophisticated research and development (R&D) activities which occupy a complete site, or even groups of sites.

Design for purpose

Many R&D facilities please aesthetically but do not function well. Their planning and execution has been ineffective. The facilities cost from 2 to 20 times as much as factory space.

Laboratories as a strategy support

To be competitive in today's world market, R&D must work quickly. Speed to market of new products is more important than development costs. First to market equals 100 percent of the market. R&D facilities should help achieve this goal.

Laboratory limitations

Laboratory work may be theoretical, abstract, creative, original, experimental, enhancement, developmental, prototype, or preproduction. Each activity may require service support.

Support services often limit research. An example is atomic physics. The capability of instruments to measure radiation by particle type limited progress. When improved support capability becomes available, this can have considerable effect, requiring planning or replanning of the affected facilities.

QA laboratories' role

The role of QA is also changing in today's world class manufacturing (WCM) environment. Local inspection and/or quality control (QC) procedures are performed by manufacturing people at the workplace. But pharmaceutical, veterinary, and food operations still need small test laboratories to check production. These will be in the manufacturing areas. There will also be the need for central, well-equipped QA laboratories. These should be minimum-cost, low-overhead, optimum-service facilities, giving rapid-response results.

QA professionals and technicians should be close to manufacturing. Their time frame must fit time-based, fast-response-to-customer strategies. They must, at the same time, meet applicable regulations and standards. All of this impacts the location, size, configuration, and layout of the QA laboratory.

R&D laboratories' role

R&D facilities are not subject to day-to-day direct manufacturing pressures. They generate new ideas, concepts, products, and processes. They test them, develop them, and bring them to fruition. R&D professionals are often loners. The laboratory and associated facilities should foster interaction between all concerned, while still respecting individual contributions.

Spectrum

The spectrum or time frame in which people work varies considerably, and this can have considerable effect on the type and location of facilities. The differences are illustrated as follows:

REAL TIME	ON TIME	ANY TIME
Q.A.	DEVELOPMENT	RESEARCH

R&D management and administration

R&D management and administration are important business success factors. They determine:

- Who works on what project
- What size the project teams should be
- What the project time frame should be
- Which laboratories and associated facilities they should use

  - Within a building
  - In which building on a site
  - At which site within a country
  - In which country

Business Plan Integration and Master Plan Development

The facilities planner should know where laboratories fit in the overall business plans. Is the present project a small, one-time job or is it part of a major objective needing careful integration and logical expansion? Business plans can have a strong bearing on facility location, configuration, and expansion.

A *facilities long-range master plan* gives effective direction for layout plans. A master plan includes:

- A mission statement for the facilities concerned: Where do they fit into the overall business picture? What is their "raison d'être"?

- An analysis of existing facilities, their condition, good features, suitability, shortcomings.
- Projections of long-term needs, possibly with dates, more likely tied to business achievement milestones.
- Short-range tactical schedule for growth projects with dates and budget estimates.

Differences between Research, Development, and QA Laboratories

There are fundamental differences between the various laboratory types.

Research laboratories conduct the basic research. New ideas are considered; new products are thought out; many different combinations may be evaluated. Research output is usually in embryo form and not yet suitable for manufacturing.

Research operations can be located almost anywhere. There are advantages in having them near development but this is not essential. In the United States much high-tech research has clustered in the West Coast's Silicon Valley and on Boston's Highway 128. Closeness to advanced-thinking universities has had much to do with this.

Development operations transition products and processes from pilot-plant to full-scale production. Proximity of the pilot plant to manufacturing is desirable. This facilitates trials and knowledge support. Without this proximity, a substantial amount of personal travel results.

An example of a multisite clustering proposal is shown in Fig. 8.1. This planning study shows how research, development, pilot plant, and R&D administration could fit on one or several sites, and their general relationships to the other corporate functions.

Development takes the embryo products from research and develops them into a form which makes them suitable for manufacturing. Development activities include pilot-plant operations, testing, field trials, government approvals, adherence to regulations, and environmental concerns.

QA laboratories are usually near manufacturing. They ensure that the manufactured products maintain the desired quality. Test samples should be readily available and close to the QA laboratory. Closeness also encourages rapid feedback and correction.

Some companies have a larger central QA laboratory for special analyses. Small satellite laboratories at manufacturing centers perform simple, basic testing.

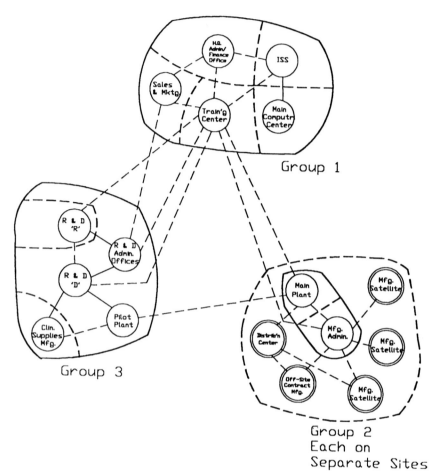

Figure 8.1   Functional group cluster focus.

Laboratories Are Expensive to Build

Laboratories by their nature are expensive to build. The least expensive "dry" laboratories—that is, those which do not have water, drains, etc.—cost about 50 percent more than an average office area. "Wet" laboratories start at 200 to 300 percent the cost of office space. The more sophisticated the laboratories, the more expensive they are. With air locks, gowning areas, air wash, ultra-high-level filtration, a lab may cost 10 to 15 times as much as an average office area. Laboratories with specialized electrical needs and of special configurations for advanced physics research can be higher-cost yet.

It is therefore important that there is minimal waste of space. Optimize expensive space and plan for logical, cost-effective future expansion.

Laboratory Office-Type Work Space

Because laboratory space is far more expensive than office space, office work areas or work space *should not* be built into the laboratory areas. Office-type work areas should be provided close to the laboratories, but outside them.

Where gowning is necessary and there is access through air locks, the minimum necessary writing space should be provided within the laboratory. Regular office space should also be provided outside for "workups" and other routine office work. This is important in any area where cleanliness is critical.

Grouping Office Work Areas

Office work areas which serve the laboratories should allow for easy interaction between the staff. Provide only partitions between workstations, using the open-plan office approach.

Provide small office "workup" areas (30 to 50 square feet) for technicians. Where practical, allow for easy rearrangement to suit restructured project teams.

Small break discussion areas foster interaction. Mini- as well as regular-size conference areas facilitate idea creation and problem solving.

Utilities and Services

Electrical equipment, computers, and peripherals can be stacked to minimize space. The stacking can be on mobile racks with appropriate umbilical cord services. Be sure that adequate cooling, ventilation, and power are available. Services which are not in constant demand can be located in service rooms at the ends of rows of laboratory benches. Examples are:

- Gas supplies in bottles or tanks
- Uninterrupted power services
- Filtration units
- Local storage needs
- Glass washing
- Cleaning service areas

Services can be located in space above the ceiling. This could accommodate:

- Air filtration units
- Water supplies
- Power and communications wiring

- Piping for special gases
- Vacuum and compressed-air services

In multistory facilities the space between floors can provide services above and below, taking drainage from the upper floor.

If ground-floor laboratories have a basement, providing drains is relatively simple since access is usually easy from the basement level. If the ground floor is built on-grade, the task is more difficult, especially if periodic reconfiguration is envisaged. One solution—during initial construction—is to run the main drains under the laboratories with service to all planned bench outlets, plus intermittent capped outlets for possible future needs.

Another option is to run the main drains in top-access tunnels in the corridors outside the laboratories with services teed off to planned bench outlets, plus some additional capped outlets. The "out-of-lab" tunnel option can help to minimize disruption when drainage problems arise. It may be necessary to run a secondary drainage system in parallel with the main drains if some effluent must be kept separate and run to a different area for monitoring and/or treatment.

An approach to providing adequate space for between-floor and below-floor services is shown in Fig. 8.2.

Flexibility and Project Teams

More and more R&D work is performed by project teams. The team works together until task completion. Because team sizes vary from project to project, there is a need to reconfigure available space and supporting services to meet the specific needs of the project.

Traditionally, laboratories have not been easy to reconfigure because of permanent walls and services. Now much greater flexibility is available.

Within an outside wall envelope, if the laboratories are configured so that the width between the outside walls stays the same, movable walls can divide sections into project team cells. Adjoining offices complete the team cell.

Two layout options are shown in Fig. 8.3. Option A, with blocks of laboratories separated by offices, makes good use of space, but office areas are long and narrow. Option B fosters more interaction in the offices, gives access to both sides of the laboratories if needed, and provides more flexibility for internal configuration.

Wet and Dry Laboratories

Where practical, separate "wet" and "dry" areas.

Wets: Include water, drains, imaging chemicals, fume hoods

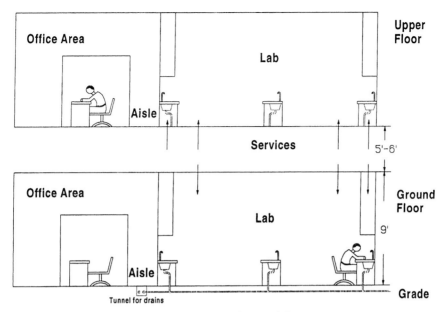

Figure 8.2 How to incorporate services in multistory lab.

Drys: Electronics, computers, mechanical, advanced physics

Wet laboratories are usually more expensive to build than dry. Floors and walls are special finishes, and floor/wall and ceiling/wall corners are usually coved to allow for easy washdown and cleaning.

Dry laboratories are often simpler to build. Where there is expected to be a great deal of wiring, it may be desirable to have either a raised or sunk "computer room"-type floor. Special heavy-duty power supplies may also require special configurations and have special service requirements.

Controlled-Condition Areas

Cleanrooms are specialized facilities with sophisticated air-cleaning equipment. They provide an ultra-clean environment. Cleanrooms are very expensive. When such conditions are necessary:

- Group activities to minimize the amount of support and service entries.

- Minimize floor space.

- Provide "gowning" area for lockers and changing, adjoining "air-wash" air-lock entries.

- Provide small "workup" areas inside for writing up notes. Provide office space for professionals and technicians outside.

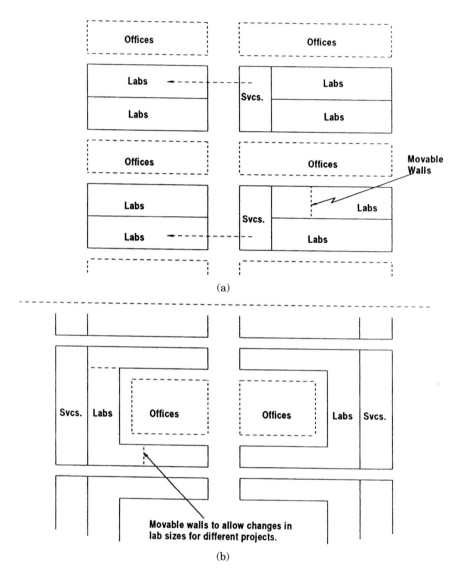

Figure 8.3 Layout options for labs/offices. (a) Option A: Block labs with adjoining offices; (b) Option B: modified "C" labs layout with grouped offices.

Special vibration damping and/or floor loading is best accomplished by locating these facilities on-grade to avoid vibration transmission through wall and floor structures from air-handling fan motors and other vibration sources.

Pilot Plants and Associated Laboratories

Pilot plants simulate manufacturing operations on a small scale. The purpose is threefold:

- Prove that the new product can be made.

- Debug before full-scale manufacturing.

- Make field-test lots prior to release to production.

Pilot-plant equipment should be easy to reconfigure. Flexibility is the key, as different projects may require different setups and combinations of equipment.

Plan to provide a staging area for holding equipment not in use. This will avoid interference with current experiments.

Pilot-plant operatives may need ready access to testing, etc. These areas should be close to the operations.

Lab-within-a-Lab

Contract research customers may require a product-focused lab-within-in-a-lab to ensure security of their "intellectual property." The layout should limit communications between projects and restrict the movement of visitors and visiting sales representatives.

Laboratory Machine Shops

Machine shops for mockups, prototypes, etc., should be close to those who will need these services. Machine shop facilities are considerably less expensive than laboratories. These services may also require security.

Access to Pilot Plants and Machine Shops

Pilot plants and machine shops should have docks with good turn-around capability for trucks. The docks should be as close as practical to these activities and to the storage areas serving them. These docks should not only provide day-to-day materials services, but also be capable of in-and-out loading of equipment.

Machine shops and pilot plants also need ground-level access. This allows for equipment moves and entry for forklift trucks and construction equipment.

Access to Laboratories

Laboratory facilities should have both dock and ground-level access for providing materials and moving equipment in and out. For multi-story facilities, the access should be as close as practicable to a service elevator.

Storage for Pilot Plants and Machine Shops

Flammables, special chemicals, and hazardous materials may need special storage facilities. Often these are on outside walls with blowout panels and special control systems.

"Controlled substances" such as drugs, or very high value materials, may also require special storage facilities. These areas will require electronic surveillance and limited access.

Storage for Laboratories

Central or distributed storage for supplies are the two options. A central storage facility allows much better materials management and control with a properly trained person in charge. Scientists and their associated technicians are not generally sympathetic to store-keeping routines, nor do they typically look after their own small, local storage areas very well.

Regulations mandate storage conditions for varying quantities and classes of hazardous materials.

Glass Wash for Wet Laboratories

Glass wash is a daily service required for most wet laboratories. A centralized service allows for the most effective equipment. But a volume/use/frequency-of-wash analysis may suggest more than one semicentralized service. This is particularly true for multistory or separate buildings.

The risk of cross contamination may dictate separate wash facilities for a particular area.

Bottled Gases—Bottles and Tanks

It may be desirable to locate bottled gases in an adjoining service area and pipe the gases to the point of use. If use is occasional and/or different gases may be required, then bring the bottles to the point of use.

Computer Services

Most R&D facilities require mainframe computer support. The facilities planner must determine with the R&D management the amount of space to allocate for this. This computer support will probably take two forms:

- A large central computer room with full *uninterrupted power supply* (UPS) and special environmental services.

- With the advent of local area networks there will be requirements for "utility closets" where computer hub concentrators and phone distribution panels are integrated. This concept would be applicable in a distributed-information environment.

The facilities planner should make every effort to ensure that adequate space is available for the information services required to support the operations.

The facilities planner also needs to be aware of the physical downsizing of central computers and the logical downsizing through distributed processing occurring at individuals' work areas.

Library Services

A good reference library is a necessity for every R&D operation. A well-maintained central library with some small satellites is often a good approach. Since the quantity of information is always growing, provide for future expansion.

Central Filing Systems

The space required for paperwork, notebooks, reference materials, microfilmed documentation, and so forth in an active R&D operation is large compared to other facilities. It must be housed, controlled, and readily accessible.

To accommodate growth, provide expansion storage or separate archive space. The facilities planner should work with R&D management to determine:

- What materials must be kept in active files
- What can be relegated to archives storage
- How much space will be needed for ongoing active filing
- Security methods

Floor loading can become critical with highly condensed filing systems.

Storage of Samples

Pharmaceutical companies are required to document and store samples of every batch or lot. Such storage may be long-term and in a humidity- and temperature-controlled environment. As operations and products grow, storage requirements increase rapidly.

Animal Operations

These are operations that use animals extensively for testing, research, and development. The following considerations apply:

- Animal operations must be kept totally separate from other R&D activities. If they are in the same building, they must be on a separate floor with locked doors between them. Access is limited to specifically authorized personnel.

- Animal operations must be self-contained. Facilities include the animal rooms, clean and dirty access corridors, cage-wash facilities, food storage and preparation, operating rooms and associated labs, and staff facilities such as locker rooms, changing areas, lunchrooms, and offices. Also included should be x-ray and "hot" (radiation) rooms.

- Air conditioning, heating, humidity control, and any other environmental controls must be separated.

- Animal operations should have their own truck dock for animals, food, and supplies.

- Trash handling and pickup must be separate.

Laboratory Facilities Layout Project Plan

Layout planning, design, and construction, whether for R&D or QA, follows the principles for planning other facilities. For example:

- The space-planning units (SPUs), activities, or cells required must be determined.

- Materials and people movement should be analyzed and incorporated in the planning process.

- Affinities need to be established, and configuration diagrams developed.

- Space needs must be calculated for both short- and long-term needs.

- Alternative layout primitives or block layouts need to be prepared and evaluated to determine the optimum layout.

- Constraints (often severe and specialized) must be considered in developing the final facilities layout plan.

The facilities layout plan should meet current requirements and anticipate future needs. It should provide for new technologies in the same locale where R&D skills and services are already established. The development of a long-range facilities master plan agreed upon with management, and updated/reviewed annually, can be a powerful tool for the facilities planner's use. Laboratory facilities planning has five phases, as shown in Fig. 8.4.

In Phase 1 the planner determines the scope of the project and locates the facility. This could be a new site, a building on a site, a floor of a building, or an area within a floor.

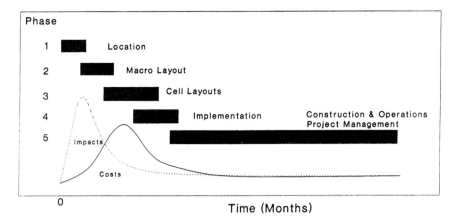

Figure 8.4 Layout planning phases with costs and impacts.

Phase 2 has the greatest impact on the organization's future. In this phase agreement will be reached on the solution in principle (the macro or block layout).

In Phase 3 the detailed layouts are made for each cell or associated work area.

The implementation planning in Phase 4 is particularly important when an existing facility is to be gutted and rebuilt. Programs continue and people need to work. There may be temporary doubling up in some areas, people may have to move temporarily into facilities located elsewhere, or other accommodations may need to be made.

Phase 5 is construction and occupancy. It also includes equipment debugging and start-up.

Activities which determine the cost occur in Phase 2.

New labs facilities example

XYZ Company is building a new R&D facility. Phase 1 of the project has been completed. The company has previously had small R&D laboratory activities in various locations on the site. The west wing of Building 2 is now available for the new R&D center. Space available is 1300 square meters, or approximately 14,000 square feet.

Planning tasks.   The *generic project plan* is shown in Fig. 8.5. It shows the tasks and their sequences for Phases 2 and 3. These tasks are:

2.01. Prepare the project schedule. Identify planning steps and determine the time frame. Agree with management.

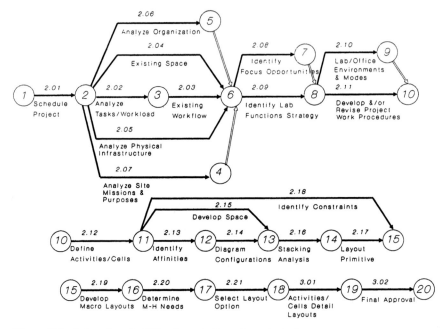

Figure 8.5   Laboratory facilities layout: Generic project plan.

2.02.–2.05. Analyze and review the existing R&D work, the workloads, work flow, space occupied, physical infrastructure, and shortcomings. This helps the facilities planner(s) to get a good understanding of what R&D does and the current problems.

2.06. Study and analyze the organization to be housed in the new facility and how it fits with the company organization.

2.07. Identify site mission(s). This helps to clarify the aims or goals.

2.08. Identify focus opportunities. It may be desirable to have a functionally focused layout with all laboratories of similar types grouped together. Alternatively it may be desirable to consider a flexible project-focused layout.

2.09. Identify the new laboratory's operations strategy. This requires the facilities planner to work closely with the R&D and company management. They present focus and operating strategy options and get agreement before proceeding with the rest of the project.

2.10. Determine the types of lab and office environments and modes to be considered in the facilities planning. The options, questions, and operating environments discussed previously in this chapter should be reviewed in depth with management.

2.11. Develop and/or revise, with management, the project work procedures for ongoing operations, particularly where they will impact on the layouts being developed.

2.12. Define the SPUs, activities, or workcells to be accommodated in the new facility. This is the first actual layout planning step. These are entered in the *affinity chart* shown in Fig. 8.6.

2.13. Identify the affinities between all activities and enter them on the chart using the A, E, I, O, U, X proximity ratings shown. These ratings range from A (absolutely necessary to be close together), in descending order through O (ordinary closeness is OK) to unimportant to X (not near, or keep apart). The reasons for the affinity determination are also shown in each case.

2.14. Prepare a *configuration diagram.* This will show the affinities established on the affinities chart in diagrammatic format, with A = four lines (like four rubber bands pulling close); E = three lines (or three bands); I = two; O = one; and X, represented by a wiggly line (a spring pushing apart).

Figure 8.7 shows the development of the configuration diagram for the project.

2.15. Determine the space required for each SPU. In this case, planning has been done to meet objectives five years ahead. As shown in Fig. 8.8, requirements have been developed in square meters (1 square meter = 10.76 square feet). Space requirements are shown net, with main circulation and access aisles added separately.

Projected space needs can be gross (including share of main aisles) or net (exclusive of aisles). Typically 12 to 15 percent of the total space is for circulation.

It is important to show what kind of space is required and what special factors should be considered. These are shown marked in Fig. 8.8.

2.16. Perform a stacking analysis where there is more than one floor (vertical stacking). When there is more than one building or separate locations on a single floor, horizontal stacking is needed. This is to determine what should go on each floor or other location to give effective space utilization and good functional balance. The configuration diagram (2.14) and possibly the layout primitive which follows in 2.17 can be used to develop appropriate clusters of those activities or cells which should stay together. Computer programs are available to use on the more complex stacking tasks.

2.17. Prepare a *layout primitive* as in Fig. 8.9. Superimpose the space for each activity area or cell on the configuration diagram. Some areas will be large, others small. This may skew the original diagram.

A simple way to make a layout primitive for smaller projects is to use squared paper and "count squares." This virtually eliminates

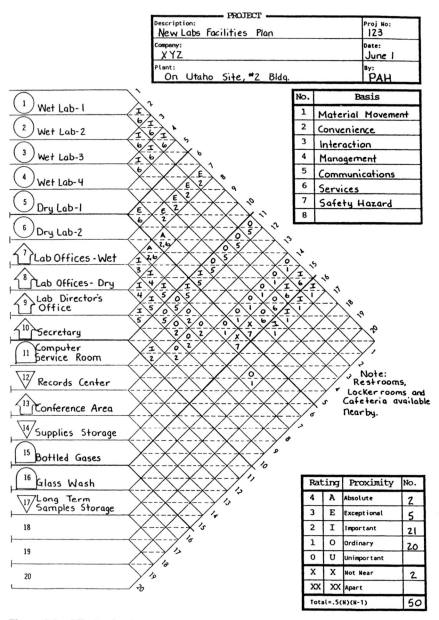

Figure 8.6 Affinity chart.

measurement. Pick a suitable scale. In this case each 1-centimeter square represents 10 square meters. The number of squares for each activity is shown in column 2 in Fig. 8.8.

Note: For this project the same scale is also used for making the macro-layout options shown later in 2.19.

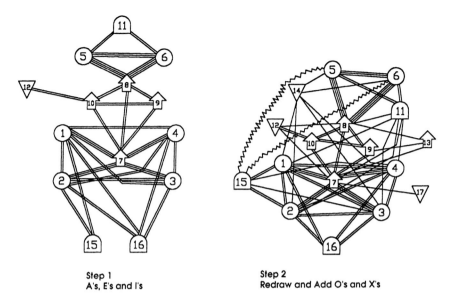

Step 1
A's, E's and I's

Step 2
Redraw and Add O's and X's

Figure 8.7 Configuration diagram.

ACTIVITY AREA SPACE SPECIFICATION

Facility New Labs in #2 Bldg.
Project 123
By PAH With
Date June 1 Sheet of

☐ Present   ☒ Plan for Year ±5

No.	No. of 1 cm Squares	Activity Name	Area in m²	Overhead Clearance	Floor Loading	Overhead Loading	Column Spacing	Foundations or Pits	Special Power	Special Environment	Water & Drains	Hazards	Special Services	Special Shape or Facility Requirements
1	12	Wet Lab-1	120	a.						✓	✓			
2	12	Wet Lab-2	120	a						✓	✓			
3	12	Wet Lab-3	120	a						✓	✓			
4	12	Wet Lab-4	120	a						✓	✓			
5	10	Dry Lab-1	100	a					✓	✓				
6	10	Dry Lab-2	100	a					✓	✓				
7	20	Lab Offices-Wet	200											
8	6	Lab Offices-Dry	60											
9	1.5	Lab Direct'rs Office	15											Include small meeting area.
10	1.5	Secretary	15											Include copier & Fax
11	2	Computer Svce. Rm.	20						✓	✓				
12	2	Records Center	20	✓										
13	3	Conference Area	30											With Audio Visual facilities
14	2.5	Supplies Storage	25											
15	1	Bottled Gases	10									✓		
16	1.5	Glass Wash	15								✓			
17	4	Long Term Samples Storage	40							✓				
18			1130											
19		Aisles 13%	150											
20														
		Totals	1280											

Notes: a. Provide 1.6m. clear above lab ceiling for overhead services.
Note: 5m. clear height available in building throughout area.

Figure 8.8 Planning space.

2.18. Constraints may force a less than ideal layout. Typical constraints are:

- Existing facilities
- Shape of the building

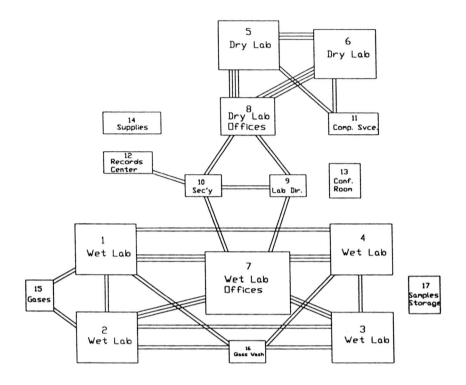

O's and X's honored, but omitted for clarity

Figure 8.9 Layout primitive.

- Column locations
- Locations of services (e.g., power, water, drains)
- Entrances and exits
- Ceiling heights
- Floor loading
- Management/user preferences
- Window locations
- Regulations
- Environmental concerns

There are, of course, many others, and their impact varies from project to project.

2.19. Develop several options of macro or block layouts. Three options are shown in Fig. 8.10. Since this is a relatively small project,

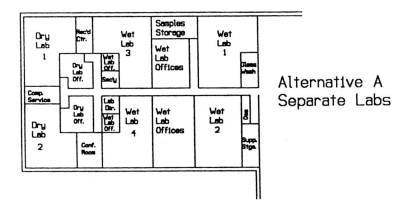

Alternative A
Separate Labs

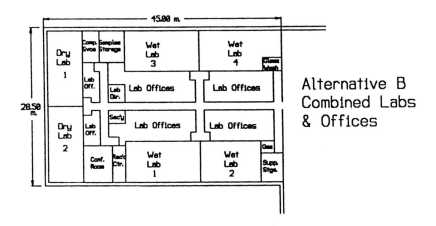

Alternative B
Combined Labs
& Offices

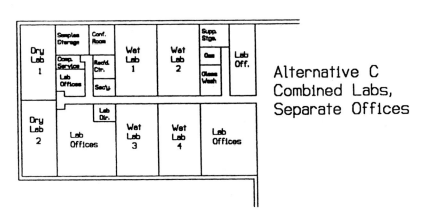

Alternative C
Combined Labs,
Separate Offices

Figure 8.10   Macro layouts for new lab facilities.

full advantage cannot really be taken of the lab/office options shown in Fig 8.3, but the principles suggested can be applied here. All three layouts honor the layout primitive in Fig. 8.9, yet they are quite different conceptually.

2.20. Determine any special materials-handling needs that may be required for each of the macro-layout options. Generally, materials-handling requirements are relatively minor in laboratories of these kinds, but they should certainly be considered.

For development and pilot-plant work materials handling becomes much more important and needs to have the same consideration as when laying out a manufacturing facility.

Special constraints apply and must be considered when handling radioactive isotopes, toxic chemicals, or other hazardous materials.

2.21. All three of the macro layouts in Fig. 8.10 are feasible, but each is different. Each has good points and some that may not be so good. All are compromises subject to the many constraints previously identified.

Undertake an evaluation study of the options. First determine that the "givens," the basic requirements, have been met by all options. Then do the qualitative and quantitative evaluation. There are many ways of doing this, such as:

- Sensory
- "PNI" (positive, negative, interesting)
- Weighted-factor analysis
- KORDANZ*

The *sensory method* relies on "I like it"; it looks good; it feels right. It stresses the importance of first-class presentation.

PNI, the "positive, negative, interesting" approach updates the pro/con method. In addition to pros and cons, it seeks points of special interest which may lead to new or enhanced options.

Weighted-factor analysis uses several qualitative factors. Each factor is given a weight (by management) reflecting its importance. Judges focus on each factor and rate the options against it. The product of weight and rating gives the score for each factor for each alternative. The highest total should be the best option.

A weighted-factor evaluation for the new XYZ Company labs facilities project is shown in Fig. 8.11. The result suggests that alternative B is the one which management prefers. This evaluation gives another opportunity for those using the facility to participate.

---

*KORDANZ is a trademark of The Leawood Group, Ltd.

———— PROJECT ————

Description: New Labs Facilities Plan		Proj No: 123
Company: XYZ		Date: June 4
Plant: On Utaho Site, #2 Bldg.	Page: Of:	By: PAH

No.	Factor	WGT	1 RTG	1 SCR	2 RTG	2 SCR	3 RTG	3 SCR	4 RTG	4 SCR	5 RTG	5 SCR	6 RTG	6 SCR	NTS
	Flexibility	9	U	0	A	36	I	18							
	Compatibility of Offices	10	I	20	A	40	O	10							
	Ease of Access	8	E	24	E	24	E	24							
	Ease of Supervision	5	I	10	E	15	I	10							
	Support Activities	7	E	21	E	21	E	21							
	Services	6	O	6	E	18	E	18							
	Honors the Affinities	10	E	30	E	30	I	20							
	Ease of Rearrangement	8	O	8	E	24	I	16							
	Totals		119		208		137								

Options:
1. A-Separate Labs
2. B-Comb. Labs & Offices   Notes:
3. C-Comb. Labs / Sep. Offices
4.
5.

Figure 8.11 Weighted-factor analysis.

KORDANZ is an advanced version of the weighted-factor method with many advantages:

- It quantifies opinion and tests concordance among evaluators.
- Effects of bias are reduced and pseudo-accuracy is eliminated.
- Qualitative and quantitative values are compared on a common basis.

- It identifies the relative importance of factors, separates factor weighting from evaluation, and generates confidence among the users through an acceptable level of agreement calculated in probability terms.

The KORDANZ approach is available as a software package for MS-DOS computers. It eliminates mathematical tedium and complexity for the user, providing a powerful and simple evaluation tool.

3.01. Now that the macro-layout or block-layout solution in principle has been decided, the detailed layout work for each activity area or cell can start. There is no point in doing detail layouts until the macro layouts have been agreed on. When doing the detailed work, minor changes in the macro layout may be suggested; these should be reviewed with management.

3.02. Completed detailed layouts should be reviewed with and approved by the laboratory management. After final approval and development of an implementation program, construction of the new facilities can begin.

State-of-the-Art Laboratory Review

There are many laboratories in the United States that have been well maintained and are in excellent condition, though they may have been built twenty to fifty years ago. Like any other facility, laboratory buildings periodically need considerable refurbishing to keep them in good shape—new roofs, major overhaul of electrical services, upgrading of other utilities, removal of walls, provision of adequate office space and storage facilities, and so forth.

In addition, to a greater extent than most business or industrial facilities, laboratories are very sensitive to changing technological, scientific, equipment, and project size needs. When competing in a world class environment one cannot afford to fall behind, or even give the appearance of falling behind, in state-of-the-art applications. While state-of-the-art "image" may be important to the outside world, it is probably more critical:

- First, in attracting the best people or employees
- Second, in retaining high-quality and scarce staff
- Third, in ensuring that projects are done as expeditiously, effectively, securely, and economically as possible.

So, for existing laboratory facilities the big questions are:

- Do the laboratory facilities meet state-of-the-art needs? If not, where do they fall short?

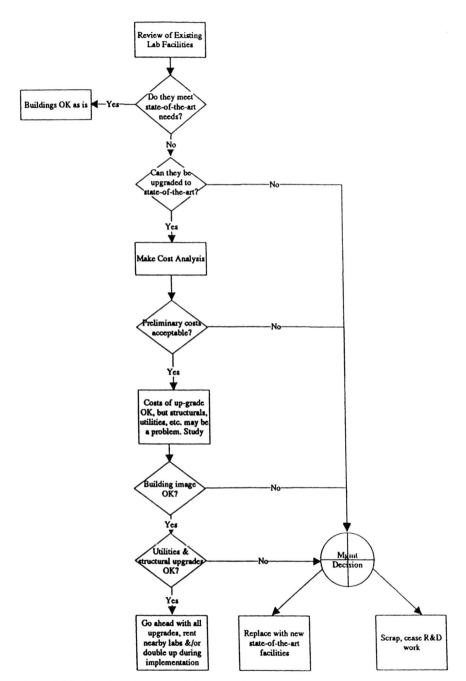

Figure 8.12 Lab facilities state-of-the-art review.

- Do the facilities need major refurbishing (utilities, structural, provision of offices, equipment, support services, etc.)?

- Can the facilities be physically upgraded to meet current and foreseeable future state-of-the-art needs? If so, what will it cost, and what do we do while the changes are being made?

Figure 8.12 gives a review of the steps to be taken.

The need to keep up with changing conditions and technologies suggests that new or rebuilt laboratories be designed to give the greatest level of flexibility to accommodate future changes. Clearly, if we knew what the future changes might be, we would make allowances for them now. But we do not; so building in flexibility will do much to extend the state-of-the-art life of laboratories.

Further Reading

Freedman, G.: *The Pursuit of Innovation,* Amacon, New York, 1988.
Hannon, N. L.: "Simplified Space Planning," Unpublished monograph, The Leawood Group, Ltd., Leawood, KS, 1986.
Hannon, N. L.: "The Well-Planned Office is a Matter of Compromise," *The Office,* March 1983, pp. 104, 106.
Tompkins, J. A., and J. A. White: *Facilities Planning,* Wiley, New York, 1984.
Wrennall, W.: "Evaluation of Management Design Options," *Manufacturing Technology International: Europe 1989,* Sterling Publications, London, 1989, pp. 50–52.
Wrennall, W.: "Kordanz Application," The Leaword, The Leawood Group Newsletter, vol. 3, Fall 1987, Leawood, KS, pp. 2–3.
Wrennall, W.: "Office Layout," *Handbook of Industrial Engineering,* 2d ed., chap. 71. Wiley, New York, 1991, pp. 1870–1901.

Designing Cleanrooms

Robert J. Ades, P.E.
Mechanical Engineer
Black & Veatch
Kansas City, Missouri

The use of cleanrooms, along with their impact on contamination management, is expanding rapidly to many areas. While the requirements for cleanrooms vary, these clean environments to produce high-technology products in the electronics, semiconductor, pharmaceutical, biotechnology, medical, and food industries have become vital.

This chapter outlines the standards that should be used by those planning a cleanroom and the design considerations for proper management of contamination.

History

Origins and development

Origins. At about the time of World War II, requirements for clean environments became apparent in the manufacture of precision optics, timing devices, and precision bearings among other items, and in the field of metrology.

Several organizations worked independently to develop specifications and standards for cleanrooms. Instrumentation and methods were developed to measure and identify contamination. Garments were developed (based on surgical clothing) to prevent particles shed from persons from contaminating the products. These early garments were ineffective because surgical clothing was primarily designed for microbiological sterility and not for low particle generation.

Some early cleanrooms involved using a pressure-tight internally smooth shell, made from a material such as polished stainless steel, with smooth radius corners and edges and pressurized with a small amount of makeup air filtered with high-efficiency particulate air (HEPA) filters. Persons inside the shell wore pressurized space suits which were designed to not allow contamination to leak out. These early cleanrooms were not successful in the battle against contamination. In fact, particle counts actually increased over time because the internal generation rate was greater than the rate of particles leaving the room in the exhaust or return airstreams. The initial costs of these first cleanrooms were prohibitive and they were difficult to use.

Laminar flow concept.   In 1961, at Sandia Laboratories in Albuquerque, New Mexico, the first *laminar flow design* was developed. The need for absolute reliability in the product required a cleanroom design which far exceeded the performance of earlier cleanroom designs.

Shortly after development of the laminar flow cleanroom concept a standard was written for cleanroom design and operation. This standard was Federal Standard 209, of which a later revision is still used, as will be seen below (see under "Federal Standard 209").

Microelectronics advancement.   Originally, electronic devices (transistors, resistors, etc.) were discrete entities which had to be connected to each other with conductors. Printed wiring boards or printed circuit boards contained the required conductors. Around 1958, technology became available to create electrical circuits containing many devices on semiconductors. These became known as *integrated circuits* (ICs). Over the years since their inception, ICs have been manufactured with more and more devices on each chip. Today, more than 1 million devices can be manufactured per circuit. With increasing component density the *line-width,* or the smallest width of a conductor or element of a device within an IC, has decreased over the years. This decreasing line-width makes the IC more susceptible to damage caused by contamination. Therefore, manufacture of ICs has been done in increasingly clean environments, and advances in cleanroom technology have primarily been focused on the needs of the microelectronics industry in recent years.

Other advances.   Between the time of the development of the laminar flow concept and Federal Standard 209, and the present, a number of significant cleanroom developments have occurred including:

- Instrumentation has changed and improved dramatically in order to keep pace with the lower particle counts of smaller and smaller particles.

- Environmental control technology was expanded to include process liquids and gases. Particle counters were developed for use in liquids. Better filtration was developed for process fluids.

- HEPA filter technology improved to keep pace with the ever cleaner cleanrooms. Filter construction, testing, and quality control were improved.

Present trends

In microelectronics, the line-width is continuing to shrink, requiring control in cleanrooms of 0.1-micron particles to less than 10 per cubic foot. This is requiring instrumentation manufacturers to improve their particle counters to be able to measure at this level. Some other trends that are occurring:

- More precise temperature and humidity control

- Better vibration isolation and elimination

- Use of robots for handling of product when practical (Robots are not necessarily cleaner than humans in appropriate cleanroom garments, but they can be used around toxic materials and for repetitive processes.)

- Better control of static electricity

Future

Integration will continue to be larger-scale, with line-widths decreasing to the molecular scale. New instrumentation technology will need to be developed to enable us to measure/count particles at this scale.

There will be increasing application of cleanrooms in the medical and pharmaceutical areas if hospitals and doctors find that this type of contamination control is cost effective and as the FDA accepts their use.

Food processing should see increased use of cleanrooms in tandem with heat/radiation as a healthier alternative to chemical preservatives.

Codes, Standards, and Practices

Codes

Codes, including building codes, fire codes, and other codes, have been written and adopted for use in the United States. Different sets of codes apply to different regions of the United States. Generally, the Uniform Mechanical Code, Uniform Fire Code, and Uniform

Plumbing Code cover the western part of the country; Southern Standard Building Code, Southern Standard Plumbing Code, Southern Standard Mechanical Code, and Southern Standard Gas Code cover the Southeast; and Basic Building Code, Basic Mechanical Code, Basic Plumbing Code, and Basic Fire Prevention Code cover the Northeast. The codes must be adopted by state law or local ordinance (whichever is applicable) to be enforceable by local authorities. Generally the codes are amended to meet local needs. The codes are similar in nature and specify many of the same requirements. Some of the requirements of the codes (as they apply to cleanrooms) are listed in this section.

Building codes. The building codes describe requirements for separation of the cleanroom from other areas in the building. The separation is to be by fire-rated walls, floors, and ceilings as applicable. The building codes specify minimum ventilation rates and requirements for exhaust ventilation. The codes specify requirements of the room in which hazardous materials are stored. The codes indicate the number of exits required in the cleanroom and separation of personnel exit corridors and hazardous material service corridors. Fire ratings of separation structures within the cleanroom—between corridors, storage areas, and other areas—are specified. The codes describe requirements for piping and tubing used to transport hazardous liquids and gases.

Fire codes. The fire codes limit the amount of hazardous materials which may be stored and used in a cleanroom and at a workstation. The fire codes require provisions be made to control excess flow of hazardous liquids and gases in piping with a fail-safe shutoff system. The codes require equipment for continuous monitoring of certain gases used in the cleanroom (e.g., hydrogen). The codes also require local exhaust at workstations. The fire codes require that only compatible exhaust streams may be combined in the same exhaust ductwork. The codes specify requirements for routing of exhaust ducts which penetrate occupancy separations. The codes specify emergency power requirements for exhaust fans. The codes require fire protection for certain exhaust ductwork carrying flammable materials. Fire protection requirements for transporting hazardous materials are specified. The fire codes require that an emergency plan be put in place and reviewed periodically by the local fire department.

Other codes. A number of other codes may need to be researched during design of cleanrooms. These include, among others, the National Fire Codes (NFC), published by the NFPA (National Fire Protection Association), and the National Electric Code (NEC), which is one Standard (Standard 70) in the NFC. The NFC comprises multiple volumes containing all the Standards, Recommended Practices, Guides, and Model Codes produced by the NFPA.

Regulations and legislation.   In addition to published codes, we have a number of regulations and a variety of legislation to protect the worker, the community, and the environment. Among these are regulations enforced by OSHA. One such regulation is the Hazard Communication Law, which requires that the manufacturer of a hazardous material provide the user with its properties on a Material Safety Data Sheet (MSDS). The Clean Air Act and Clean Water Act were passed to control dumping of toxic wastes in waterways and in the atmosphere. These acts are enforced by the U.S. Environmental Protection Agency (EPA). In recent years, more aggressive environmental legislation has been passed, which forces hazardous material users into recycling programs. Legislation has been passed concerning leaking underground storage tanks (USTs) containing hazardous materials. All new USTs require double containment, among other provisions such as monitoring and leak detection, to protect the environment. International regulations are banning the use of certain chlorofluorocarbons (CFCs) which deplete the earth's ozone layer.

Standards

Standards have been developed which include standards setting the performance requirements for cleanrooms and cleanroom components, and establishing fire protection requirements.

Cleanroom performance and testing standards.   The most relevant cleanroom standard for performance and testing is produced by the federal government. Other cleanroom standards are mentioned for reference.

    Federal Standard 209.   The federal government first published Federal Standard 209 in 1963. Over the years, 209 has been revised several times. Currently, Revision D (209D) is applicable and is in widespread use. Revision E has recently been published.*

Essentially, 209 covers only aerosol particulate contamination, which will be discussed in the section titled "Contamination." The standard defines cleanroom *classes,* the class essentially being determined by the number of 0.5-micron airborne particles per cubic foot of air in the cleanroom. The class defines the cleanliness of the cleanroom. The standard describes measurement and monitoring of parti-

---

*This chapter does not reflect the publication of Revision E.

Class	Measured particle size (microns)				
	0.1	0.2	0.3	0.5	5.0
1	35	7.5	3	1	NA†
10	350	75	30	10	NA
100	NA	750	300	100	NA
1,000	NA	NA	NA	1,000	7
10,000	NA	NA	NA	10,000	70
100,000	NA	NA	NA	100,000	700

†NA = not applicable.

Figure 9.1 Class limits in particles per cubic foot of size equal to or greater than particle sizes shown (in microns).

cle concentration. The standard also indicates how measurement data is to be handled statistically.

Cleanroom classes. The definition of cleanroom classes in accordance with Federal Standard 209 is summarized Fig. 9.1.

If the particle counts, measured and statistically treated in accordance with 209, are less than what is listed in Fig. 9.1 for a particular class, then the cleanroom can be certified for that class. For example, if a cleanroom particle count for 0.3-micron particles is 300 or less (after statistical manipulation of data) at all locations where counts were taken in the cleanroom, then the cleanroom can be certified as *Class 100*. For Class 100 and greater cleanrooms, particle counts of only one size particle need to be taken. For less than Class 100 cleanrooms, the sizes and the number of different sizes need to be specified by the user. A graphical representation of the information shown in Fig. 9.1 is shown in Fig. 9.2.

Particle sizes not listed in Fig. 9.1 can be interpolated from Fig. 9.2. Other classes besides the ones in Figs. 9.1 and 9.2 can be defined by drawing a line on Fig. 9.2 intersecting the desired class at the 0.5-micrometer particle size and extending the line parallel to the other class lines. Note that the class limit particle concentrations shown in Figs. 9.1 and 9.2 are defined for class purposes only and do not necessarily represent the size distribution to be found in any particular situation.

Measurement and monitoring. In order to certify that a cleanroom meets a classification, multiple particle counts must be measured at different locations for specific lengths of time. Federal Standard 209 specifies how many samples must be taken, locations of samples, and sample volume. The operating condition of the cleanroom is specified by the user. The cleanroom may be *as built* (no personnel or production equipment in the area), *at rest* (no personnel in the area), or *operational* (normal production operation in the area). Other condi-

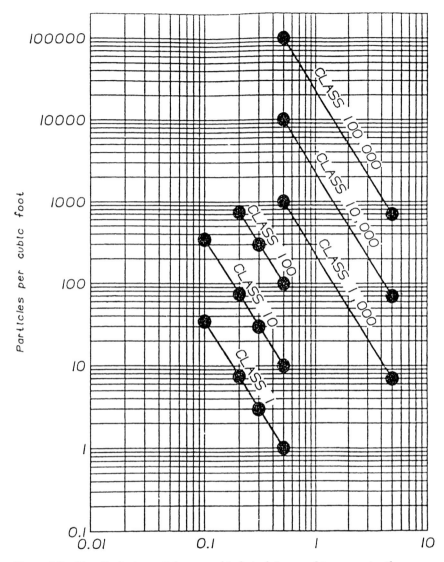

Figure 9.2 Class limits in particles per cubic foot of size equal to or greater than particle sizes shown (graph).

tions of the cleanroom are recorded, including air velocity, air volume change rate, temperature, humidity, and others.

After the cleanroom has been certified, the user may require monitoring of particle concentrations during operations. Federal Standard 209 describes standards for monitoring operational cleanrooms.

Federal Standard 209 also describes methods and equipment for measuring aerosol particle contamination. Generally, for particle sizes 5 microns and larger, a manual sizing and counting method or an optical counting instrument is used. For particle sizes 0.1 to 5 microns, an optical particle-counting instrument is used. Limitations of both particle-counting methods are described.

Statistical treatment of data.   A statistical analysis of measured particle counts is specified by 209. Two criteria must be met if the cleanroom is to be certified for a specific classification: (1) the average of the particle counts at each location must be at or below the class limits, and (2) the mean of the averages must be at or below the class limit with a 95% confidence limit. These criteria may allow some of the samples taken to exceed the class limit particle concentration, but as a whole the counts are below the class limit. The statistical analysis attempts to reduce the effects of random errors (lack of precision).

Other cleanroom standards.   Other standards exist for classifying cleanrooms or materials in cleanrooms. Military Standard 1246 specifies product cleanliness levels. The American Conference of Governmental Industrial Hygienists (ACGIH) produces their *threshold limit values* (TLVs) for various hazardous materials. The American Society of Heating, Refrigeration, and Air Conditioning Engineers (ASHRAE) produces a set of handbooks which contain a wealth of information on general HVAC design and some limited information on cleanroom design. The United Kingdom has its own standard, *Environmental Cleanliness in Enclosed Spaces* (BS 5295), paralleling Federal Standard 209. Another alternative to Federal Standard 209 is the U.S. Air Force Technical Order T.O. 00-25-203, *U.S. Air Force Contamination Control of Aerospace Facilities.* The National Environmental Balancing Bureau (NEBB) produces *Procedural Standards for Certified Testing of Cleanrooms.*

Fire protection standards.   Fire protection standards include those produced by the NFPA, among others. Many of these standards have been adopted throughout the United States. The NFPA produces *Standard for Protection of Cleanrooms,* NFPA 318, which is a good reference for life safety considerations. NFPA 231 and 231C apply where bulk storage and rack storage are used. NFPA Standard 30 applies where flammable liquids, gases, and volatile solids are used and stored.

HEPA filter standards.   In the United States, HEPA filters are normally tested using the *thermal di-octyl phthalate (DOP) method,* which is described in the U.S. Military Standard MIL-STD-282. In this method, a smoke cloud of DOP droplets which are about 0.3 micron in diameter is generated by vaporizing DOP by heating and condensing into a cloud, and the cloud is fed to a HEPA filter. The upstream and downstream concentration of DOP smoke is measured. The filter penetration and the filter efficiency are calculated from these measurements.

HEPA filters are tested for flammability in accordance with Underwriters Laboratory Standard 586. ASHRAE Standard 52 specifies tests for dust-holding capacity.

Practices

In addition to codes and standards, guidelines and practices are published to assist cleanroom designers and operators.

IES recommended practices.   The Institute of Environmental Sciences (IES) produces a series of Recommended Practices which are useful in design of cleanrooms. *Cleanroom Design Considerations* (IES-RP-CC-012) is a good guideline for use when designing cleanrooms. Other IES Recommended Practices include *Testing Cleanrooms* (IES-RP-CC-006) and *Recommended Practices for HEPA Filters* (IES-RP-CC-001).

Other practices.   The American Society for Testing and Materials (ASTM) publishes standard practices for particle measurement, filter leak testing, and instrument use and calibration.

Contamination

Proper cleanroom design requires definition, a thorough understanding, and proper identification of contaminants in the cleanroom or in cleanroom process materials.

Types of contamination

In order to control contamination in a cleanroom, it is extremely important to define and identify the types of substances which are to be considered "contamination."

Contamination will be defined as any substance (solid, liquid, vapor, or gas) which is undesirable in any material or on any surface in the cleanroom. Materials often present in cleanrooms include atmospheric air, process gases, and process liquids. Surfaces in the cleanroom often will include surfaces of product (e.g., integrated circuit), surfaces of production equipment, furniture, walls, floors, ceilings, and surfaces of facilities equipment. Contamination in atmospheric air or process gases may be aerosols (solid particles or vapors in gases) or other gases. Vapors originate from solid or liquid surface interfaces with gases.

Contamination in liquids may be solid particles, other liquids, or dissolved gases or vapors. Contaminating materials which dissolve in liquids may be difficult to remove, presenting a monumental challenge to the individual trying to provide a high-purity process liquid for use in the cleanroom.

Contaminating materials may be categorized as either organic or inorganic substances. Organic substances usually are naturally occurring materials but may be manufactured materials in some cases. All organic compounds contain carbon and may be in solid, liquid, or gaseous form. Organic substances may be viable or nonviable. Examples of organic substances are plastics (solid), alcohols (liquid and vapor), and propane (gas). All of the listed examples may be found in an ordinary cleanroom. Bacteria is an example of a viable, complex, naturally occurring organic substance which may be considered contamination in many cleanrooms and process materials.

Inorganic substances may be naturally occurring or human-made solids, liquids, or gases. These substances may be elements (e.g., silicon), compounds (e.g., oxides of iron or other metals), or mixtures of elements (e.g., metal alloys).

Solid particles.   Solid particulate contamination may be carried by fluids (liquids or gases). Solid particles in atmospheric air generally are small, ranging in size from visible (around 50 microns) to microscopic (smoke particles as small as 0.01 micron in size). For comparison purposes, the human hair is between 30 and 200 microns in diameter. Solid particles in liquids may be much larger than particles in atmospheric air because of the higher viscosity and density of the liquids (resulting in larger drag forces).

A generally accepted rule of thumb is that approximately spherical particles greater than 1 micron will settle out of still or slow-moving air relatively quickly under the force of gravity. Particles between 0.1 and 1 micron will settle more slowly and may be carried indefinitely by the slightest air movements. Less than 0.1-micron solid particles will have a tendency to remain suspended in still air permanently because of Brownian motion; the particles are constantly bombarded from all sides by air molecules, causing them to bounce around and resist the effects of gravity.

Solid particles in air may be considered contamination in cleanrooms because they may be deposited on the material that is to be kept clean. For example, in microelectronics, the IC must be kept microscopically clean to prevent damage by deposition of particles on the IC, causing short-circuiting or other types of failures. Deposition of particles may be caused by contaminated air or process fluids coming in contact with the IC and the particles in the contaminated fluid settling on and attaching to the IC because of gravity and other forces. Once a particle is attached to a surface, it may be difficult to remove without damage to the IC. Therefore it is essential that the surface of the IC not be allowed to become contaminated—this is accomplished by maintaining a microclean environment and using only high-purity production materials.

Gases and vapors. Gases and vapors may be considered contamination in some cleanrooms. Gaseous contamination may be a component in a mixture of gases such as oxygen in atmospheric air, or could be dissolved in a liquid. To truly be a gaseous contaminant, the gas must react directly with the product from the gas phase (e.g., oxidation of metallic surfaces). The gaseous contamination may be carried in a moving fluid or may be transported by diffusion (transport of substance from a higher concentration to a lower concentration) in a still fluid.

Vapors may react with the product directly from the vapor phase or they may condense on cool surfaces, forming a layer of condensed liquid on the surface (e.g., oil vapors condensing on a cool surface).

Liquids. Liquids may be considered contamination in some instances. The liquid contamination could be broken up in a mist (small particles in liquid state) and carried by a gas. The mist will either vaporize into the gas or will be carried by the gas in much the same way solid particles are. The mist particles may settle out of the gas by gravity and reform on a surface as a liquid.

Liquid contamination may also be carried by another liquid. Liquid contamination may be insoluble in the carrying liquid and will either float or sink in the still liquid depending on the specific gravities of the contaminant and the carrying liquid (e.g., water will sink in xylene or gasoline). Liquid contamination which is soluble in the carrying fluid will not float or sink and will be difficult to remove (e.g., water in isopropyl alcohol).

Sources of contamination

As important as identifying the types of contamination in a cleanroom is the identification of the sources of contamination. No attempt to reduce a particular type of contamination in a cleanroom will be effective without determining the source of the contamination.

External sources. Many of the contaminants in cleanrooms or in cleanroom process materials have a source external to the cleanroom. The contaminants are carried into the cleanroom from outside the building in fresh air (outdoor air) makeup airstreams. Fresh air is required to replace process exhaust air and air lost by exfiltration and relief dampers. A certain quantity of fresh air is required by building codes and other codes and standards for human respiration.

Much of the contamination is removed from the airstream by filtration systems, but a small amount will pass through and will contaminate the cleanroom atmosphere.

Process fluids will carry a small amount of contaminants into the cleanroom from external sources by the nature of the less than 100 percent efficiency filtration systems.

Natural causes. The natural environment contributes great quantities of contamination in solid and liquid particulate, liquid, vapor, and gas forms. The environment surrounding the facility to a great extent determines what kinds of contaminants are present. In sandy or desert areas, oxides of silicon are likely to be a problem; near the ocean, sea mist will carry dissolved salt which may deposit on ICs after the water has evaporated.

Pollen produced by reproducing plants will be present in most areas and constitutes a contamination produced by biological sources. Dead plant and animal cells may become airborne when they dry up. Bacteria and molds may be present in the atmosphere or may grow in other fluids. Decaying animal or vegetable matter may release methane gas into the environment.

Industry. Industrial activity contributes large amounts of contaminants to the atmosphere external to the cleanroom. Burning of fossil fuels contributes soot (carbon particles) and fly ash to the atmosphere, increasing average particulate contamination levels.[2] Foundries produce airborne particles consisting of metallic compounds.

Burning fossil fuels produces large amounts of gaseous contamination including carbon dioxide and sulfur dioxide. Exhaust from automobiles contributes ozone and carbon dioxide, among other gases, to the environment. CFCs are released from refrigeration systems and contribute to gaseous contamination of the atmosphere. Many areas such as Denver and Los Angeles may hold contaminants in because of terrain (mountains) and prevailing winds.

Facilities. Facilities which are used to modify or produce cleanroom air or production fluids may contribute to contamination. Moving equipment (such as rotating or reciprocating pumps, fans, valves, dampers) will have wearing surfaces. The material from these surfaces may manifest itself in the form of airborne or fluid-transported contamination. Duct, piping, and other equipment will contain contaminants during its installation which may not be removed by initial cleaning prior to being put into service. These contaminants may "break loose" during the life cycle of the equipment and contribute to overall contamination.

Production materials. Even "electronics-grade" bottle-source gases will contain trace amounts of contaminants. No production material can be made 100 percent pure at any cost. The higher the desired level of purity, the higher the premium paid for the material.

Internal sources. Much of the contamination present in cleanrooms can be traced to internal sources (i.e., sources within the boundaries of the cleanroom). Highly effective filtration systems may prove useless in controlling contamination from these sources. Contamination control may involve taking measures at the source of the contamination.

Personnel. Personnel present a double-edged sword of contamination problems. People produce an extraordinary amount of contamination and people are often in close proximity to the product. Covering personnel with "bunny suits" will result in a dramatic reduction of contamination liberation if used properly.

Skin flakes are the predominant contribution to cleanroom contamination by humans. Breathing produces water vapor and carbon dioxide. Coughing and sneezing will expel contaminants at high velocity. Cosmetics and deodorants compound the problem of personnel contamination. Smoke particles continue to be emitted from a cigarette smoker even after a period of time has lapsed after smoking.

Equipment and facilities. As described above under "Facilities," equipment and facilities with moving parts will generate particles and possibly oil vapors from lubrication. Coatings such as paint may out-gas for years after their application. Equipment moved into cleanrooms may have been stored in dirty areas, and if proper decontamination procedures are not followed, the equipment will contribute to contamination in the cleanroom. Walls, floors, ceilings, and other facilities must be properly cleaned prior to commissioning the cleanroom.

Improper work practices. Personnel can contribute to contamination problems by not following proper work procedures. Examples include the following: not dressing or wearing cleanroom apparel correctly; wearing cosmetics where not allowed; not cleaning up spills promptly and appropriately; not handling product properly or with the correct equipment.

Detection and measurement of contamination

While defining problem contaminants and isolating their sources is important, detecting and measuring contaminants in the cleanroom is no less important. In fact, it is the detection and measurement of contamination that helps us determine the true contamination problems in a given cleanroom and in turn guides us to the source.

Visual techniques.   The human eye is the most fundamental instrument used for detection and measurement of contamination. Using a smooth, flat plate and strong light, a trained and experienced inspector with good eyesight should be able to resolve large particles (greater than 50 microns). All visual techniques involve observing stationary particles on a plate or smooth surface of a membrane filter medium and manually counting. A grid can be etched on the observation plate to aid in counting particles. By placing the observation plate in a region of interest in the cleanroom and allowing particles to settle on it for a measured length of time, the observer will be able to determine the "fallout rate" per unit area. This measurement does not tell anything about the small particles which do not settle, but they may not be of interest anyway.

Optical microscope.   An obvious extension to the naked-eye visual technique is the use of an optical microscope. The optical microscope will enable the operator to easily resolve 1-micron particles. By drawing a measured quantity of air through a membrane filter and observing the filter surface under a microscope, the observer can measure the particle count of a particular volume of air.

Electron microscope.   Scanning electron microscopy (SEM) will allow much higher magnification, and therefore much smaller particles can be resolved. Use of the electron microscope requires a high degree of training and skill and its use for everyday monitoring may be limited.

Light-scattering techniques.   Because visual techniques require manual counting of particles on a plate or filter of some sort, there is a high risk of operator error in their use. An automated technique which does not involve transport of a sample is much more desirable.

Light-scattering techniques utilizing the *Mie scattering theory* are often used. A white light or laser light beam is aimed at a collection of particles and the scattered light is collected by a detector. Computer programs relate the scattered-light intensity and scattering angles to the Mie theory to yield information on particle sizes and counts. Laser-based airborne particle counters using Mie theory are reasonably accurate to about 0.1 micron. To increase the range of light-scattering counters, supersaturated vapor from an organic material is condensed around the particles (the particles act as nuclei) to increase their size. These systems are appropriately known as *condensation nuclei counters* (CNCs) and can be used down to 0.01-micron particles.

UV-induced emission of electrons.   When energetic ultraviolet (UV) light strikes a material, electrons may be emitted. This is known as the *photoelectric effect.* The emitted electrons may be collected by an electrode near the material bombarded by UV light. This system may be used to gain qualitative information on contamination on surfaces.

Corona discharge. This system involves the use of a corona discharge which pulses when particles cross the discharge path and affect conduction. This system will detect particles on the order of nanometers, much smaller than can be detected by scattering techniques.

Gas and vapor detection. In the cleanroom many chemicals are used which may give off toxic or hazardous vapors. Detection of these vapors is important, not only for contamination control but for personnel safety as well. Explosive, corrosive, toxic, and hazardous gases may also be present in the cleanroom for process requirements. Each cleanroom may require one or more of the detection devices described below.

Hydrogen detectors. Hydrogen detectors utilize a simple electrical bridge arrangement along with hydrogen-sensitive elements to determine the amount of hydrogen in air. The units generally output the percent of the lower explosive limit of hydrogen in air (4 percent). The units will have local alarms and provision for transmitting alarm signals to a central panel, where automatic action is undertaken to shut down the hydrogen system and purge with inert gas.

Solvent vapor detectors. Solvent vapor detectors are similar in operation to hydrogen detectors. Each element is calibrated to detect one of many organic solvent vapors, including xylene, acetone, and alcohol, among others. These units are placed in locations where vapors are likely to collect if a spill occurs.

Chlorine monitors. Chlorine gas is toxic and extremely corrosive; thus, monitors must be installed where the gas is used. Automatic action should be undertaken to shut down and purge the chlorine system when a leak is detected.

Toxic and hazardous gas detection systems. Generally these systems employ a central monitor with gas tubing carrying samples from remote sites. A number of types of monitors are employed. One kind uses chemically sensitive paper which is read optically. Another type uses flame photometry. Another unit uses gas chromatography in tandem with a mass spectrophotometer. A final type uses photoionization.

Smoke detectors. Fire codes require that smoke detectors be installed in the return duct or plenum to the air-handling units. The sensors may utilize photocell technology to measure blockage of light by smoke particles or may utilize an ionizer to charge the particles which can then be measured electrically. An alarm signal from a smoke detector can be used to automatically shut down equipment and to activate smoke purge systems.

Cleanroom Types

The type of cleanroom can be described by the layout and the airflow pattern. The types of cleanrooms will be defined as follows:

Nonunidirectional

Unidirectional

Mixed flow

Mini-environment

The type selected is always based on a perceived need, facility operation, level of cleanliness required, and cost. Cleanrooms are expensive to build and operate. It is important for the user to specify only the type and level of cleanliness absolutely necessary.

Nonunidirectional

Nonunidirectional cleanrooms, as the name implies, are those for which no particular attention is given to the airflow direction. In a typical nonunidirectional cleanroom, air flows from partial-coverage HEPA filters normally located in the ceiling to returns located in the raised floor or side walls. Figure 9.3 shows a typical nonunidirectional airflow arrangement. Due to the layout of the filters and the location of the returns, the airflow in the space is normally turbulent.

Unidirectional

Unidirectional cleanrooms have one-directional airflows. This is also commonly known as *laminar flow.* Air flows vertically downward or horizontally. In a typical vertical unidirectional cleanroom, air flows from 100 percent coverage filters of the high-efficiency particulate air (HEPA) and ultralow particulate/penetration air (ULPA) types which are located in the ceiling, to a perforated raised floor. Figure 9.4 shows a typical unidirectional airflow arrangement.

Mixed flow

Mixed-flow cleanrooms have a combination of nonunidirectional flow and unidirectional flow in the same room. The concept is to provide higher-velocity air at the process than in the remainder of the room. The theory is that the different velocities will cause the particles to

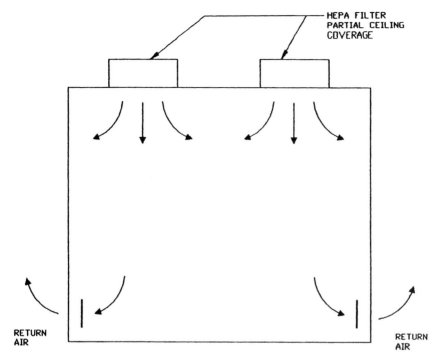

Figure 9.3   Nonunidirectional cleanroom.

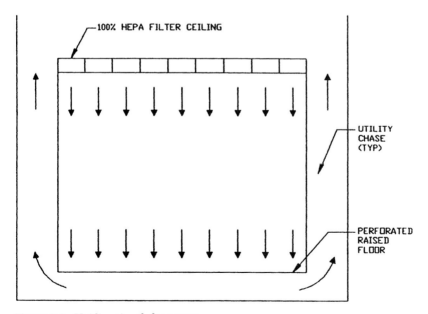

Figure 9.4   Unidirectional cleanroom.

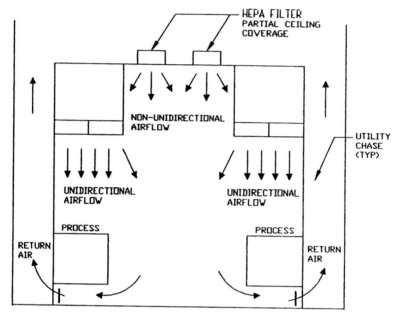

Figure 9.5   Mixed-flow cleanroom.

be transported away from the process. The unidirectional, higher-velocity air is provided over the process and the nonunidirectional, lower-velocity air is to be provided in the remainder of the room. Figure 9.5 shows a typical mixed-flow cleanroom arrangement.

Mini-environment

A *mini-environment* is a manufacturing area in which the process is physically isolated from the surrounding area.

The mini-environment concept has existed since cleanrooms were first applied but is receiving renewed interest, particularly in the semiconductor industry. As the semiconductor industry moves toward smaller and smaller line-widths, mini-environments are being considered as a way to meet the particle, temperature, and humidity requirements. The mini-environment shows great potential to lower capital and operating costs while providing improved particle contamination management.

In a modern Class 1 cleanroom similar to the unidirectional cleanroom shown in Fig. 9.4, the process is not physically separated from the remainder of the room. In this arrangement the product is exposed during the process to the particles generated from cleanroom activities. Studies have shown that most of the particles present in a modern Class 1 cleanroom are generated from cleanroom activities.

Current mini-environments consist of a *clean hood,* sometimes known as a *laminar-flow hood,* or an enclosure around the process. One concept currently receiving much attention is the use of a specially designed system known as a *standard mechanical interface* (SMIF). The SMIF system consists of SMIF transport bores, SMIF interfaces, and sometimes SMIF transport systems.

Design Practices

Planning

The planning of a cleanroom facility requires participation of all the team members who will be responsible for the project. The team normally includes:

Designers	Safety group
Users	Environmental group
Facilities planners	Contractor
Process group	

The team must establish why the cleanroom is being built and what the budget and schedule will be. The contamination management goals for particles, gases, and vapors, and for biological and ionic contaminants must be established. Life safety relative to the functions to be performed in the cleanroom must be considered. The energy conservation goals must be developed. Use of isolation technology must be considered. The clean construction protocol program must be specified.

The careful planning of the design and construction of a cleanroom is critical to the success of the function of the cleanroom.

Materials of construction

No perfect cleanroom material exists. This statement is a generally accepted conclusion throughout the cleanroom design field. The perfect cleanroom material would need to excel at a variety of different property characteristics including non-out-gassing, low particle shedding, electrostatics, mechanical strength and toughness, ability to resist abrasion, ability to be cleaned, resistance to bacterial growth, appearance, chemical composition compatible with product or process materials, and others. One material may excel at one or more of the characteristics, but no materials excel at all characteristics. The cleanroom designer must choose the characteristics that are most relevant and important in the particular cleanroom being designed and choose materials excelling at these characteristics.

Walls. Walls must have the ability to resist abrasion and must be easily cleanable. Walls must be able to withstand impacts without chipping or perforating. Cleanroom walls can be constructed of a core consisting of wood, gypsum, or honeycombed aluminum with a skin or coating of polymeric material. Sometimes hollow-core walls are used with a steel skin coated with an enamel paint. Painted coatings may out-gas and should be considered carefully. Hard plastic laminates may be used but tend to be more expensive. If wood or gypsum cores are used, great care must be taken when cutting and sealing penetrations. Walls may be static-dissipative or static-retentive depending on requirements. Glass is used for windows and rates well in most property characteristics.

Floors. Floors must be able to withstand heavy loads and constant abrasion without breaking, particle shedding, or excessive deformation. Tile-type floors are unacceptable because the gaps between the tiles will collect dirt and are difficult to clean. Generally, a vinyl floor covering is bonded to a concrete floor. The floor may be static-conductive and grounded to prevent accidental static discharge, which may destroy products. Floors should not be treated with wax because the wax will flake off and become a contaminant.

Ceilings. Ceilings do not need to be as durable as floors and walls because they are not exposed to the same abuse and abrasion. However, ceilings still need to have smooth, cleanable finishes which do not collect dust and have attractive, light-reflective finishes. The acoustic ceilings which are often used in office-type spaces are not acceptable. Ceilings for cleanrooms are usually modular with a grid system and standard spacing so that HEPA filters, lights, and tiles are interchangeable. The grid is usually inverted-T sections or special HEPA filter ceiling grid made from aluminum or coated steel with gaskets to seal against air leakage. The tiles are clipped down to force them to seal against the gaskets against the positive pressure of the room. Tiles must be designed to allow penetrations such as process piping and fire protection sprinklers. The modular design allows the ceiling components to be easily rearranged or replaced as requirements change.

Air filtration

The goal of the cleanroom is to be able to provide a space for activities requiring nearly absolute freedom from particulate and other forms of contamination. Air filtration is one of the essential means through which this goal may be reached.

Gas and vapor removal. Of primary concern in the cleanroom is removal of particles. However, increasing attention is given in the cleanroom to removal of gases and vapors. Water vapor can be removed in the air-conditioning system by cooling the air below its dew point. Gases and vapors which are generated within the cleanroom (e.g., from out-gassing of materials, fabrication processes, leaks from production material systems) may be removed directly by dedicated exhaust systems if the source is known. The gases and vapors may be diluted by exhausting a certain amount of cleanroom air (possibly in addition to dedicated exhaust) and introducing makeup air from outside.

If these measures are not sufficient for removal of organic vapors or the source of organics is the makeup air, activated-charcoal filtration should be used. Activated-charcoal filters consist of beds of a dry, porous form of carbon which has a strong affinity for hydrocarbon vapors. After some time, the carbon beds will become "loaded" with organics and they can be regenerated by heating the carbon beds or passing steam through them to drive off the contaminants.

Care must be used with these filters because they can become a source of contaminants.

Particulate contamination removal. The most efficient and cost-effective means for removing particles from air is with filters. However, other technologies have emerged which enhance filtration or offer alternatives. Particulate filters may be either fibrous or membrane. Fibrous filters are used exclusively for environmental filtration in cleanrooms (exception: membrane filters used to sample and test air using microscopy).

Particulate filtration theory. A brief discussion of particle filtration theory is necessary to understand the mechanisms which fibrous filters use to remove particles from air.

Five major mechanisms are responsible for the removal of particles throughout the entire particle size spectrum, as illustrated by Fig. 9.6.

- *Gravity.* Particles greater than approximately 5 microns will settle out of the airstream under the force of gravity. The larger the particle, the more rapidly the particle will settle out of the airstream.

- *Direct interception.* A particle follows a streamline which passes (within one-half of the particle outside dimension) next to the outside dimension of the filter fiber. Any size particle can be intercepted by this mechanism.

- *Inertial impaction.* More massive particles, by the nature of their higher inertia, do not strictly follow the fluid streamlines and will tend to continue on a straight path. If this path is near a filter

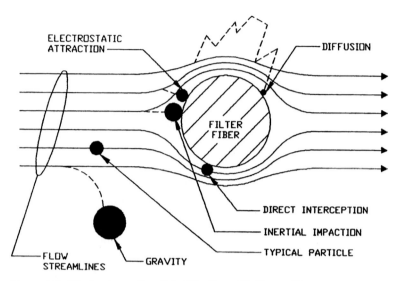

Figure 9.6 Mechanisms of particle filtration with fibrous filters.

fiber, the particle will be intercepted. This mechanism applies only to particles larger than about 0.5 micron.

- *Diffusion.* Small particles are "bounced around" by the molecules of the air and take a random path rather than following the streamlines. This phenomenon (Brownian motion) has been discussed earlier in this chapter. The more time the particles spend residing in the filter the more likely they are to be intercepted by the filter fibers. This is one reason a thicker filter medium will be more efficient. Only particles less than about 0.5 micron are intercepted by this mechanism.

- *Electrostatic attraction.* Electrical charges on either the particle, the fiber, or both will cause an attractive or repellent force between. Unless both items are charged with the same polarity, the force will be attractive. Any size particle can collected with this mechanism. As we will see later, this principle can be used to enhance the collection efficiency of filters.

Because some mechanisms are more effective for larger particles (inertial impaction, gravity) and diffusion is more effective for smaller particles, each filter has a *most penetrating particle* depending on air velocity, particle density, and fiber diameter. The most penetrating particle size for most high-efficiency filters is between 0.1 and 0.3 micron in diameter. The efficiency of a filter (percent of particles which are intercepted and retained) is always associated with a particle size, which should be near the size of the most penetrating particle.

After removal of the particles, the filters must retain them on the surfaces of the fibers. The particles are retained by the fibers by surface chemistry, van der Waals' forces, electrostatic attractions, and combinations of these forces.

Low- and medium-efficiency filters. These types of filters are used as prefilters in a cleanroom. By liberal use of inexpensive prefilters the life expectancy of the expensive HEPA or ULPA filters can be extended. However, each prefilter has associated with it a pressure drop and excessive use of them will result in increased capital investment (larger fan and motor) and increased energy expenditure. These filters should always be used on the makeup air to the cleanroom and may be required on the return air from the cleanroom.

The prefilters may be "staged" by using coarse filtration to capture large particles (leaves, insects, etc.) and using successively finer filters to prepare the air for cleanroom use. The low- and medium-efficiency filters may be rough-fiber filters similar to filters in home central heating and air-conditioning systems or may be bag filters similar to woven cloth filters used in vacuum cleaners. These filters may be as efficient as around 95 percent on 0.3-micron particles.

High-efficiency particulate air (HEPA) filters. The HEPA filters are the "workhorses" of the cleanroom. Their efficiencies range from 99.97 to 99.9999 percent on 0.3-micron particles. It is for this reason that these filters are sometimes referred to as *absolute filters*. HEPA filters are constructed of a single sheet of filter medium folded many times, with the pleats separated from each other by corrugated anodized aluminum elements called *separators*. The medium and separators are housed in a frame, usually made from anodized aluminum. By pleating the media a large amount of filter surface area can be compacted in a small volume. The filter medium is a collection of randomly oriented fibers of varying lengths and diameters compressed into mats. The filter medium consists of both submicron-size fiberglass fibers for high-efficiency collection of submicron particles and larger-diameter fibers for strength.

Ultralow particulate/penetration air (ULPA) filters. ULPA filters are more efficient than HEPA filters (up to 99.99999 percent efficient for 0.12-micron particles) and more expensive. One type of ULPA filter uses a media of expanded polytetrafluoroethylene (PTFE).

Electrostatic collection techniques. As mentioned previously, the principle of electrostatic attraction can be used to collect particles or enhance the efficiency of filters. One method uses a corona discharge upstream of the filter medium to charge and agglomerate the particles and a HEPA filter with an applied electric field to collect the particles. Another method uses *electrets* (permanently polarized material) to collect aerosols from the air.

Testing filters. A number of different filter-testing techniques exist to determine efficiency, flammability, dust-holding capacity, and so forth, and to locate leaks in filters. The standard test that has been used in the United States for some time for rating HEPA filters is the thermal (or hot) DOP method.

The *hot DOP method* involves the creation of monodisperse (single-size) particles of DOP using a vaporization condensation process. The liquid DOP at room temperature and pressure is heated and vaporized. The air with DOP vapor is mixed with cool air in a proportion such that the DOP condenses into droplets of about 0.3 micron. The DOP droplets are carried by a fan-pressurized airstream which passes through the filter. The concentration of DOP both upstream and downstream of the filter is measured with light-scattering photometers and these measurements are what determine the efficiency of the filter. The hot DOP test is suitable only for use in the factory for rating the filter. It does not indicate where a pinhole leak might have occurred.

The *cold DOP test* involves creating polydisperse (multiple-size) particles of DOP using a nebulizer (mechanical agitator). This test is suitable for use in the field to locate pinhole leaks in HEPA filters. The cold DOP method does not create a high enough concentration of small DOP particles for efficiency testing of the filters. A measured concentration of DOP is introduced into the air upstream of the filter. A scanning probe which draws air for measurement by a light-scattering photometer is passed over the entire surface of the filter, and any leaks detected are marked for repair at a later time.

DOP has come under criticism in recent years because it may be carcinogenic, is not entirely monodisperse, and may itself become a contaminant because of its out-gassing properties. One suggestion is to use corn oil instead of DOP.

HEPA/ULPA filter installations. HEPA filters may have air supplied to them through individual flexible ducts to each filter or through a common plenum duct where the entire space above a number of filters is pressurized. Individually ducted HEPAs may be sealed using latex sponge rubber or neoprene sponge seals and clamps (to resist the room positive pressure). Where filter plenums are used, a gel seal is often employed. For this seal, the HEPA frame has a ridge which extends down from the frame and fits in a trough in the ceiling grid. The trough is filled with a non-out-gassing gel material, thus providing an airtight and contamination-proof seal. Some companies use a negative-pressure seal where the space around the filter frame is under negative pressure relative to the cleanroom and the plenum, thus preventing contaminated air from leaking around the filter and into the cleanroom.

Air velocity

Air velocity in the cleanroom is an important parameter which must be selected during the design, and will affect the performance and cost of the cleanroom. The *face velocity* of a HEPA filter is the average velocity, computed by dividing the volume flow rate of air passing through the filter by the face area of the filter. It is usually expressed in feet per minute (fpm). The face velocity of a HEPA filter is not equal to the velocity of every point across the filter and may vary as much as 20 percent. To maintain laminar flow from a HEPA filter, the face velocity should be maintained between 50 and 150 feet per minute. This is only important for laminar-flow designs, and velocities outside this range may be chosen for nonlaminar cleanrooms. At velocities less than 50 feet per minute the air may not have enough momentum to maintain its streamlines. At velocities greater than 150 feet per minute the flow becomes turbulent. It may be noted that lower velocities will make a filter more effective at collecting and retaining smaller particles because the particle will spend more time residing in the medium and the diffusion collection process will be more effective. However, more surface area and therefore more filters will be required, resulting in greater capital expenditure. There is a distinct relationship between velocity and volume air changes per unit time in a cleanroom, and they are related by the geometry of the room and the amount of HEPA coverage. If a cleanroom with 80 percent HEPA coverage and a 10-foot ceiling has a face velocity of 100 feet per minute, the number of air changes per hour is 600. The general formula is:

$$\text{Air changes per hour} = \frac{\text{face velocity} \times \text{HEPA coverage} \times 60}{\text{ceiling height}}$$

The *average air velocity* of a cleanroom is the velocity computed by taking the total volume air flow through all the HEPA filters and dividing by the plan area of the cleanroom. The average air velocity will be less than the face velocities of the HEPA filters except in the case of 100 percent filter coverage.

Recommendations.   The IES offers recommendations of velocity ranges for various classifications of cleanrooms. These should be interpreted as guidelines or rules of thumb; they are not mandatory. Figure 9.7 is a summary of the IES guidelines. Note that the velocities listed are the average air velocities of the room as calculated in the previous section and not the face velocities of the filters.

Class	Average air velocity (fpm)
100,000	5–8
10,000	10–15
1,000	25–40
100	60–90
10	70–90
1	80–100

Figure 9.7   IES recommended airflow velocity guidelines.

Temperature

Cleanrooms are usually maintained at required temperatures with tight tolerances for both personnel comfort and product/process requirements. Depending on the classification of the cleanroom, personnel may be dressed in cleanroom apparel which does not allow heat and moisture to escape very well. Generally, the higher the cleanroom class, the more cleanroom garment coverage required. Because the operator will be wearing her normal clothing under stuffy cleanroom garments, cleanroom temperature setpoint will probably be lower than that of other areas. Cleanroom temperature setpoints are often in the range of 68 to 72°F for Class 1,000 and better, and 70 to 74°F for Class 10,000 or Class 100,000 cleanrooms.

Sometimes process or product requirements will require temperatures outside the range of human comfort. The process/product requirements will overrule the comfort requirements almost without exception. Cleanrooms where certain photoresist chemicals are used may need to be held at 68°F or lower.

Temperature control.   Often in cleanrooms (particularly for semiconductor manufacture) the temperature of the room is not as critical as the ability to hold a constant temperature within a tolerance. Large swings in temperature can contribute to human discomfort but, more important, temperature must be held to a tight range for product/process requirements. Changes in temperature can contribute to product damage during photolithography processes and can cause particle generation during expansion or contraction of substrate materials. Control of temperature to $\pm 1°F$ is common and control may be as tight as $\pm 0.1°F$ for certain photolithography processes.

In order to control temperatures to tight tolerances, zones must be limited in size. For control of zones with evenly distributed heat generation to $\pm 1°F$, width should be no greater than 16 feet and length should be no greater than 40 feet. Tighter tolerances and concentrated heat loads will require smaller zones.

Control of temperature must be separated from control of humidity in order to maintain the tight tolerances required for most cleanrooms. Temperature control must be accomplished at both the makeup and recirculation air units. Figure 9.8 shows an example of a temperature control schematic for a typical cleanroom zone. Many variations of this schematic are possible and it is not to be considered to be a universal solution for temperature control. However, it does serve to illustrate some general principles of cleanroom temperature control design. During cold outdoor conditions the air entering the makeup air unit needs to be heated at the preheat coil to prepare the

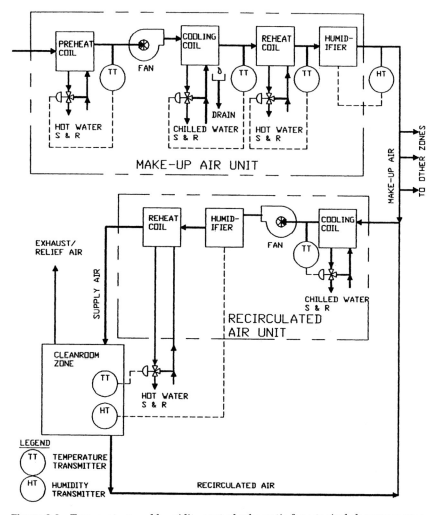

Figure 9.8 Temperature and humidity control schematic for a typical cleanroom zone.

airstream for humidification. The cooling coil in the makeup air unit is designed to cool the airstream and remove moisture during warm outdoor air conditions. The reheat coil in the makeup air unit is designed to heat the airstream to prepare it for controlled humidification. Leaving the makeup air unit is an airstream of constant temperature and humidity.

In the recirculated-air unit, the cooling coil removes heat from the mixed airstream consisting of makeup air and recirculated air. The heat originates primarily from process loads and fan heat. The humidifier adds moisture (if required) to the airstream to meet humidity requirements in the zone. The reheat coil adjusts the final temperature of the supply airstream to meet temperature requirements of the zone.

Temperature measurement. In order to meet the temperature tolerances discussed above, the instrumentation used to measure temperature must be accurate. Bimetallic-type thermostats such as used in residential heating and cooling systems are insufficient to maintain the precise temperature setpoints in cleanrooms. Several types of temperature elements are used for temperature transmitters installed in ducted air streams and in cleanroom zone air spaces, including thermocouple and resistance temperature devices (RTDs). The outputs from the transmitters are sent to an electronic control system such as discussed under "Controls" below for processing to provide output signals for controlling devices and monitoring. Temperature transmitters should be installed at various points in the system for monitoring and alarming in addition to those installed for control.

Humidity

The level of humidity in a cleanroom affects both human comfort and products/processes. Humidity above 50 percent may cause corrosion on some metal surfaces. High humidity may also cause condensation and encourage microbial growth. On the other end of the scale, low humidity levels will allow static electricity to build which may attract particles and contaminate surfaces. A very small static electric discharge can be damaging to delicate ICs, and yield reduction is possible because of low humidity.

Humidity levels for human comfort may range from 30 to 70 percent. Process and product requirements will vary but generally they will require humidity levels between 30 and 50 percent. Often a control point of 40 percent is specified.

Humidity control. Control of humidity in air involves both removal of moisture (dehumidification) and addition of moisture (humidification). Dehumidification may be accomplished by cooling the air to below its dew point so that water condenses out. This may be done with a chilled-water coil down to about 42 to 43°F dew point using 38°F chilled water. If the space temperature is held to 72°F, it is not possible to reduce the space humidity below 33 to 34 percent. If lower humidity is required, it must be accomplished by using a direct-expansion (DX) refrigerant system, a brine coil, or a chemical desiccant dehumidifier. Desiccant dehumidifiers contain a highly hydrophilic chemical compound which absorbs moisture from the air on contact. After a period of time the desiccant will not take on any more moisture and will need to be regenerated. Some desiccant dehumidifiers regenerate by heating the desiccant to drive off moisture. Others use a liquid desiccant which is continuously being regenerated. Desiccant dehumidifiers may be unacceptable because they have potential to introduce contamination into the airstream.

Humidification can be achieved by several methods including nebulizing water by vibration, mechanical agitation or spraying through a nozzle, passing the airstream through a saturated medium, or direct injection of steam. Steam injection is the most effective means for introducing moisture into air and is the preferred method for cleanroom use. One major problem with steam injection is the vaporization of water treatment chemicals used in steam generators, which can be a source of contamination. One suggestion is to use deionized (DI) water for steam generation. However, hot DI water and DI steam are extremely aggressive to metal surfaces and are unacceptable in a normal HVAC system. A compromise might be to use reverse osmosis (RO) water but it can still dissolve surfaces in boilers. A way around this problem is to use glass-lined boilers or use a system with a reboiler which boils clean water (RO water) in a vessel by passing steam through tubes immersed in the clean water. These systems are readily available but are costly.

The humidity tolerance requirements vary from cleanroom to cleanroom and are generally ±2 percent to ±5 percent. As with temperature, specifying a tolerance smaller than what is required by process/product considerations is a costly mistake. Some processes involving chemical reactions require precise control of humidity in order to be consistently successful.

Refer to Fig. 9.8. In this example of humidity control for a typical cleanroom zone two humidifiers are employed. Also, a cooling coil in the makeup air unit is used to dehumidify using chilled water. A brine or DX coil could be used in its place depending on how low a dew point is desired. One humidifier and the cooling coil in the makeup air unit control the humidity of the makeup airstream. The cooling coil is controlled by a temperature transmitter which senses the dew-point temperature of the air downstream of the coil. The air leaving the makeup air unit is controlled to a precise humidity level prior to mixing with the recirculated-air stream. The humidifier in the recirculated-air unit is controlled by the zone humidity transmitter which is needed if there are different humidity setpoints in each zone. Dehumidification will not be required in the recirculated-air unit because all wet processes are normally exhausted and a large amount of makeup air is required (which has already been dehumidified). Control of humidity in a cleanroom is more influenced by variations in outdoor conditions than by variations in humidity generation or adsorption in the cleanroom. Therefore, most of the adjustment will be made in the makeup air unit.

Control of humidity can be challenging when zones are set at different relative humidities, or the cleanroom is surrounded by spaces at higher or lower relative humidities. Moisture can pass through walls and ceilings because of diffusion of water vapor driven by the humidity gradient. Vapor barriers may be necessary. Velocities at relief openings need to be high to prevent diffusion of water vapor upstream. Buffer zones may need to be created around zones of precise humidity requirements.

Humidity measurement.   Humidity in air can be measured using a wet-bulb/dry-bulb psychrometer in which the dry-bulb air temperature is compared to the wet-bulb temperature (measured by covering the bulb of a thermometer with a wick moistened with water and passing air over it) and the humidity is determined from a psychometric chart. This is not acceptable for an automatic control system and other humidity measurement devices have been developed. Some materials change dimensions with changes in humidity (e.g., hair) and this mechanical motion can be transduced into an electric signal. Another device measures the change in conductivity of a material depending on the amount of water absorbed.

Pressure

Air pressure is used in cleanrooms as a means of contamination control. If cleanroom building materials could be sealed absolutely airtight and no doors or openings in the cleanroom envelope existed, pressurization would not be required. However, in the real world, building materials cannot be made perfectly airtight and personnel and materials need to enter and exit the cleanroom by some means, making room pressurization a necessity. By pressurizing a cleanroom at a higher pressure than surrounding areas without contamination control or cleanrooms of higher classification, particles are not able to diffuse through cracks and openings in envelope materials, and the outward flow of air experienced when doors are opened prevents most particles and air laden with particles from entering.

Recommendations.   The IES offers recommendations for levels of cleanroom pressurization. IES recommends a pressure difference between cleanroom space and uncontrolled space of 0.05 inch w.c. Less pressure will allow excessive particle migration. Higher pressure will make doors hard to open or close and may create noise problems with high flows of air through small openings. Air pressure difference between two cleanrooms, one of a higher classification than the other, should be about 0.02 inch w.c. with the cleaner space at the higher pressure. The pressures of multiple-classification cleanrooms should be staged, with the cleanest spaces having the highest pressures and each successively less clean space having a lower pressure than the preceding.

Air locks may be used to prevent loss of air and pressure in cleanrooms. *Air locks* are small rooms with two doors that allow passage from a space at higher pressure to a space at lower pressure without appreciable loss of air from the higher-pressure space. Air locks should be maintained at the pressure of the lower-pressure space and may have interlocks on the doors to prevent both from being open at once.

Pressure control.   When product requirements dictate or when budgets allow the use of pressure control, one can choose from a variety of means to achieve dynamic control of pressure. Variable-speed drives may be used on makeup air fans to furnish additional air when doors are opened allowing air loss from the space. Relief dampers which are mostly open during normal conditions and will close down to maintain pressure when doors are opened may be used. Since all spaces with human occupants will require makeup air, the relief dampers do not require an additional amount of air to be supplied unless process exhaust in space equals or exceeds the makeup air. Pressure transmitters measure pressure differentials between adjacent spaces and transmit signals to a central controller. The controller transmits signals to the relief dampers or variable-speed drives to modulate position or speed to maintain pressure differential.

Pressure measurement.   Differential pressures in cleanrooms should be monitored even if not dynamically controlled. Changes in pressure may indicate need to change operating strategies, or to maintain equipment, and may provide information related to reduction in product yield. Pressure may be measured with sensitive wall-mounted gauges utilizing a diaphragm and mechanical movement. Pressure may be recorded with a chart recorder or digital data acquisition system. For electronic measurement, pressure transmitters with diaphragm sensors and strain-gauge transducers will have required accuracy to measure and transmit pressures for control and monitoring.

Controls

Modern cleanrooms have large numbers of parameters to be monitored, controlled, and alarmed. With the reduction in cost of microprocessor-based controllers, these are the units of choice for cleanrooms. Pneumatic controls still have limited uses (e.g., valve and damper actuators) but they have virtually been replaced by microprocessor-based controls. A local area network (LAN) or a peer-to-peer communication network should be used for larger cleanroom application. The network will link up stand-alone controllers. A computer equipped with supervisory control and data acquisition (SCADA) should be tied in to the network. Tasks performed by the SCADA will include alarm, historical trend, graphic display, control, troubleshooting, and various other functions.

Microprocessor-based controllers include the distributed control system (DCS), the programmable logic controller (PLC), and the direct digital control (DDC). These systems basically perform the same control functions. Field instrument input signals (discrete and analog) are converted into digital signals. Output signals from the controller are capable of emulating virtually all conventional signals used. When making a decision on the type of controllers to be used, major issues examined should include the following:

- Future and immediate control objectives
- Existing communication system and control system
- Control and communications requirement
- Complexity and the type of process
- The size of the control system (number of I/Os)
- Personnel's knowledge and comfort with the control system
- Selected controller's experience for the particular control task
- Manufacturer and local vendor support
- Software programming difficulties
- Support documentation
- Cost/payback analysis

Vibration

Another form of contamination not previously discussed which can be quite damaging in a cleanroom used for microelectronics fabrication is vibration.

Sources of vibration.   Vibration is generated by both natural and human-made sources. Natural sources include seismic activity in the crust of the earth, tidal forces, and winds, among others. Cleanrooms are, by necessity, located in the vicinity of roads with vehicular traffic. Large trucks can impart vibration into the road which is transmitted through the earth to the cleanroom facility. Construction and heavy industry near the cleanroom facility may use equipment which generates large-amplitude–low-frequency vibration. Inside the cleanroom facility, vibration is created by equipment such as chillers, air handlers, and compressors. Even some production equipment which may be near the cleanroom itself, such as vacuum pumps and packaging testing equipment, may contaminate the environment with vibration. Personnel walking and carts being pushed through the hallways can contribute to vibration contamination.

Effects of vibration.   Many cleanroom-based processes are not vibration-sensitive. However, many high-level photolithography processes require alignment of a mask and wafer within high tolerance. Any vibration may cause the light projection boundaries to become blurred and the product may be rendered useless. Testing of ICs may require the use of minute probes physically touching conductors on the chip. Vibration can cause the probe to move relative to the conductor and this may damage the IC. Scanning electron microscopes require vibration-free environments.

Reducing effects of vibration.   Reducing the effects of vibration ideally starts in the planning and design stages of the facility. After the vibration criteria have been established, decisions can be made by the designers which will affect the performance of the cleanroom.

Site selection.   Obvious characteristics need to be observed initially when proposing to construct a facility with cleanrooms on a site. Is the site in an area with a large amount of seismic activity or near the ocean with tidal waves? Is there a superhighway passing near the site? One needs to confer with zoning officials to see what kind of future construction will occur in the area around the site. Ground vibration testing needs to be undertaken by specialists at different locations on the site to determine baseline vibration levels and transmissibility of the soil.

Plant layout. Within the facility, the locations of the vibration-sensitive process equipment and the utilities equipment (chillers, compressors, etc.) need to be separated as much as possible. Ideally, the utilities equipment should be in a completely separate building from the cleanroom facilities with vibration isolation in all piping between the buildings. The best place, generally, to locate vibration-sensitive equipment is on a ground-level slab-on-grade or basement floor. If vibration-producing equipment needs to be in the same building with vibration-sensitive equipment and areas, the vibration-producing equipment should be located in the basement, with the vibration-sensitive areas located on other floors.

Foundation and structure. It may or may not be beneficial to set foundations on bedrock (where it occurs close to surface), depending on whether the bedrock is likely to be excited by vibration sources. Sometimes a massive foundation with drilled piers will be useful in dampening vibration, but care must be exercised to prevent installing a foundation with a natural frequency the same as what is present in the ground. Essentially, a vibration analysis must be performed using the site test data to determine the optimal foundation design. A stiff floor such as a thick concrete waffle slab with closely spaced columns is preferred for all upper floors. A slab-on-grade floor should utilize a thick (8- to 12-inch) concrete slab. The structure should be reinforced concrete and not steel.

Vibration isolation. The final step in producing a vibration-free design is isolating the source and recipient of the vibration.

Isolating sources of vibration. All vibration-producing equipment should be installed on vibration isolators. Vibration isolators come in various forms including high-deflection springs, air-filled bellows, inertia bases with isolators, and others. Again, putting as much distance as feasible between vibration-producing equipment and vibration-sensitive equipment is most effective in reducing transmitted vibration. All piping and ductwork connections to equipment should have flexible connections. Piping and ductwork should be hung using vibration-isolating hangers.

Isolating vibration-sensitive equipment. Equipment which is highly sensitive to vibration should be installed on foundations separate from the building foundations with its own separate columns if not on-grade. All connection to the building foundation and facilities should be through flexible coupling. Piping and other facilities should not be supported from the separated foundations. As an added precaution the equipment may also be installed on some type of vibration-isolating supports.

Detuning. Another vibration reduction strategy which should be used during design and continued through construction, start-up, and occupation of the facilities is detuning. During design the entire system of soil, foundation, structure, and mechanical equipment should be analyzed to prevent designing resonating systems. Masses of foundations and floor slabs should be selected so they are not in resonance with the soil. Rotating speeds of elements in mechanical equipment should be selected to avoid resonance with the ground, floors, and foundations. Vane axial fans are preferred over centrifugal fans because their higher speeds do not produce low-frequency vibrations. Avoid the use of reciprocating compressors and chillers because they are inherently unbalanced and produce low-frequency vibrations.

Noise

Noise can be a significant problem in cleanrooms. Not only can high levels and certain types of noise be a nuisance to personnel, but noise can affect sensitive processes by changing from sound energy to vibratory energy in the product or tool.

Design criteria. Often noise is characterized by its *A-weighted sound pressure level,* which does not set limits on a frequency-by-frequency basis. For cleanrooms a *noise criterion* (NC) *curve* should be used which defines sound levels in each frequency range. Currently most class 1 to 100 cleanrooms are designed to meet NC 55 to NC 65.

Sources of noise. Because of the large volume of air moved through the cleanroom, large recirculation fans are required. These fans are usually located near the cleanroom out of necessity. Vacuum pumps are noisy but need to be near the process in order to be most effective.

Reducing noise. When specifying equipment, the acceptable radiated sound power must be limited. Ductwork must be designed to limit velocity and prevent excessive wall vibration. Acoustical enclosures should be used around noisy equipment. Consideration must be given to cooling requirements of equipment in enclosures. Silencers, lined ducts, and plenums should be used to attenuate the sound carried by the supply air.

Fire protection

Fire protection for cleanrooms requires professional fire protection engineering, coupled with a thorough knowledge of the cleanroom operation and experience with state-of-the-art detection and suppression systems. Building codes such as the Uniform Building Code, the

Uniform Fire Code, and NFPA Standard 318 are all excellent guides, but detection and suppression systems for cleanrooms, especially semiconductor cleanrooms, demand tailor-made systems to fit the unique needs of each facility.

Preaction systems.   Automatic sprinklers are required in cleanrooms by most building codes, fire codes, and NFPA Standard 318, but to prevent any possibility of accidental water damage due to a leaking fitting or a damaged sprinkler head, these should be preaction sprinkler systems.

NFPA Standard 318 mandates wet-pipe sprinkler systems for cleanrooms, but a supervised preaction system should be considered based on engineering judgment. With a supervised preaction system the piping is empty except for a small amount of compressed air for supervision. If a leak should occur in the system and bleed off the compressed air, a trouble alarm is activated. System water is held back by a preaction valve (actually a deluge valve). When the detection system is activated, the preaction valve is tripped, piping fills with water, but no water is discharged until a sprinkler head fuses. Only a signal from the detection system can automatically trip the preaction valve.

Detection.   When combustion occurs in a cleanroom, rapid detection is of major importance. Reliance on the fusible link of a sprinkler head as a detection system is inadequate due to the time lag between ignition and activation—depending on the intensity of the combustion, 19 to 35 minutes is possible. Smoke detection response time can be from 11 to 18 minutes, and by this time the combustion products have contaminated the area.

The sensitivity of a well-maintained ionization detection system should be considered, and for chemical workstations using flammable liquids, the use of optical (UV and/or IR) for millisecond detection of combustion is imperative.

Local detection and suppression.   In addition, it is highly recommended that wet-chemical workstation optical detectors automatically activate a Halon suppression system designed for the individual workstation.

Detection and suppression of combustion prior to activation of the sprinkler system discharge minimize the extent of contamination, water damage, and business interruption. The sprinkler system should be considered as a disaster backup to prevent thermal damage of other equipment, and a localized fire becoming a structural fire should it overcome the suppression efforts of personnel and/or dry-agent suppression systems.

Energy conservation

Energy conservation has often been overlooked during the design of cleanrooms. This is unfortunate because cleanrooms use such inordinate amounts of energy for cooling, heating, and air movement. It has been estimated that cleanrooms use up to 45 times as much energy per square foot as office buildings.

Much of this energy is used to power fans which maintain positive pressure and proper velocities and air changes. Energy savings here can be realized by optimizing the airflow to the actual product requirements. In other words, do not overdesign the cleanroom beyond the requirements of the process or product. Additionally, cleanroom airflows can be cut back at night and during idle periods. Only a minimum amount of air is needed to maintain cleanliness when the room is not occupied and production has ceased. Use of backward-curved plug fans will minimize energy consumption because these fans do not create wasted velocity pressure.

Energy is also used in cleanrooms to heat and cool airstreams and to humidify and dehumidify airstreams. There is room for much innovation in this area using free-cooling, energy recovery, cogeneration, and dual-temperature chilled-water systems, among other concepts. Chillers should always be run with minimum condenser water temperatures. Free cooling can be accomplished through the use of plate-and-frame heat exchangers coupling the cooling-tower water with the chilled water during periods of low wet-bulb temperatures.

Clean construction protocol

It is important in the design of a cleanroom to consider the requirements of a clean construction protocol. The design must permit the construction to follow a clean construction schedule. The design of a critical cleanroom component such as the recirculation air-handling unit should be such that the units do not have to be installed before they can be protected from contamination.

A typical clean construction protocol has a defined number of steps or levels. Figure 9.9 shows a typical clean construction protocol. Each level is an increasing level of cleanliness. It is important to define only the minimum number of levels which will permit construction of the facility to the cleanliness required. Increasing the number of levels unnecessarily may increase the construction time and cost.

Certification

After construction of a cleanroom, the room must be certified to operate in accordance with the standard for which it was designed. For

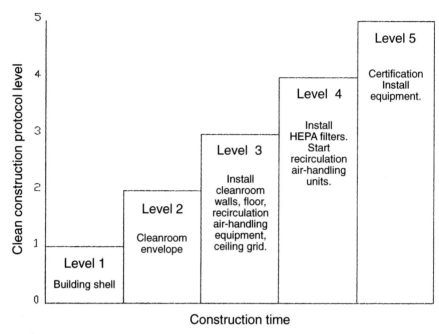

Figure 9.9 Construction time.

illustrative purposes, Federal Standard 209D will be referred to in this section as the standard of choice.

The first step in certification is to specify the environmental test conditions such as temperature, humidity, and pressure. These conditions should be the conditions for which the cleanroom was designed to operate. The cleanroom operational conditions should be specified and recorded (as built, at rest, or operational). The environmental conditions and use parameters should be recorded at the time of the test. These will include air velocity, activities by personnel, and others, in addition to temperature and humidity.

The method of particle counting must be determined. Manual counting can be used for particles 5 microns and larger. For particles less than 5 microns an optical counting instrument must be used.

The locations and numbers of sample locations are specified by 209D for both unidirectional and nonunidirectional airflow. 209D requires that at least two sample locations be chosen for each cleanroom zone and that at least one sample be taken at each location. No less than five samples must be taken in each cleanroom zone.

Federal Standard 209D specifies the minimum air volume to be drawn for each sample. For the lower classifications (Class 1, 10, and 100) larger sample volumes are required than for the higher-classification rooms (Class 1,000, 10,000, and 100,000). After collection of all of the required samples, the data must be treated statistically in accordance with 209D. If the data indicate conformance with the 209D requirements, the cleanroom is certified and a report is prepared to document the certification.

Certification should be performed by a third party employed by the owner, with no ties to either the contractor or the designer. This ensures no bias in the measurement and analysis of data. After initial certification, the cleanroom should be recertified periodically (particularly after people and equipment have been brought in and before full-scale production). The cleanroom should be monitored and should use the same procedures as were used for certification during operation to ensure clean environmental conditions and limited loss of product.

References

1. American Conference of Governmental Industrial Hygienists: *Industrial Ventilation, A Manual of Recommended Practice,* 19th ed., ACGIH, Cincinnati, OH, 1986.
2. Kozicki, M. N., S. A. Hoenig, and P. J. Robinson, *Cleanrooms: Facilities and Practices,* Arizona State University, Tempe, 1990.

Further Reading

Agnew, B.: *The Laminar Flow Clean Room Handbook,* 3d ed., Agnew-Higgins, Garden Grove, CA, 1968.

American Society of Heating, Refrigeration, and Air-Conditioning Engineers, Inc.: *1987 ASHRAE Handbook, HVAC Systems and Applications,* ASHRAE, Atlanta, 1987.

American Society of Heating, Refrigeration, and Air-Conditioning Engineers, Inc.: *1988 ASHRAE Handbook, Equipment,* ASHRAE, Atlanta, 1988.

Austin, P. R.: *Design and Operation of Clean Rooms,* rev. ed., Business News Publishing, Detroit, 1970.

Austin, P. R.: *Contamination Control Seminar, 1990 Seminar Series,* Contamination Control Seminars, Livonia, MI, 1990.

Bishop, D. E.: "DRI-TEST™...The New Way to Challenge HEPA Filters," *Proceedings, 9th International Symposium on Contamination Control,* September 26–30, 1988, Institute of Environmental Sciences, pp. 179–182.

Brown, W. K., and C. A. Lynn: *Fundamental Clean Room Concepts,* SF-86-06 no. 2, *ASHRAE Transactions 1986,* vol. 92, Part 1.

"Cleaning Up Clean Room Design": *Engineering News Record,* April 12, 1984, pp. 20–24.

Eidam, H.: "Low-Energy Consumption Clean Room Equipment with Laminar Flow," *Proceedings, 9th International Symposium on Contamination Control,* September 26–30, 1988, Institute of Environmental Sciences, pp. 131–136.

Federal Standard, Clean Room and Work Station Requirements, Controlled Environment, FED-STD-209D, General Services Administration, Fort Worth, TX, 1988.

Filter Units, Protective Clothing, Gas-Mask Components, and Related Products: Performance Test Methods, MIL-STD-282, U.S. Government Printing Office, Washington, D.C., 1956.

Grout, R. V., and C. B. Williams: "Vibration Design Considerations for Semiconductor Manufacture," *Proceedings of the Institution of Mechanical Engineers, Clean Room*

Design, January 29, 1986, Mechanical Engineering Publications, Ltd., pp. 13–15.

Hardy, T. K., and R. H. Shay: "Assessing Gas Cabinet Manifold Purge Techniques," *Solid State Technology*, PennWell Publishing, 1987.

High Efficiency, Particulate, Air Filter Units, UL Standard 586-85, Underwriters Laboratories, Inc., Northbrook, IL, 1985.

King, J. G.: *The History of Clean Rooms*, SF-86-06 no. 4, *ASHRAE Transactions 1986*, vol. 92, Part 1.

Mizogami, S., Y. Kunimoto, and T. Ohmi: "Ultra Clean Gas Transport from Manufacturer to Users by Newly-Developed Tank Lorries and Gas Storage Tanks," *Proceedings, 9th International Symposium on Contamination Control*, September 26–30, 1988, Institute of Environmental Sciences, pp. 352–359.

Rose, T. H.: *Noise and Vibration in Semiconductor Clean Rooms*, SF-86-06 no. 3, *ASHRAE Transactions 1986*, vol. 92, Part 1.

Silver, W., and E. A. Szymkowiak: "Vibration on Various Floor Types in Microelectronic Fabrication Facilities," *Proceedings, 9th International Symposium on Contamination Control*, September 26–30, 1988, Institute of Environmental Sciences, pp. 187–191.

Standard for Protection of Cleanrooms, NFPA 318, National Fire Protection Association, Quincy, MA, 1991.

Tolliver, D. L.: *Handbook of Contamination Control in Microelectronics, Principles, Applications and Technology*, Noyes Publications, Park Ridge, NJ, 1988.

10

Production and Capacity Planning

D. J. Medeiros, Ph.D.
Associate Professor
Department of Industrial Engineering
Penn State University
University Park, Pennsylvania

Production planning is the activity of developing schedules for manufacture, which specify products, quantities, and timing. The capacity planning activity utilizes these schedules to plan changes which affect the maximum output of products, such as adding or removing facilities, machines, and labor hours. These two activities are often carried out in an iterative fashion. A proposed production level may be unrealistic from a capacity planning viewpoint, therefore requiring a change in the production schedule. The revised production schedule would again be compared to the capacity plan, and the process repeated until satisfactory plans were obtained.

Production and capacity planning are carried out at many levels within the firm. These plans can be classified into three groups: *strategic* (or long-term), *tactical* (or medium-term), and *operational* (or short-term) planning.

At the strategic planning level, production plans are based on forecasts of customer demand. These plans are utilized for making significant changes in capacity which require a long lead time and a large capital investment, such as development of a new manufacturing facility or acquisition of major equipment items. Plans may extend over a time period of five years or more.

At the tactical planning level, production plans are specified in more detail over a shorter time horizon. These plans include major product families; sources of demand include customer orders, forecasts, planned service parts, planned inventories, and others. Capacity plans focus on decisions such as acquisition of smaller items of equipment, subcontracting, and changes in workforce levels.

At the operational (detailed) planning level, the production schedule includes planned production of each end item by time period (such as a week). Capacity planning seeks to effectively utilize available capacity, and may include planned overtime or undertime, alternate job routing, and so forth.

The different levels of planning are related; decisions made at a higher level constrain the lower-level plans, and the lower-level plans, when aggregated, should match the higher-level plans.

Strategic Planning

Production and capacity planning are part of the firm's overall business plan. These high-level plans extend over several years. Planned output is expressed in aggregate units (tons, dollars, labor hours); market forecasts and product life cycles are important considerations. The production plan is used primarily as a basis for developing an appropriate capacity (or resource) plan. Decisions related to the capacity plan are long-term and large in scope. These may include location and development of a new manufacturing facility, shutdown or conversion of an existing facility, acquisition of equipment with high capital costs or long lead times, or other major changes in the methods of production.

A common method for converting planned output into capacity requirements is the use of *capacity factors* based on historical performance of the firm. This approach is effective for planning at the strategic level, assuming that there are no major changes in the product mix—i.e., the product mix of the proposed plan and product mix at the time during which the performance data were collected must be similar.

Production and capacity planning at this level are not often standardized. Needs vary by industry and type of market, goals frequently are difficult to quantify, and risks are high. Typically, a company will develop its own specific models based on its particular circumstances and needs. General approaches include economic models, decision tree models, optimization models, and computer simulation models, as well as ad hoc methods. Most of these models are used to select from alternative strategies or to provide performance information concerning a particular strategy. Computer-based production planning and control systems can be utilized to provide information to assist in the planning process, but are not typically used in developing the plan.

Tactical Planning

Once the strategic or business plan has defined the major resources available for manufacturing, an *aggregate production plan* is developed. This aggregate plan will extend over a medium-length time horizon (typically 12 months, although this will vary based on product lead time). Individual products are aggregated into families or major product groupings, and planned output is measured in consistent units such as tons, labor hours, cubic feet, or dollars. The planning time unit is selected based on the needs of the company; often a month is used as the planning unit. Sometimes longer time units are used near the end of the planning horizon.

The objective of aggregate production planning is to determine the quantity of each product family to produce in each time period. These quantities should be consistent with the strategic plan after conversion of units and summing over several time periods to match the units and timing of the strategic plan.

Production quantities are determined based on demand (from firm orders, forecasts, planned service requirements, etc.) and on the cost of production under various scenarios related to capacity. Costs that may be considered include subcontracting costs, cost of adding equipment, overtime/undertime costs, hiring/layoff costs, inventory or holding costs, and backorder or shortage costs.

Often, the aggregate planning problem is solved by developing planned production levels, estimating the cost of the proposed plan based on required changes in capacity, and repeating the process until a satisfactory plan is achieved. The process of evaluating plans to determine cost can be incorporated into a computer-based spreadsheet, allowing many alternatives to be examined in a brief period of time.

Another approach to aggregate planning is optimization models; these have been an active area of research for many years. A simple formulation incorporating subcontracting and overtime is shown below.

$$\text{minimize } Z = \sum_{i=1}^{N} \sum_{t=1}^{T} r_i X_{it} + o_i V_{it} + S_i B_{it} + C_i I_{it} \qquad (10.1)$$

$$I_{i,t-1} + X_{it} + V_{it} + B_{it} - d_{it} = I_{it} \qquad \begin{aligned} i &= 1,\dots,N \\ t &= 1,\dots,T \end{aligned} \qquad (10.2)$$

$$I_{it} \geq w_{it} \qquad \begin{aligned} i &= 1,\dots,N \\ t &= 1,\dots,T \end{aligned} \qquad (10.3)$$

$$h_i X_{it} \leq k_{1t} \qquad i = 1,\ldots,N; \; t = 1,\ldots,T \qquad\qquad (10.4)$$

$$h_i V_{it} \leq k_{2t} \qquad i = 1,\ldots,N; \; t = 1,\ldots,T \qquad\qquad (10.5)$$

$$h_i B_{it} \leq k_{3t} \qquad i = 1,\ldots,N; \; t = 1,\ldots,T \qquad\qquad (10.6)$$

$$X_{it}, V_{it}, B_{it} \geq 0 \qquad \begin{array}{l} i = 1,\ldots,N \\ t = 1,\ldots,T \end{array}$$

where X_{it} = the number of units of product family i produced in period t using regular time

V_{it} = the number of units of product family i produced in period t using overtime

B_{it} = the number of units of product family i produced in period t using subcontracting

I_{it} = the inventory of product family i at the end of period t

r_i = the cost of producing one unit of product family i using regular time

o_i = the cost of producing one unit of product family i using overtime

s_i = the cost of subcontracting one unit of product family i

c_i = the cost of holding one unit of product family i in inventory for one time period

d_{it} = the demand in units for product family i in time period t

w_{it} = the desired inventory of product i at the end of time period t

h_i = the capacity required to make one unit of product family i

k_{1t} = the capacity available on regular time in period t

k_{2t} = the capacity available on overtime in period t

k_{3t} = the capacity available by subcontracting in period t

Equation (10.1) minimizes the cost of producing part families plus the holding cost. The formulation assumes that production and holding costs per unit do not change over the planning horizon, although it could easily be modified to include such changes. Equation (10.2) is a balance equation; it states that ending inventory equals beginning inventory plus production minus demand. Equation (10.3) specifies the required level of inventory of each family at the end of each time period; this can be zero or some positive value. Equations (10.4),

(10.5), and (10.6) state that capacity cannot be exceeded for regular time, overtime, and subcontracting, respectively.

This model would typically be used in a rolling horizon fashion. The model would be solved, and the beginning period or periods would be implemented. Once the results from using the plan in the first time period are available, these are input to the model as initial conditions. The model is again solved, but with the past time period dropped and another time period added at the end of the planning horizon. Although the model is solved for the entire planning horizon, only the first portion of the solution is implemented.

Some possible modifications to this formula include back-ordering, adding or removing capacity, penalizing idle capacity, and incorporating nonlinear costs. Hax and Candea[1] describe and compare several types of optimization models for aggregate planning which include many of these factors.

In some cases—particularly if the plant operates on a just-in-time (JIT) scheduling philosophy—a more appropriate objective function might include smoothing production levels (reducing variability from one period to the next) in addition to, or instead of, cost minimization.

Operation Planning

It is at the operational level where most of the work in computer-based production and capacity planning has been done. At this level, production and capacity planning is usually done using a *manufacturing resources planning* (MRP II) *system.* MRP II systems are large computer software packages which are typically divided into modules such as bill of materials, purchasing, inventory, master scheduling, final assembly scheduling, and so forth. A firm implements modules as needed to support its particular manufacturing environment. The MRP II system both plans production and manages information related to production.

Master production scheduling

The first step in operational planning is to develop a *master production schedule* (MPS). This schedule is a disaggregation of the production plan. It includes end items to be manufactured, as well as quantities and timing for those end items. (For assemble-to-order manufacturing, the MPS will include major options, subassemblies, and components rather than actual end products.) The time frame for the MPS is chosen so that it exceeds cumulative lead time (material procurement plus manufacturing lead time). Inputs used to produce the plan include forecasted demand, customer orders, on-hand inventory, and safety stock requirements. Considerations include smoothing production requirements across time periods, minimizing setup requirements, and remaining within production and capacity limitations established by the aggregate plan.

The MPS may be produced using "trial and error" methods. After developing a trial schedule, a *rough-cut capacity planning* (RCCP) *module* is used to determine if the MPS is feasible. RCCP can be performed either using historical data to estimate capacity factors or using a *bill of capacity* (BOC) for each end item. The BOC contains the capacity required to produce one unit of the end item. At this level of planning, it is usual to consider critical operations, facilities, or machines in the BOC rather than all capacity requirements. The RCCP module computes capacity requirements per time period for the trial MPS. Analysis of these requirements allows identification of potential bottlenecks if the trial schedule were to be implemented. This information is used to revise the MPS, and the process is repeated until a feasible schedule is obtained.

Mathematical programming methods may be utilized to disaggregate the aggregate production plan into an MPS. Hax and Candea[1] present several disaggregation algorithms based on a "knapsack" formulation. Smith[2] discusses the use of linear and integer linear programming algorithms. Bedworth and Bailey[3] describe a three-step procedure which is useful if high setup costs are incurred when changing from one part family to another. The procedure first identifies items which are in short supply and schedules replenishment of all parts in those part families. Next, the production quantity of each family is found using a procedure that minimizes setup and holding costs (family disaggregation). Finally, the individual item quantities are found by attempting to have each item in the family reach its safety stock level at the same time (item disaggregation). The Bedworth and Bailey family and item disaggregation algorithms are shown below.

Family disaggregation algorithm.   Set $ß = 1$, $P^1 = x^*$, and $z^1 = z$, for iteration 1

Step 1: Compute for all $i \in z^1$:

$$y_i^ß = \frac{\sqrt{S_i \sum_{j \in i} (k_{ij} D_{ij,t})}}{\sum_{i \in z^ß} \sqrt{S_i \sum_{j \in i} (k_{ij} D_{ij,t})}} * P^ß$$

Step 2: For any $i \in z^1$

$$\text{if } LB_i \leq y_i^\beta \leq UB_i$$

$$\text{set } y_i^* = y_i^\beta$$

For other families, go to Step 3.

Step 3: Divide other families into two groups:

$$z_+^\beta = \left\{ i \in z^\beta : y_i^\beta > UB_i \right\}$$

$$z_-^\beta = \left\{ i \in z^\beta : y_i^\beta < LB_i \right\}$$

Compute

$$\Delta^+ = \sum_{i \in z^\beta} \left(y_i^\beta - UB_i \right)$$

$$\Delta^- = \sum_{i \in z^\beta} \left(LB_i - y_i^\beta \right)$$

Step 4:

If $\Delta^+ \geq \Delta^-$, let $y_i^* = UB_i$ for all $i \in z_+^\beta$

If $\Delta^+ < \Delta^-$, let $y_i^* = LB_i$ for all $i \in z_-^\beta$

Set $\beta = \beta + 1$, $z^{\beta+1} = z^\beta - \{$all families for which y_i^* has been found$\}$, $P^{\beta+1} = P^\beta - y_i^*$ (for all i scheduled in iteration β)

If $z^{\beta+1} = \emptyset$, then stop; otherwise go to Step 1.

where x^* = the production requirement from the aggregate plan
S_i = the setup cost to produce family i
K_{ij} = a conversion factor for units of item j in family i to the aggregate production unit
$D_{ij,t}$ = demand for item j in family i during period t
LB_i = lower bound for production of family i
UB_i = upper bound for production of family i
z = those families selected for production (because some items will reach the safety stock level)

Item disaggregation algorithm

Step 1: For each family i being produced, determine the number of periods N for which

$$y_i \leq \sum_{j \in i} K_{ij} \left[\sum_{n=1}^{N} D_{ijn} + SS_{ij} - I_{ijt-1} \right]$$

Step 2: Calculate

$$E_i \leq \sum_{j \in i} K_{ij} \left[\sum_{n=1}^{N} D_{ijn} + SS_{ij} - I_{ijt-1} \right]$$

Step 3: For each item in family i, calculate the production quantity

$$y_{ij}^* = \sum_{n=1}^{N} D_{ijn} + SS_{ij} - I_{ijt-1} - \frac{E_i D_{ijN}}{\sum_{\forall j \in i} K_{ij} D_{ijN}}$$

Step 4: If $Y_{ig}^* < 0$ for any item g, set $Y_{ig}^* = 0$, remove item g from the family, subtract $K_{ig} D_{igN}$ from the denominator in Step 3, and repeat Step 3

The result of these algorithms will be a list of items to produce in each time period, along with the quantity of each. This will constitute the MPS. Again, a capacity analysis could be performed using RCCP as described above.

If a firm is utilizing JIT manufacturing concepts, the goal of master production scheduling is to smooth and balance production of end items to ensure a smooth flow of product through the manufacturing facility. This approach is often called *uniform plant loading.* Since JIT emphasizes setup minimization, cellular production, reduction of safety stock, and production to demand rather than forecast, the disaggregation techniques cited above may be less applicable in the JIT environment. The MPS will be arranged to produce the same number of end items each day for some time period (often a month or longer) and production of different end items will be balanced (often, some of each item will be produced each day). Capacity planning is greatly simplified in such an environment since the work load does not vary greatly.

Material requirements planning

Material requirements planning (MRP) is a method for planning the production purchase of subassemblies, components, and raw material used in manufacturing the end items as planned in the MPS. Bills of material describe the components and manufacturing sequence for the end items, and an item master list describes, for each component, the inventory level, lead time, cost, manufacturing time, and related information.

Using the bill of materials, along with component availability and lead time, the MRP system plans purchase or production orders for components as required to support the MPS. Lot-sizing algorithms and lead times used within the MRP system will affect the capacity of the facility to produce to the MPS requirements. The time-phased production requirements from MRP can be input to a *capacity requirements planning* (CRP) module. The output from the CRP module is projected load at each workcenter for each time period. Typically, this is done without considering capacity constraints at the workcenter; however *load leveling* can be employed to correct any major problems. Load leveling attempts to move work orders from one time period to another, subject to material availability and the due date of the work order. If load leveling cannot correct an overload, the planner may choose to change workcenter capacity (for example, by planning overtime), or to change the MPS by moving some end-item requirements into an earlier or later time period.

Implementation

Once the planning phase is completed, implementation of the plans occurs on the shop floor. If a JIT system is used, implementation will involve utilizing a "pull" system to move work through the facility. End-item requirements from the MPS are released to the final assembly operation; final assembly in turn creates requirements for new components or subassemblies by utilizing existing work-in-process, and this is passed back through the product structure. Small imbalances in capacity are dealt with on the shop floor; large imbalances do not occur due to the nature of the master production scheduling process.

If an MRP system is used, a shop floor scheduling system or module can be utilized to schedule work in the facility. Shop floor scheduling systems typically involve simulating the flow of work orders through the facility to identify bottlenecks or other capacity constraints. These systems allow the user to try various scheduling rules, alternate routings, temporary changes in capacity, and so forth, until a feasible plan of action is obtained. Shop floor systems also support related activities such as order release, order tracking, data collection, production reporting, and closeout of completed orders. Such systems may be part of the MRP II package or stand-alone computer software that receives schedules from the MRP system and reports production data to it.

Summary

Capacity planning systems are used at many levels in the planning and control of production. These systems report required capacity and allow the planner to make changes as required. Methods of changing capacity include:[4]

Long-range changes

Facilities and land

Capital equipment

Workforce

Medium-range changes

Make/buy decisions

Process plans (machine routing)

Subcontracting (over long periods)

Tooling

Workforce (reallocate)

Short-range changes

Overtime/undertime

Subcontracting (for specific orders)

Part routings

Workforce (temporarily move)

Computerized production and capacity planning systems have grown in importance with the complexity of modern manufacturing and increasing availability of reasonably priced computers and software. These systems do not perform capacity planning; rather, they aid the planner by displaying the effects of his or her decisions on available and utilized capacity. At the highest planning levels, few systems are available to assist the planner, but as the plan becomes more detailed and the difficulty of effect prediction increases, computerized systems can be effectively utilized to ensure that a feasible and realistic production schedule is available for the manufacturing facility.

References

1. Hax, A. and D. Candea, *Production and Inventory Management,* Prentice-Hall, Englewood Cliffs, NJ, 1984.
2. Smith, S. *Computer Based Production and Inventory Control,* Prentice-Hall, Englewood Cliffs, NJ, 1989.
3. Bedworth, D. and J. Bailey, *Integrated Production Control Systems,* 2d ed., Wiley, New York, 1987.
4. Lankford, R., Olson, C., Revere, J., and S. Smith, *American Production and Inventory Control Society (APICS) Certification Program Study Guide: Capacity Management,* APICS, Falls Church, VA.

Further Reading

R. T. Lubben: *Just-in-Time Manufacturing: An Aggressive Manufacturing Strategy,* McGraw-Hill, New York, 1988.
J. Orlicky: *Material Requirements Planning: The New Way of Life in Production and Inventory Management,* McGraw-Hill, New York, 1975.

Evaluating and Selecting Facility Design Options

William Wrennall
President, The Leawood Group, Ltd.
Leawood, Kansas

Introduction

Evaluation of facility design options aids selection. The selection methods of this chapter apply to many facility design tasks.

Evaluation and selection help in determining the preferred site, space, building, and material-handling plan. This chapter concentrates on nonfinancial methods.

The KORDANZ* method developed by Knott,[1,2] based on Kendall[3] and Nelson,[4] receives special attention. This is because it is suitable for facility design evaluation as well as job evaluation.[5] This is convenient, because facility design as part of operations restructuring often results in a need for job regrading.

Chapters on site selection and simulation supplement this chapter.

The reasons for generating alternatives

The space-planning process described in Chap. 4, "Factory Layout and Design," generates layout options. This leads to the question: "Why generate options?" Some managers just want one answer. However, many management situations and problems do not have one correct answer or solution. In such cases, it becomes necessary to develop sound alternatives, evaluate them, and make a choice.

*KORDANZ is a trademark of The Leawood Group, Ltd.

Progressive senior management, forever cognizant of uncertainties, values sound and participative contributions to decision making. Faced with uncertainties, it favors a range of options. Thus management is not deprived of room to maneuver.

Developing Design Options

Options may result from:

- Varying constraints of the layout primitive
- Mirror images
- Combination or splitting of space-planning units (SPUs)
- Single or multistory stacking
- Different or multiple buildings
- Building constraints

More fundamental differences in the layout design process can create options—for example,

- Use of different processes
- Advanced technology
- Proven technology
- Different SPUs

In the strategic analysis portion of the factory layout process the designer will consider several ways to develop the SPUs. The design can incorporate cellular manufacturing concepts, group technology (GT) cells, classic Detroit assembly lines, flexible manufacturing systems (FMSs), focused factories, central or point-of-use material deliveries. This gives rise to options.

From the above, the planner can generate several site, factory, office, or material-handling plans. Varying the constraining of the layout primitives gives additional design options.

Limit options to three to five sound alternatives. Too many and similar options will cause unnecessary confusion and invite criticism.

Presenting alternatives allows management to choose. It avoids one-option rejection and negativism. The selection process can be creative and innovative. This releases new information, constructive criticism, and input. Thus further options emerge. The new inputs do not negate previously generated alternatives. They may be selected as they are, slightly modified, or enhanced as a result of the selection team's catalytic action.

A major benefit of this selection process is acceptance. The selling is done. The choice is not made by the designers. Colleagues now claim ownership of the selected version.

Who Should Evaluate?

The designer, the design team, the management, or an empowered team may evaluate and select. The designer or design team should do a preliminary evaluation. This serves as a rehearsal and quality check. A cross-functional team of users and managers then evaluates and selects the preferred option. For an expeditious decision, the evaluating team should be empowered to authorize approval of its choice. This supports a time-based competitive strategy.

Evaluations from those affected by a decision are readily accepted by them. The participative procedure earns commitment. The release of new and needed information results in better design options.

Facility Design Objectives

Structuring the evaluation and design process forces the identification of key factors. It is important to think through what is wanted, and how it will be recognized and attained.

Acceptable options must meet certain primary objectives. It is advisable to list feasible evaluation factors and submit them to management for approval. Managers often have not considered how they will recognize a good design from a not so good option. Submitting a preliminary list saves management time.

The facility designers gain an advantage by knowing what is considered "good" by management before the project is started. The design plan, for acceptance, must be consistent with the company culture, management biases or preferences, and, often, unstated policies.

This process of criteria validation verifies the project assumptions and avoids the wasted effort associated with unclear project definition.

Examples of desirable facility features are:

- Meets forecast capacity needs
- Consistent with company image
- Supports operations strategy
- Allows for new product introductions
- Provides for rapid material velocity
- Improves working conditions
- Easy to install

The Selection and Evaluation Process

The selection of a facility design can be based on a range of evaluative processes.

The first hurdle is qualification. Does the new or redesign meet the basic design criteria? These criteria can be operational management requirements; they may be corporate-project-specific, or company-wide design decisions or regulations.

The designer does not have a choice. This is illustrated in Fig. 11.1. Qualifying criteria are the minimum design requirements that have to be met for the design to be a contender.

Given Requirements

- ◆ Area
- ◆ Power
- ◆ Time
- ◆ Etc.

Evaluation Factors

- ◆ Unit Cost
- ◆ Support
- ◆ Fit Plans
- ◆ Etc.

Figure 11.1   Site selection process.

Examples of qualifying criteria are:

- The plant will be able to produce 500 windows per eight-hour shift.

- The facility should be capable of converting raw materials and purchased parts into products within four hours.

- Layout rearrangements must not reduce output.

- Safety and fire codes must be met.

Evaluation then differentiates between feasible options.

Methods to Assess Design Options

For qualified design candidates the evaluation methods can be qualitative, quantitative, nonanalytical, or analytical. Qualitative methods are judgmental. Quantitative methods measure differences on a numerical scale.

Nonanalytical methods of evaluation may use qualitative or quantitative measures. The evaluation will make holistic decisions. Analytical methods will examine parts, features, or factors. The evaluation will add the parts and give a total evaluation.

Frequently the capital costs of alternatives are similar or are not significant discriminators. Tangible capital costs can be measured or estimated, but some operating costs may be difficult to determine. These other factors are no less significant—indeed they may be strategically vital.

The less quantitative methods and evaluation of intangibles have received less publicity than economic analysis procedures. The less quantitative methods are often most important in establishing and selecting facility designs for the future success of the firm.

Evaluation discriminates between alternatives and helps decide what option to adopt. The benefit of formalizing the process is increased confidence and readier acceptance later. Formalization minimizes the risk of accepting a nonfunctioning design. The process also discourages autocratic and overcentralized decision making.

An important step is the design of the evaluation criteria. An understanding of what is to be achieved and how it will be judged by your peers before the work is performed can only enhance the productivity of the layout design process.

Accounting for intangibles in the decision-making process

Many factors in evaluation and selection are difficult to quantify. These factors are called *intangibles*. The fact that they are intangible does not make them less real.

Experience aids anyone who makes decisions that involve intangibles. Yet, the problem exists—how to separate prejudice from experience? The evaluation methods that follow are designed to lessen the impact of such prejudices.

Evaluation Methods

For facility design evaluation the following methods are useful:

- Sensory
- Intangibles

  PNI

  Ranking

  Weighted-factor analysis

  KORDANZ

- Material flow

  Iso-cost/distance intensity

  Quantified flow diagrams

  Simulation

- Affinity analysis
- Decision tree
- Fluidity

Sensory method

The *sensory method* draws on instincts, experiences, and prejudices. Do we like the look, smell, taste, or feel of it? Does it seem right? Miñoza-Gatchalian confirms that this method is particularly important in the food, fashion, and cosmetic industries.[6] It is qualitative and nonanalytical.

The sensory method requires decision makers experienced in facility designs. These individuals pick the best option based on their experience. The designers and their colleagues will see what they like. It may satisfy a hidden agenda that cannot be publicized for policy reasons.

The sensory method has obvious weaknesses. How, for example, to tell when a decision comes from prejudice? Still, when experienced individuals perceive strengths and/or weaknesses, do not ignore it. They are probably more right than wrong. What looks good on paper may be good visually in operation.

PNI

PNI stands for **P**ositive, **N**egative, and **I**nteresting. Decision makers rate options with respect to these categories. The idea comes from de Bono, an authority on innovative and "lateral" thinking.[7]

In a meeting environment, PNI works like this:

- Decision makers consider each alternative
- For each they are asked which features are:

   Positive

   Negative

   Interesting

- Go around the room to exhaust P, N, or I comments; set a time limit for commentary
- From the consensus, determine the preferred option

Interesting features are neither good nor bad, but pique curiosity, involvement, and innovation. These are the most difficult to generate. Take the necessary time to find interesting features. They can help you develop novel and more creative alternatives.

On the surface, PNI may seem too simple, little more than a pro-con listing that brings people no closer to a decision. Used correctly, PNI is a mind-focusing tool. Instead of people simply blurting out their prejudices, PNI forces them to focus on a particular aspect of the problem. By doing so, PNI produces more useful results than a pro-con list. It also stimulates and releases creative thinking.

Ranking

A simple *ranking* can evaluate design options. This is an example of a qualitative, nonanalytical method of evaluation. It is a fast method. It can be analytical if the ranking is done by factor.

Weighted-factor analysis

The *weighted-factor method* is analytical. The factors can be quantitative or qualitative.

Figure 11.2 is an example of a weighted-factor analysis. To use the procedure, fill in the project block in the upper right-hand corner, then list the alternatives under "Options" at the lower right. Next:

1. Ask decision makers what factors they will and should consider to make their decision. List these under "Factor."
2. Have decision makers weigh each factor with respect to each other on a scale of 1 to 10. Place these numbers under "Wgt." Note that

Description:		Proj. No.
New Office Expansion		1144
Company:		Prep By:
Additives Int.		KB
Location:	Page:	Date:
Central Offices	1	7-May

			Ratings / Weighed Scores										
			1		2		3		4		5		6	
No.	Factor	Wgt.	R	WS	R	WS	R	WS	R	WS	R	WS	R	WS
1	Ease of Installation	5	1	5	4	20	2	10	4	20				
2	Acoustics	7	3	21	2	14	4	28	2	14				
3	Ease of Utility Installation	5	2	10	3	15	2	10	4	20				
4	Use of Space	8	2	16	4	32	2	16	3	24				
5	Employee Convenience	9	2	18	2	18	2	18	3	27				
6	Simplifies Workflow	10	2	20	2	20	3	30	3	30				
7	Communication	7	1	7	3	21	2	14	4	28				
8	Flexibility	6	1	6	3	18	2	12	4	24				
	Totals:			103		158		138		187				

RATING VALUES:		OPTIONS:	
4	A = Almost Perfect	1	V.P.s near President
3	E = Especially Good	2	Sales near Reception
2	I = Important (good)	3	V.P.s with their departments
1	O = Ordinary (fair)	4	President in N.E. corner office
0	U = Unimportant	5	
*	X = Not Acceptable	6	

Figure 11.2 Weighted-factor analysis.

these *need not be* sequential; for example, there could be two 9's and three 7's.

3. Then decision makers consider the factors in each option, rating factors on the A, E, I, O, U, X scale, with numerical equivalents of 4, 3, 2, 1, 0, Invalid.

4. A score—"WS"—for each factor comes from multiplying the weight by the rating.

5. When the rating and scoring is complete, total the scores for each option. The highest score gives the preferred design.

In Fig. 11.3 are shown different block (macro) space plans from a lay-out project. Figure 11.2 was a completed weighted-factor analysis for these space plans. Notice that factor 7, "Communication," has a weight of 7. Following factor 7 across the three options, note that it received a different rating for each layout. The ratings are O (1) for number 1, E (3) for number 2, I (2) for number 3, and A (4) for number 4 option. Note, too, the difference in score that these ratings created.

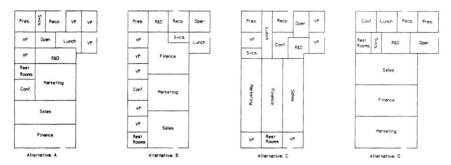

Figure 11.3 Alternative layouts.

Weighted-factor analysis is not a precise technique. Option 4 is the preferred layout, with a total score of 187. However, the difference in totals for numbers 2 and 3, while significant, is not huge. A 10-point difference (or less) means that the layouts have similar merit.

The weighted-factor method has weaknesses:

- Factor weighting is subjective and the weighting scale is linear.
- There is bias in the judging.
- There is no measure of evaluation consistency.
- Where group discussion precedes rating, the opinions of a strong personality may influence judges.
- The method does not create or demand a deep understanding.

And yet, the method does have benefits:

- It helps determine decision parameters (i.e., factors and their weight).
- It creates significant discussion and narrows the range of options.
- Numbers give it a semiquantitative nature.

Weighted-factor analysis is among the simplest and most effective means of deciding between layout options, because:

- It brings in many people, often creating its own consensus.
- It is often unnecessary to go beyond this method.

The KORDANZ method

KORDANZ is a computerized form of weighted-factor analysis, which overcomes many of the negatives of traditional weighted-factor evaluation. It is a multifactor method with the following enhancements:

1. The evaluation scale is relative, rather than absolute.
2. It eliminates pseudo-accuracy.
3. There is a common basis to compare qualitative and quantitative values.
4. The relative importance of the separate objectives of the options is reflected in the final evaluation.
5. It calculates level of agreement in probability terms.

The advantages of KORDANZ are:

- Quantifies opinion
- Tests concordance among evaluators
- Reduces effect of prejudice
- Makes prejudgment difficult
- Encourages participation of specified management levels
- Reduces evaluation to binary decisions
- Captures knowledge of informed evaluators
- Minimizes time required by evaluators
- Separates factor emphasis from evaluation
- Generates confidence among users
- Identifies the relative importance of factors
- Clarifies and polarizes the purpose
- Minimizes emotion
- Is analytical

Figure 11.4 flow-charts the KORDANZ process. The first step is to decide on the factors. Some examples are shown in Fig. 11.5.

KORDANZ achieves factor ranking by forced comparisons between factor pairs. Ties are not allowed. The number of comparisons are $n(n - 1)/2$, where n is the number of factors. The ranking is performed by a number of judges.

The method tests the consistency of each judge and tests for inconsistencies between them. From the rank data for each factor a ratio is calculated. The *ratio*, or *emphasis coefficient*, is the weighting given to each factor by the evaluators. By matrix analysis the system tests for inconsistencies of judgment. Factors ABC are ranked in pairs by a judge who prefers:

A to B

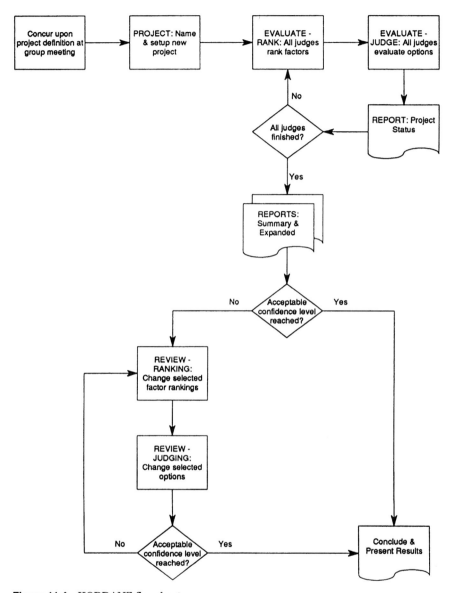

Figure 11.4 KORDANZ flowchart.

B to C

C to A

The system highlights this inconsistency by an asterisk in the report.
Figure 11.6 depicts inconsistent and consistent judging.

FACTORS	SOME CONSIDERATIONS
Cost of implementation	Construction and setup costs Downtime Equipment and personnel costs
Layout flexibility	Ability to expand or contract Minimize rearrangement Use of areas by various shifts
Material handling effectiveness	Simplified flows Types and bulk of materials Entry and exit points within processing Movement intensity and timing
Space utilization	Utilization of s.f. by class of space Use of cubic space Use of aisles, etc.
Supervision and control	Communication enhanced Line-of-sight management Placement of computerized controls
Support service integration	Support service areas Movement and intensity amongst serviced areas Communication and information sharing
Environment	Environmental impact Safety Workstation ergonomics
Ease of implementation	Schedule and timing Production interruption Availability of implementation resources

Figure 11.5 Evaluation factors.

The inconsistencies may occur because:

- The judge cannot distinguish between factors.
- There is no distinction between the factors.
- The judge is a poor evaluator.
- There is deliberate distortion by the judge.

Judges should know about their inconsistencies. Remove causes of ambiguity such as poor factor definition. This step is the quality control of the process.

An inconsistency:

- Raises doubts and suspicions

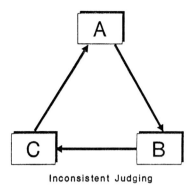
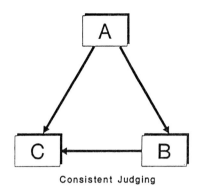

Inconsistent Judging   Consistent Judging

Figure 11.6 KORDANZ judgments.

- Reflects randomness of judgment
- Is the only way tied values can occur with forced comparisons

Inconsistencies between judges are common. They will have an effect in addition to the inconsistency of individual judges. This creates either an acceptable or unacceptable level of agreement. If acceptability or concordance is not achieved, it may be desirable to redefine the factors in clearer terms.

Differences of opinion are inevitable within a pooled group. It may be worthwhile to examine these differences of opinion and motivation. This can be done by testing the level of concordance between judges.

The factors, each with a coefficient of emphasis, are now ready for evaluation. Random pairs of alternatives are compared for each factor and by all judges. The sum of the factor evaluations combined with the weights for each alternative give a total utility or score for each option evaluated.

The requirements of the evaluation are:

- Independent evaluator responses
- Capture of evaluator knowledge
- Avoidance of bias
- Acceptable concordance
- Evaluations that gain support

A KORDANZ site selection example.   KORDANZ was successful in selecting a new site for an aerospace company. The results of the evaluation are given in Fig. 11.7.

Scenarios and factors are the means of identifying the preferred manufacturing site. The factors were far more than conventional wis-

```
------------------------------------------------------------------------
Project: PHOENIX          * Kordanz - Summary  Report *        Page: 1
Thu Sep 27          Evaluation by Forced Binary Decision       14:33
------------------------------------------------------------------------

                    Composite Ranking of Factors

                    Factor                Weight
                    --------------------  ------
                    OPERATIONAL NEEDS        8
                    TECH & ORG. SUPPORT      7
                    ENHANCED TECHNOLOGY      6
                    FIT LONGRANGE PLAN       4
                    BTD RESPONSIVENESS       4
                    SECURITY                 4
                    COMPATIBLE PRODUCTS      2
                    COMPETITIVE COSTS        1
                    COMMUNITY IMPACT         0


                    Composite Ranking of Options

                    Option                Score
                    --------------------  -----
                    NEAR SITE                80
                    PRESENT SITE             72
                    FAR/NEAR MANUFACTURE     49
                    FAR SITE                 15


                    Level of Confidence: 99.00

                An acceptable confidence level was obtained.
```

Figure 11.7   KORDANZ summary report (site selection).

dom and comforting philosophical statements.

It is important in scenario design that:

1. The scenarios be realistic

2. They modify the decision maker's assumptions of reality

3. There is willingness to face uncertainty

4. There is understanding of the causes of uncertainty

Scenarios that merely quantify the obvious do not help. Scenarios should assist managers with strategic insights which go beyond their initial spectrum of consideration. This change in perspective is fundamental to developing scenarios that incorporate managers' deep concerns.

In factor development, separate the predetermined from the variable. Site-planning literature is full of checklists. These checklists usually are lists of prerequisites. Prerequisites do not differentiate between options.

In this project the review, modification, and approval of scenarios were done at a meeting with director-level managers. A preliminary list was the basis for discussion and approval. The group decided that two of the twelve factors were not mutually exclusive. They revised and approved eleven factors.

After factor approval, the next step in the KORDANZ procedure calculated factor weights. The directors saw pairs of factors in random sequence. They chose the most important. Draws were not allowed. Responses were tested for consistency and the factor weights calculated.

In early discussions cost was the single most important factor. The directors' responses, surprisingly, resulted in a factor weight of only 16 percent for cost.

After factor weighting, site alternatives were judged by a group of project managers. They used the approved scenarios and factors. They did not know the factor weights.

The cost factors were separated because a different group calculated cost comparisons for each scenario. Cost can be quantified; many other factors cannot.

In this case a total score for each site scenario was calculated from the managers' forced comparisons and the directors' weights for the factors.

Sensitivity analysis determined the accuracy of cost estimates needed to discriminate between leading contenders. In this example a 50 percent lower cost for the second-ranked candidate was needed to make it a serious contender. This saved time in cost calculation, since great accuracy was not necessary.

When the participants and top management saw the results, their response was: *"Aha, we are surprised."*

The KORDANZ approach is available as a software package for DOS computers and is described by Wrennall.[8,9] The software eliminates mathematical tedium and complexity. It provides a powerful and apparently simple process. KORDANZ is not limited to site selection. It has a wide range of applications and is easy to use.

Figures 11.8 and 11.9 are from a KORDANZ summary report. A list of factors and weights comes from the judges' forced comparisons. The composite ranking of the options is also generated. Review the level of confidence of the weighting before proceeding. In this example the level of confidence is 95.00% and so judging of the options can proceed.

The material flow method

Material flow analysis (MFA) can help decide between site or factory layout options. It has the advantage of getting away from intangibles, because material flow can be quantified.

There are three ways to evaluate a layout relative to material flow considerations:

- Iso-cost/distance intensity
- Quantified flow diagrams

```
------------------------------------------------------------------------
Project: OFFICE              * Kordanz - Summary Report *
Page: 1               Evaluation by Forced Binary Decision          14:30
------------------------------------------------------------------------

                    Composite Ranking of Factors

                    Factor                   Weight
                    --------------------     ------
                    Honors Affinities          20
                    Minimum Noise              16
                    Fit with Other Depts       13
                    Ease of Supervision        12
                    People Convenience         10
                    Best Use Of Space           8
                    Rearrangement Ease          4
                    Ease of Installation        1



                    Composite Ranking of Options

                    Option                    Score
                    --------------------     -------
                    Mgr. Lower Center          454
                    Mgr. Upper Right           169
                    Mgr. Lower Left            133

                    Level Of Confidence:   95.00

              An acceptable confidence level was obtained.
```

Figure 11.8 KORDANZ summary report (location of manager's office, p. 1).

- Computer simulation

Using iso-cost/distance intensity.   To develop an *iso-cost/distance intensity plot,* produce a graph with distance along the horizontal axis and *flow intensity units* (FIUs) along the vertical axis. An example of such a graph is given in Fig. 11.10.

Select a pair of blocks, cells, or SPUs from the macro layout for evaluation.

- Figure the distance between the pair.
- Determine the FIUs between them.
- Plot this point on the graph.
- Repeat for each possible SPU pairing.

Plotting a similar graph for each layout has two uses:

1. It helps to improve the layouts.
2. Calculations derived from these graphs will help evaluators to decide on the layout with the best material flow.

Iso-cost/distance intensity graphs lead to improved layouts. They show those site units or SPUs that have a high material flow between them but are far apart.

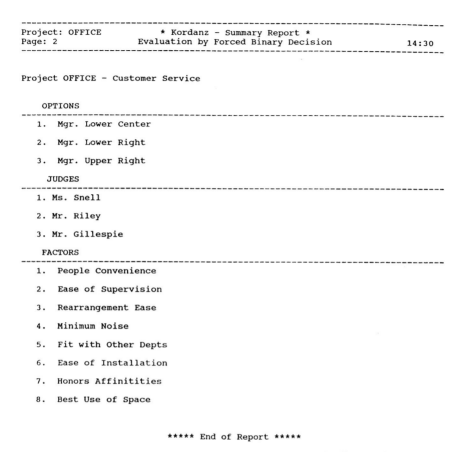

```
-----------------------------------------------------------------------
Project: OFFICE            * Kordanz - Summary Report *
Page: 2                 Evaluation by Forced Binary Decision      14:30
-----------------------------------------------------------------------

Project OFFICE - Customer Service

    OPTIONS
-----------------------------------------------------------------------
   1.  Mgr. Lower Center

   2.  Mgr. Lower Right

   3.  Mgr. Upper Right

    JUDGES
-----------------------------------------------------------------------
   1. Ms. Snell

   2. Mr. Riley

   3. Mr. Gillespie

    FACTORS
-----------------------------------------------------------------------
   1.  People Convenience

   2.  Ease of Supervision

   3.  Rearrangement Ease

   4.  Minimum Noise

   5.  Fit with Other Depts

   6.  Ease of Installation

   7.  Honors Affinitities

   8.  Best Use of Space


              ***** End of Report *****
```

Figure 11.9   KORDANZ summary report (location of manager's office, p. 2).

Most points follow the pattern shown in Fig. 11.10. A point far away from the general group, as in Fig. 11.11, indicates something wrong with the layout.

The point in question represents two areas with a flow of nearly 150,000 FIUs and a distance of 400 feet. The two should have been closer together.

The iso-cost/distance intensity plot is a diagnostic and refinement tool. Using this graph, blocks that are relatively close but with a low material flow are prime candidates for relocation. The plot also helps in material-handling analysis.

In physics, work = force × distance. Material handling has a similar concept called *transport work*. Transport work = intensity of flow × distance.

Iso-transport work curves are joined points of equal transport work. An example of an iso-transport work curve is shown in Fig. 11.12.

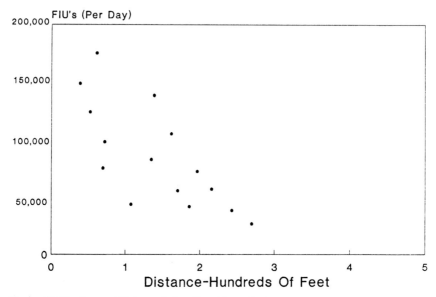

Figure 11.10 Iso-cost/distance intensity—Example 1.

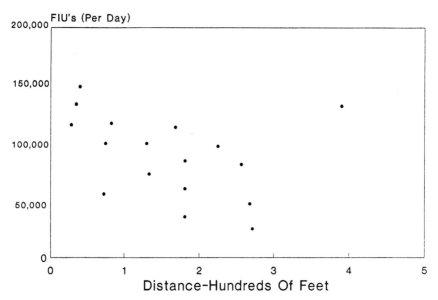

Figure 11.11 Iso-cost/distance intensity—Example 2.

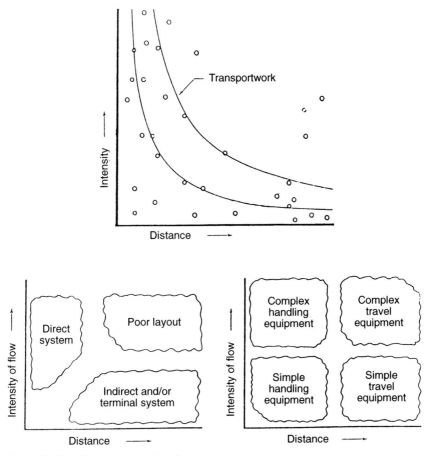

Figure 11.12   Distance-intensity plot.

Spaceplan* generates a series of curves. Layout refinement drives the curves toward the origin.

To evaluate alternative layouts using transport work analysis:

- Calculate the transport work for each path and sum for all paths.

- Compare the layouts with respect to transport work.

- The lowest number gives the preferred layout.

The "cloud" diagrams in Fig. 11.12 illustrate the use of the plots to select material-handling systems and equipment.

*Spaceplan is a trademark of Prime Computervision.

Quantified flow diagrams. *Quantified flow diagrams,* or *location diagrams,* are visualizations of routes and/or quantities of material movements in a macro layout.

This visual display also highlights material movement complexity, route material intensity, and distances traveled as a result of the macro layout.

To create location diagrams:

- Create a diagram similar to a configuration diagram.
- Between the centroids of the SPU pairs use bars or thickness of lines to rate the material flow.

Figure 11.13 shows a location diagram.

In a study of material flow between SPU pairs, short, thick A and E (high-flow) lines indicate their location is good. On the other hand, long, thick lines between SPUs with high material flows suggest that something is wrong with the layout.

Obvious route simplification or reduction in backtracking or distances moved can be more powerful than attempts at numerical comparisons. Quantified flow (or locational) diagrams lead to a layout with the efficient and effective material flow. These diagrams are also helpful in deciding material-handling methods. Short, thick lines or high-volume short distances suggest *handling equipment.* Long, thin lines or low-volume long distances indicate an *indirect transport system.*

Computer simulation. A facility is one piece of a complex system of random inputs and variables that are difficult to quantify. *Computer*

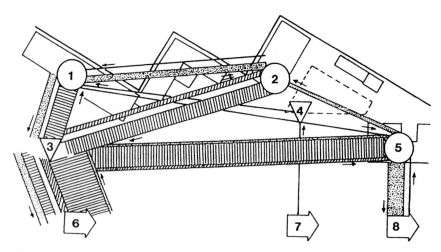

Figure 11.13   Location diagram.

simulation is one method that can anticipate for and demonstrate the probabilistic effects these inputs and variables can create.

Computer simulation can be either numeric or animated and results from mathematical modeling of material movements under varying conditions. With high investment in process and support equipment there is a need to determine if the system is sound before capital has been expended. A computer model can establish to what extent the system will (1) function, (2) provide the specified results, and (3) benefit from modifications.

Simulation can economically provide information as to what will happen over a period of time. An animated display tests the validity of the simulation model and presents a live display of what the system will do under varying operating conditions. The results are often very different from predictions based on simplistic arithmetic calculations, "experience," or conventional wisdom.

For example, the management of a plastics extrusion plant wanted to use a Kanban system of inventory control. Kanban, they felt, would allow them to respond more effectively to changes in product mix.

At any one time, the plant had about $12 million of finished goods inventory in its warehouse. This inventory equaled about 55,000 units of finished production. Computer simulation incorporated a Kanban system into the existing plant. The simulation showed that the plant received many small orders which Kanban handled effectively. But every second or third day a large order came in. These large orders created havoc with the Kanban system.

The planners running the simulation then modified the Kanban system to accommodate the expected schedule of orders. The resulting operation required that the plant now only store 3500 units of finished production, as opposed to the 55,000 they had been carrying. Warehouse space was freed up. Plus, the plant greatly reduced delivery time.

Affinity analysis

A simple but effective means of evaluating a layout is *affinity analysis*.

To perform an affinity analysis, examine each layout option, determining how many affinities between the SPUs have been honored. The layout where the A and E affinities are between neighboring areas and X-related areas are apart will be a good layout.

Decision trees

Decision trees are evaluation diagrams that resemble a tennis play-off diagram.

They have the following advantages:

- Are easy to set up and use
- Are easy to communicate
- Help in the identification of all possible outcomes
- Can assign probabilities based on a consensus of opinion for various events

Decision tree analysis has significant benefit when little "hard" information is available but "guesstimates" abound. Forecasts of cost, size, shape, etc., are coupled with assigned probabilities to make the tree. Decision trees are helpful for both site selection and facility design.

In the following example, a manufacturer is deciding the location for a new plant. Decision criteria involve occurrences that are two to five years in the future. Plant construction must begin within the next six months to meet the sales demand.

Two sites are available in the general geographic area. Two independent future events will affect product cost. The first is the location of a new interstate highway. Depending on its location, the highway will add or reduce freight costs. These cannot be passed to the customer.

The other event is the construction of a facility for an important raw material. The facility site selection for raw material processing is years away. Current capacity is adequate. The location for the new raw material facility is either in the "plains" or in the "valley" (Fig. 11.14). The associated costs and likelihoods of occurrence are shown on Fig. 11.15.

The decision tree of Fig. 11.16 reflects this situation. Relative probabilities were assigned and the best (least annual cost) site was selected (Fig. 11.17). In this scenario it was site S with an annual cost of $1,019,000. The eight outcomes ranged from $885,000 (Interstate S. to E.N.E. and raw material facility in the valley) to $1,192,000 (Interstate S.W. to N.N.E. and raw material facility in the plains). Since these two alternatives are possible with the same site selection (S), it was determined to be very sensitive to assigned probabilities. Because of this, the alternative probabilities were reassigned.

In the second and final scenario (Fig. 11.18), 50/50 probabilities were assigned for each outcome. When this was evaluated, the least cost was now site R at $1,003,000 per year. The selection of site R would result in an annual savings of $36,000 over site S. Decision tree analysis provides the user with the eventual costs of each outcome and the likelihood of occurrence.

Fluidity

Nyman uses the term *fluidity* to cover *flow, density,* and *velocity* as evaluative features in cell design.

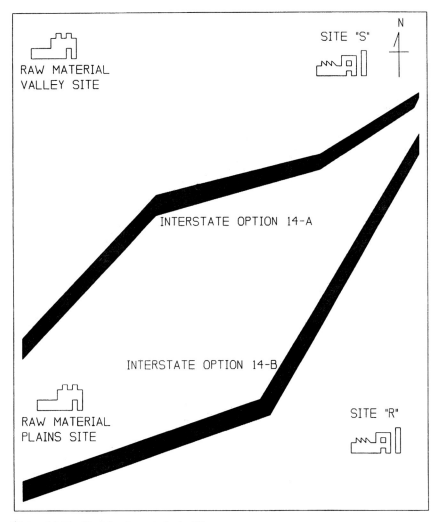

Figure 11.14   Decision tree analysis: Map.

Flow is the minimum path that a part must travel in a manufacturing plant to accumulate its value-added content.

Assessment of flow time and distance is made by evaluation of:

- Proximities
- Operation distances more than 6 feet apart
- Backtrack frequency
- Flow path complexity

Outcome	Site R probability	Site S probability
Interstate		
S. to E.N.E.	630,000 (0.6)	490,000 (0.6)
S.W. to N.N.E.	555,000 (0.4)	762,000 (0.4)
Raw material		
Plains	460,000 (0.7)	430,000 (0.7)
Valley	360,000 (0.3)	395,000 (0.3)

Figure 11.15   Decision tree analysis: Alternative Table "A."

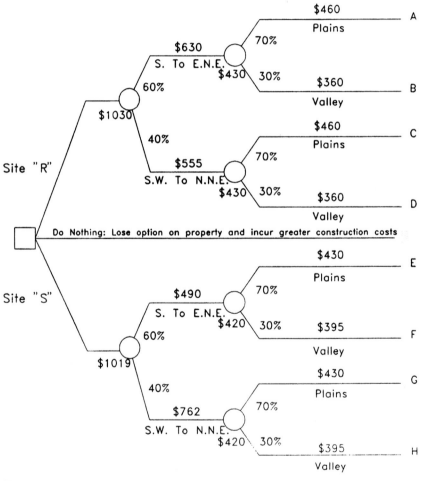

Figure 11.16   Decision tree for site selection.

Outcome	Site	Interstate	Raw material	Cost (000)
A	R	S. to E.N.E.	Plains	1090
B	R	S. to E.N.E.	Valley	990
C	R	S.W. to N.N.E.	Plains	1015
D	R	S.W. to N.N.E.	Valley	915
E	S	S. to E.N.E.	Plains	920
F	S	S. to E.N.E.	Valley	885
G	S	S.W. to N.N.E.	Plains	1192
H	S	S.W. to N.N.E.	Valley	1157

Decision Tree with 60/40 and 70/30 Probabilities

Cost per year of Site R = $1,030,000
Cost per year (volatile) of Site S = $1,019,000

Decision Tree with 50/50 and 50/50 Probabilities

Cost per year of Site R = $1,003,000
Cost per year of Site S = $1,039,000

Figure 11.17   Decision tree analysis: Table "B."

- Positive movements

Density defines the relative closeness of operations. Density examines:

- Machine spacing
- Automation opportunities
- Communication effect
- Inventory elimination
- Decision-making focus

The *density ratio* is defined as:

$$\text{Density ratio} = \frac{\text{sum of equipment areas}}{\text{gross cell area}}$$

Nyman considers that a cell density ratio less than 35 percent is generally too low. Density ratios of 50 percent or more are possible.

Velocities of materials, ideas, designs, and transactions are targets for improvement in world class companies. Facility design aims to increase material velocities by reducing travel distances. Time-based competitors reduce physical and mental time paths by integrated restructuring.

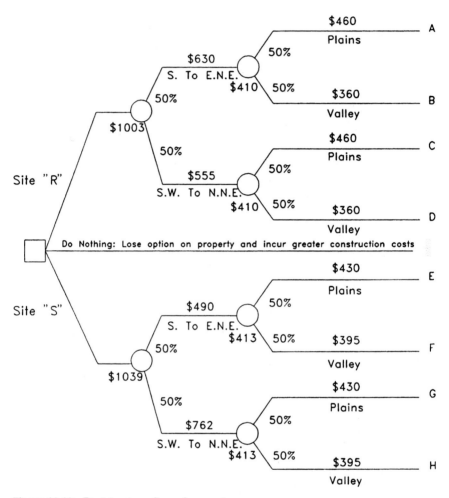

Figure 11.18   Decision tree: Second scenario.

$$\text{Velocity} = \frac{\text{Value–adding time}}{\text{lapsed time from order to customer receipt}}$$

The acceleration of value adding will be characterized by:

- Rapid problem elimination
- Continuous material flow
- Real-time decisions
- Customer satisfaction

- Negative inventories
- High return on assets employed
- Increased market share

Summary

It is unnecessary to use more than a few evaluation techniques on any given project. Choose methods depending on the key issues for distinguishing between the options. Key issue examples could be material-handling consequences of the layout, ease of implementation, quality effect, and communications. The overall objectives of the project should also have been met, supposing that with hindsight they are still valid.

The evaluation process will release a lot of information and understanding of facility plans. Participation will spread this knowledge and ownership will follow resulting in acceptance of the selected decision.

References

1. Knott, K., "Forced Comparisons and Youden Squares as the Basis of Improving Job Ranking in Job Evaluation," *International Journal of Production Research,* Taylor & Francis, New York, 1983.
2. Knott, K., "An Approach to the Evaluation of Design Using Ranking Theory," IMSE Working Paper 84-102, Department of Industrial and Management Systems Engineering, Pennsylvania State University, State College, PA, 1984.
3. Kendall, M. J., *Rank Correlation Methods,* Charles Griffin, London, 1975.
4. Nelson, C. L., A Methodology for Choosing between Competing Material Handling Systems, unpublished ms., Department of Industrial and Management Systems Engineering, Pennsylvania State University, State College, PA, 1985.
5. Otis, J. L., and R. H. Lenkart, *Job Evaluation,* 2d ed., Prentice-Hall, Englewood Cliffs, NJ, 1954.
6. Miñoza-Gatchalian, Miflora, *Sensory Evaluation Methods,* College of Home Economics, University of the Philippines, Philippines, 1989.
7. de Bono, Edward, *Letters to Thinkers,* Harrap, London, 1987.
8. Wrennall, W., "Evaluation of Management Design Options," reprinted from *Manufacturing Technology International,* Sterling Publications Limited, London, 1989.
9. Wrennall, W., "KORDANZ Software Now Available," The Leaword, The Leawood Newsletter, Leawood, KS, vol. 8, Spring 1990.

Further Reading

Bell, K.: *KORDANZ User Manual,* Version 2:30, The Leawood Group, Ltd., Leawood, KS, 1992.
Knott, K., P. C. Cohen, and A. Ferraco: "Method for Evaluating Flexible Manufacturing Systems," *MTM Journal,* Fairlawn, NJ, 1985.
Nyman, L. R. (ed.): *Making Manufacturing Cells Work,* Society of Manufacturing Engineers, Dearborn, MI, 1992.

12

Queueing Methods for Facility Evaluation

Dr. Mark C. Springer
Associate Professor
Department of Management
College of Business and Economics
Western Washington University
Bellingham, Washington

Dr. Paul K. Makens
Director of Quality Management and Technology
Quality Resource Group
Hospital Corporation of America
1 Park Plaza
Nashville, Tennessee

In designing a manufacturing or service facility, a facilities planner is frequently concerned with evaluating the performance of alternative layouts prior to beginning construction. The planner may well be interested in the systemwide throughput of the proposed facility, the existence of bottleneck workcenters, or the buildup of inventory or customers at certain stages of the production or service process. One popular technique for evaluating a facility layout is computer simulation; software packages such as SIMAN/CINEMA[1] enable a facilities planner to "build" the proposed facility on a computer and track its performance through simulated time. Although great strides have been made in developing user-friendly simulation packages, the planner may still consider simulation models too time consuming to build and use in the early stages of facility design. In this phase, the planner is interested in relatively quick feedback for the purpose of winnowing out large numbers of undesirable facility layouts. Fredericks[2] suggested that analytical models are more appropriate for such rough-cut analyses of a large variety of alternative layouts. After the range of promising layouts has been narrowed down, a detailed simulation can then be constructed to aid in the final stages of layout design.

Queueing theory provides one increasingly popular avenue for analytically modeling service and manufacturing systems. In everyday speech, a *queue* is simply a waiting line where people, materials, work orders, or vehicles "queue up" and wait to be served. Queueing theory is therefore the science of waiting, and provides exact and approximate formulae for the performance evaluation of *queueing systems.* A queueing system may consist only of a single isolated queue, which in queueing theory terminology usually refers not only to the line of people or things waiting to be served but also the server(s) who must serve the waiting entities. More realistically, a queueing system may be comprised of several interacting queues; this is referred to as a *queueing network.* In a queueing network, each individual queue may send and receive entities to and from other queues. Queueing networks can be used to represent a service or manufacturing system; each queue in the network represents a workcenter or workstation processing customers or materials and subsequently forwarding them to another workcenter. Thus queueing networks may be used as an alternative to simulation in evaluating the feasibility or desirability of a proposed layout.

This chapter will proceed as follows. First, the simplest queueing system will be considered: a single isolated queue. This will introduce basic queueing terminology and apprise the reader of the information which may be obtained by applying queueing theory. Queueing networks will be considered next, and the differences between several existing models will be discussed. Finally, the application of queueing network software packages for the evaluation of manufacturing systems will be briefly discussed.

The Isolated Queue

The simplest queueing system is comprised of a single isolated queue which receives entities from an outside source; once entities are served, they leave the queue. Although this is an extremely simple system, many service systems may be modeled by a single queue: an automated teller machine, a car wash, a toll booth, and so forth. In each of these systems, customers arrive, wait in line, are finally served, and then continue on their way. A facilities planner designing one of these systems may be interested in determining the system throughput, the average number of customers waiting, the likelihood that a server is idle, or the chance that the system becomes flooded with customers and newly arriving customers are turned away. All of these questions may be answered using queueing theory.

Queue parameters

Before performance measures can be obtained, however, it is necessary to identify the *parameters* which determine the configuration of an isolated queue; even a queueing system as simple as a single isolated queue has a variety of possible forms. As shown in Fig. 12.1, a single isolated queue is characterized by its arrival process, the service process, the number of servers at the queue, and the queue buffer capacity. The *arrival and service processes* determine how quickly entities enter the system and how quickly they are served by each server; the *number of servers* at a queue can range from one to infinity; the queue *buffer capacity* refers to the number of entities which can fit in the queueing system, including those entities being served.

Analytically, the simplest queues have arrival and service processes which are each characterized by a single number, namely the rate at which customers enter the system and the rate at which each server processes them. The *arrival rate* is commonly denoted by the Greek letter λ (lambda) and the *service rate* per server is denoted μ (mu). The *traffic intensity* of the queue, denoted ρ (rho), considers how fast customers are being served relative to their arrival. If there are *M* servers in the queue, each one of which serves entities at the rate of μ, then the traffic intensity is simply given as the ratio of the system arrival rate to the maximum system service rate:

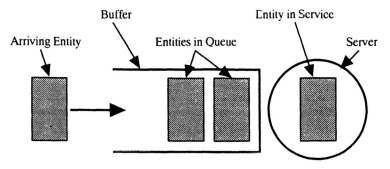

Figure 12.1   An isolated queue.

$$\rho = \frac{\lambda}{M \times \mu} \qquad (12.1)$$

In a queue with unlimited waiting room and any variability in the arrival or service process, clearly ρ must be less than one or the number of customers waiting will grow to infinity.

In addition to knowing the arrival and service rates, it is also necessary to know the nature of the arrival and service process: do the interarrival times and the service times never vary, or do the arrival and service rates simply represent an average of otherwise variable interarrival and service times? If the interarrival or service times are nonvarying, they are termed *deterministic*; if they vary from entity to entity, the corresponding arrival or service process is termed random, or *stochastic*. With stochastic arrival and service processes, large (but finite) lines can build up in front of a queue even though the traffic intensity is well below one; invariably, a server will temporarily slow down just as a large influx of customers arrives.

The variability of an arrival or service process may be measured by a quantity known as its *squared coefficient of variation,* or SCV, which is simply the *variance* of the interarrival or service time distributions multiplied by the square of the corresponding arrival or service rate. The variance, denoted σ_a^2 for the arrival process and σ_s^2 for the service process, is a statistical measure of the variability of a random variable. The expressions for the SCV of the arrival process, denoted c_a^2, and of the service process, denoted c_s^2, are given below:

$$c_a^2 = \sigma_a^2 \times \lambda^2$$

$$c_s^2 = \sigma_s^2 \times \mu^2 \qquad (12.2)$$

When the SCV for an arrival or service process is zero, the process is deterministic; when the SCV is one, then the process is *Poisson* and the interarrival or service times of the Poisson process possess an *exponential* distribution. Although the exponential distribution indicates a large degree of variability, the SCV for an arbitrary arrival or service process may well increase beyond one. Figure 12.2 provides examples of three different service time distributions, each with the same service rate of one entity per minute but with different service time SCVs ranging from 0.1 to 2.0.

Poisson processes and the corresponding exponential distributions are important in queueing theory since they exhibit what is termed the *memoryless* property. In a Poisson arrival process with exponential interarrival times, the expected time until the the next entity

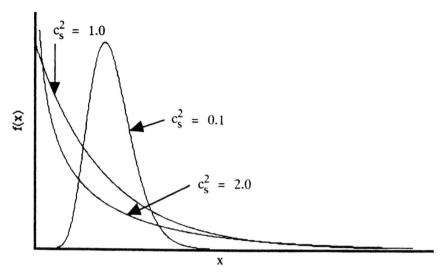

Figure 12.2 Service time distribution of increasing variability.

arrives is independent of the arrival time of the most recent entity; similarly, in a Poisson service process the expected time until the server finishes serving the current customer is independent of how long the server has been serving the customer. While this assumption may not be satisfied in many situations, it is frequently close enough to reality to justify using the analytically tractable formulae which can be derived for Poisson systems. Non-Poisson arrival and service processes, which are denoted *general* arrival and service processes, frequently do not possess exact formulae for the performance measures the facilities planner is interested in; as shall be shown below, there are a variety of approximations which can be used in place of exact results. It should not be assumed that Poisson or exponential queues are useless abstractions, however, as Poisson processes have been shown to be good approximations to processes that result from the aggregation of a large number of independent processes; this frequently is the case in arrival processes, where customers arrive from several different sources.

In addition to the arrival and service processes, it is also necessary to specify the number of *parallel servers* in the queueing system. Each of the M servers at the queue is equally skilled and possesses the same service time distribution; if the second of two servers in a two-server queue is not busy, the next arriving entity does not wait for the first server to become idle but proceeds immediately to the second server. Readers searching for an application may think of a group of tellers in a bank.

Many isolated queues have limited space available for waiting entities. If there is space for no more than a total of K entities waiting *and* being served, the *buffer capacity* of the queue is K. Entities arriving at the queue when it is full are said to be *blocked*; they are turned away and are lost to the queueing system. Entities cannot wait for space to become available since there is no place to wait; neither do they leave temporarily and return later. Thus an entity blocked is an entity lost to the system.

Other parameters for the single isolated queue exist which have not been discussed here. It is possible to model a queue with a *finite population* of entities or customers; we have implicitly assumed that the population of customers is, for all practical purposes, infinite. *Balking,* which is a more sophisticated form of blocking, has not been considered; a balked entity is one who arrives, observes the number of waiting customers, and thereupon decides not to enter the queue even though space is available. The *queueing discipline,* which describes how customers are selected for service from those waiting to be served, will be assumed to be *first-come–first-served* (FCFS); other methods are available for selecting entities for service, such as the *shortest processing time* (SPT) rule. The reader interested in these extensions is referred to one of the standard works on queueing theory.[3,4]

As a means of shorthand to simplify the following discussion on performance evaluation, Kendall's notation will be used to characterize the precise configuration of a queue: *A/B/C/D* represents a queue with a type *A* arrival process, a type *B* service process, *C* different servers, and a finite buffer capacity of *D*. Poisson arrival and service processes are denoted by "M" and general processes are denoted by "G". The capacity parameter is customarily omitted from the notation when it is set to its default value of infinity. Thus an M/G/2/4 queueing system is a single isolated queue with Poisson arrivals, two general servers, and a total available buffer capacity for four entities; an M/M/1 queueing system is a single-server queue with Poisson arrivals, exponential service times, and infinite buffer capacity.

Performance measures of isolated queues

Once the parameters of an isolated queue have been defined, it is possible to evaluate the performance of this queueing system using exact or approximate formulae. Expressions exist which calculate the average percentage of server idle time, the average customer waiting time, the average queue length, the average probability of a customer being blocked, and so forth. For queues with Poisson arrival and service processes, an exact analysis is possible; for many others, it is necessary to rely on approximations. It should be noted that all of the performance measures under consideration here are *steady-state* averages of system performance; i.e., they estimate the performance of the system over time after it has "settled down" following an initial *transient,* or start-up phase.

Two of the queue performance measures are the probability of the queueing system being empty, denoted by p_0, and the probability of the queueing system being full, denoted by p_k. The probability p_0 may be thought of as the percentage of time that all of the servers are idle, while p_k represents the proportion of entities which are turned away because there is no room in the queue. Of course, p_k will be zero for those queueing systems with infinite buffer capacity. For all other single queues, however, the larger p_k becomes the lower the queue's *throughput* becomes. Let μ_r denote the average throughput of a single isolated queue. Then the throughput of the queue is simply that portion of the arrival process which is not blocked by a full queue:

$$\mu_r = \left[1 - p_k \right] \times \lambda \qquad (12.3)$$

Note that in a queueing system with infinite capacity the throughput is simply the arrival rate, since all entities which enter the queue must also depart from it after being served.

Another key concern in any queueing system is the amount of time which an entity spends waiting or waiting and being served. The average time spent by an entity in the system, i.e., the average time waiting and in service, is denoted W_s; the average waiting time only is represented as W_q. Since the total time in the system is simply the time spent waiting and the average service time, the following relationship exists between W_s and W_q:

$$W_s = W_q + \frac{1}{\mu} \qquad (12.4)$$

This is convenient since if one knows one of the two performance measures the other can be readily determined.

Two closely related performance measures are concerned with the average number of customers either in the system or simply waiting. The notation is similar to that used for waiting times, with L_s representing the average number of entities either waiting or being served, and L_q indicating the average number of entities waiting. Intuitively, it appears as if the time spent by an entity waiting or waiting and in service should be related to the average number of entities waiting or waiting and in service. In this instance, intuition is completely justified: Little's law provides for the following relationship between the waiting time and queue size measures:

$$L_q = \mu_r \times W_q$$

$$L_s = \mu_r \times W_s \qquad (12.5)$$

Note that by using Little's law and the relationship between W_q and W_s, it is necessary to know only one of the performance measures L_q, L_s, W_q, and W_s in order to be able to calculate all four. In all following discussions, therefore, only formulae for L_s will be provided.

Other performance measures may also be of interest to the facilities planner. In particular, one might want more information on waiting times, queue sizes, and so forth than simple averages. In that case, the probability distributions for waiting times and queue sizes would be of interest. For the queueing systems discussed below, analytical formulae for such probability distributions are generally obtainable if an analytical expression is available for the average-based performance measures discussed above; otherwise, the probability distributions must be approximated.

Queues with Poisson arrival and service processes

In order to obtain some working formulae for the performance measures discussed above, consider first the isolated Poisson queue with a single server and infinite buffer capacity. Queueing theory shows that the probability of the server being idle is simply one minus the traffic intensity ρ:

$$p_0 = 1 - \rho \qquad (12.6)$$

This equation shows that it is necessary for the traffic intensity to be less than one (or, equivalently, for λ to be less than μ) for the system to reach steady state; if ρ is one or greater, the number of entities in the system will continue to grow without bounds as entities arrive as fast or faster than they can be served.

The traffic intensity ρ can also be used to determine the average number of entities in the system:

$$L_s = \frac{\rho}{1 - \rho} \qquad (12.7)$$

The remaining system parameters, namely L_q, W_s, and W_q, can then be found using Little's law as discussed in the previous section. For example, suppose that an automated teller machine has been designed to process 30 customers an hour, while its location near a busy intersection would result in 20 customers an hour demanding its services. If we assume infinite waiting space (at a busy intersection?) and Poisson arrival and service processes (a more reasonable assumption), the teller would be idle $1 - {}^{20}\!/_{30} = 33$ percent of the time, while $({}^{2}\!/_{3})/(1 - {}^{2}\!/_{3}) = 2$ customers on average would be waiting or being served. As with all infinite-capacity systems, the probability of the teller being full is zero, and hence the throughput of the teller is simply the arrival rate of 20 customers per hour. The average number of customers waiting, the waiting time per customer, and the average combined waiting and service time per customer are then found to be 1.33 customers, 0.07 hours, and 0.10 hours, respectively. The details of the calculations are given in Fig. 12.3.

The usefulness of the queueing formulae for facility evaluation may be seen by making the example a bit more realistic. Suppose that the planner had to decide how many parking spaces to locate next to the teller machine. The more parking spaces, the more expensive the setup for the teller machine. However, installing too few parking spaces will cut down on the utilization of the teller machine and result in unhappy customers. The planner's decision can be facilitated by treating the teller machine as an M/M/1/K queueing system; the planner knows the arrival rate, the service rate, and the number of servers, but needs to evaluate alternatives for the buffer capacity K.

$\rho = \dfrac{\lambda}{\mu} = \dfrac{20}{30} = \dfrac{2}{3}$	$P_k = 0$
$P_0 = 1 - \rho = 1 - \dfrac{2}{3} = \dfrac{1}{3}$	$\mu_r = \left(1 - P_k\right) \times \lambda = (1 - 0) \times 20 = 20$
$L_s = \dfrac{\rho}{1 - \rho} = \dfrac{\frac{2}{3}}{1 - \frac{2}{3}} = 2.0$	$W_q = W_s - \dfrac{1}{\mu} = \dfrac{1}{10} - \dfrac{1}{30} = \dfrac{2}{30}$
$W_s = \dfrac{L_s}{\mu_r} = \dfrac{2}{20} = \dfrac{1}{10}$	$L_q = \mu_r \times W_q = 20 \times \dfrac{2}{30} = \dfrac{4}{3}$

Figure 12.3 Calculation of performance measures for M/M/1 queue.

In a queueing system with finite buffer capacity, the traffic intensity may be greater than or equal to one, since this will simply result in large numbers of entities being turned away. The expression for the probability of the system being empty, therefore, is given as follows:

$$
p_0 = \begin{cases} \dfrac{1 - \rho}{1 - \rho^{K+1}} & (\rho \neq 1) \\[2em] \dfrac{1}{K+1} & (\rho = 1) \end{cases} \tag{12.8}
$$

Note that a separate formula must be used when the arrival and service rates are equal. This also holds true for the probability that the queue is full:

$$
p_k = \begin{cases} \dfrac{(1 - \rho) \times \rho^K}{1 - \rho^{K+1}} & (\rho \neq 1) \\[2em] \dfrac{1}{K+1} & (\rho = 1) \end{cases} \tag{12.9}
$$

Finally, the average number of entities waiting and in service is given below:

$$
L_s = \begin{cases} \dfrac{\rho}{1 - \rho} - \dfrac{(K+1) \times \rho^{K+1}}{1 - \rho^{K+1}} & (\rho \neq 1) \\[2em] \dfrac{K}{2} & (\rho = 1) \end{cases} \tag{12.10}
$$

Given this information it is possible to calculate all of the performance measures for the automatic teller machine example. Suppose that the planner was initially considering having only two or three parking spaces available, since (after all) the average number of customers in the uncapacitated system was only two. The various performance measures for these two scenarios and for a buffer capacity of four are given in Fig. 12.4. Note that even with three parking spaces, more than 10 percent of the customers would be turned away. Clearly, some rethinking is necessary on the planner's part.

| Performance | K | | |
Measure	2	3	4
p0	0.4737	0.4154	0.3839
pk	0.2105	0.1231	0.0758
μ_r (Customers)	15.7895	17.5385	18.4834
L_s (Customers)	0.7368	1.0154	1.2417
W_s (Hour)	0.0467	0.0579	0.0672
L_q (Customers)	0.2105	0.4308	0.6256
W_q (Hours)	0.0133	0.0246	0.0338

Figure 12.4 Performance measures for the M/M/1/K queue with increasing buffer capacity.

Finally, consider a single M/M/M queue with M identical servers and infinite buffer capacity. If the automated teller machine is to be conveniently located in the parking lot of a large shopping center, it is possible to assume an infinite buffer capacity. However, demand in the shopping center will be considerably greater than at the previously discussed busy intersection; how many machines should be installed at this newly proposed location? If the new arrival rate is expected to be 55 customers per hour and the service rate per machine of 30 customers per hour remains unchanged, one clearly needs at least two machines to ensure that the traffic intensity, which is now defined as the ratio of the arrival rate to the total service rate, is less than one. To determine precisely how many machines are necessary, it is necessary to define acceptable levels for one or more of the performance measures and see how the different scenarios compare with each other.

For systems which satisfy the traffic intensity requirement, the probability of all M servers being idle is given below. The exclamation mark indicates a factorial, e.g., $M!$ is shorthand for $M \times (M - 1) \times (M - 2) \times \cdots 2 \times 1$.

$$p_0 = \frac{1}{\sum_{i=0}^{M-1} \frac{(M \times \rho)^i}{i!} + \frac{(M \times \rho)^M}{M! \times (1 - \rho)}} \qquad (12.11)$$

The formula for the average number of entities in the queue and in service is also somewhat complex:

$$L_s = M \times \rho + \left[\frac{(M \times \rho)^{M+1}}{M! \times (1 - \rho)^2 \times M} \right] \times p_0 \qquad (12.12)$$

Performance	M	
Measure	2	3
p_0	0.0435	0.1398
μ_r (Customers)	55	55
L_s (Customers)	11.4783	2.4134
W_s (Hour)	0.2087	0.0439
L_q (Customers)	9.6449	0.5800
W_q (Hours)	0.1754	0.0105
W_q (Minutes)	10.5217	0.6328

Figure 12.5 Performance measures for the M/M/M queue with an increasing number of servers.

As with the M/M/1/K queue, the values for L_q, W_s, and W_q may be found for the M/M/M queue by using Little's law.

Consider now the automatic teller example discussed above. Suppose that management wanted to ensure that, on average, a customer would have to wait no longer than two minutes, or one-thirtieth of an hour. How many machines would need to be installed at the shopping center? Figure 12.5 shows the average waiting time, along with the other performance measures, for queueing systems with two and three machines. Note that if only two machines are installed at the shopping center, the average waiting time per customer is over ten minutes. Three machines must be installed if management wishes to reduce waiting time to less than two minutes.

Performance measures may also be calculated for other single isolated queues with Poisson arrival and service processes. The M/M/M/K system, which combines multiple servers and finite capacity, may be appropriate for several facility decisions. Models for balking, finite arrival populations, and other configurations also exist. The reader who is interested in these more complex systems is referred to one of the standard texts listed earlier.

General queues

When the arrival and service processes are not both Poisson, the equations describing the performance of the queue generally become more difficult. One exception to this situation is the M/G/1 queue, a single-server queue with Poisson arrivals and a *general* service process. Although the arrival process is still characterized by an arrival rate λ and an implied interarrival time SCV of one, both the service rate μ and the service time SCV c_s^2 must now be specified. Let the traffic intensity be defined as for any single-server queue, i.e., as the ratio of the arrival rate to the service rate. The probability of the server being idle is still dependent only on ρ as in Eq. (12.6) from the previous section. The average number of entities waiting and in service, however, depends not only upon the arrival and service rates but also upon the service time SCV c_s^2:

$$L_s = \rho + \frac{\rho^2 \times (1 + c_s^2)}{2 \times (1 - \rho)} \qquad (12.13)$$

The above expression is referred to as the *Pollaczek-Khintchine expression* after the researchers who developed it.

As with other single isolated queues, Little's law may be used to find L_q, W_s, and W_q once L_s is found. Note the similarity of the above formula for L_s to the expression for that performance measure given for the M/M/1 queue; when c_s^2 is set equal to one, the above expression reduces to that for the M/M/1 queue. This is not a coincidence; when c_s^2 is set equal to one, the service process is Poisson and thus the M/G/1 queueing system is just an M/M/1 system. As the service time SCV increases, however, the average number of entities in the system—and by implication the average waiting time and other related performance measures—also increases. As c_s^2 decreases, however, so does average queue size and waiting time; clearly, less service time variability is always preferable, as it will improve customer service.

Consider once again the automatic teller machine example discussed in the previous section. Most of the variability in the service time distribution will be derived from the customer who is using the machine; the automatic teller itself, being a machine, should contribute very little to the variability. Customer confusion and mistakes, however, can contribute significantly to the variability of the service time. Suppose that although the service rate will remain at 30 customers per hour, and the arrival process is Poisson with a rate of 20 customers per hour, management is considering a simplified user interface which experiments have shown will significantly reduce service time variability. What impact on average waiting time would reducing c_s^2 from 1.0 to 0.2 have? Figure 12.6 shows that such a decrease in c_s^2 will reduce the average waiting time by nearly one-half; clearly reducing variability may be as advantageous as increasing the number of servers. In addition, Fig. 12.6 illustrates that *doubling* the service time SCV will *increase* waiting time by one-third.

Once one is confronted with not only a general service process but a general arrival process, exact solutions for the performance measures become unobtainable. There are no exact formulae for any of the quantities discussed above for the G/G/1 queue, the G/G/M queue,

Performance	Service Time SCV		
Measure	0.2	1.0	2.0
p_0	0.3333	0.3333	0.3333
μ_r (Customers)	20	20	20
L_s (Customers)	1.4667	2.0000	2.6667
W_s (Hour)	0.0733	0.1000	0.1333
L_q (Customers)	0.8000	1.3333	2.0000
W_q (Hours)	0.0400	0.0667	0.1000

Figure 12.6 Performance measures for the M/G/1 queue with increasing server variability.

or the G/G/1/K queue. Researchers have therefore focused on developing simple approximations for these queueing systems which perform well under a wide variety of parameter settings. The approximation of L_s for the G/G/1 queue developed by Kraemer and Langenbach-Belz[5] has been shown to perform well compared to other approximations in an experiment conducted by Shanthikumar and Buzacott.[6] Essentially, the Kraemer and Langenbach-Belz approximation extends the well-known Pollaczek-Khintchine formula for the mean waiting time in an M/G/1 queue by incorporating the SCV of the interarrival times into the Pollaczek-Khintchine expression. In addition, an empirically derived "correction factor" is included:

$$L_s = \rho + \frac{\rho^2 \times (c_a^2 + c_s^2)}{2 \times (1 - \rho)} \times g(\rho, c_a^2, c_s^2) \qquad (12.14)$$

The correction factor, $g(\bullet)$, is dependent upon the value of the SCV for the arrival process:

$$g(\rho, c_a^2, c_s^2) = \begin{cases} e^{-\frac{2 \times (1 - \rho)}{3 \times \rho} \times \frac{(1 - c_a^2)^2}{c_a^2 + c_s^2}} & c_a^2 \leq 1 \\ e^{-(1 - \rho) \times \frac{c_a^2 - 1}{c_a^2 + 4 \times c_s^2}} & c_a^2 \geq 1 \end{cases} \qquad (12.15)$$

Springer and Makens[7] similarly investigated several alternative approximations for the G/G/1/K queue and found that the approximation of Gelenbe[8] consistently outperformed four other approximations. Before determining the performance measures, it is first necessary to calculate the parameter γ (gamma):

$$\gamma = \frac{2 \times (\lambda - \mu)}{\lambda \times c_a^2 + \mu \times c_s^2} \tag{12.16}$$

Once γ has been determined, the empty and blocking probabilities, as well as the average number of entities in the system, may be determined using the following formulae:

$$p_0 = \begin{cases} \dfrac{1 - \rho}{1 - \rho^2 \times e^{\gamma \times (K - 1)}} & (\rho \neq 1) \\[4ex] \dfrac{1}{2 + \dfrac{2 \times (K - 1)}{c_a^2 + c_s^2}} & (\rho = 1) \end{cases} \tag{12.17}$$

$$p_k = \rho \times p_0 \times e^{\gamma \times (K - 1)} \tag{12.18}$$

$$L_s = \begin{cases} \dfrac{p_0 \times \rho \times \left[\dfrac{1}{\gamma} - \dfrac{1}{2} \right] + p_k \times \left[K - \dfrac{1}{\gamma} - \dfrac{1}{2} \right]}{\rho - 1} + p_k \times K & (\rho \neq 1) \\[4ex] \dfrac{K}{2} & (\rho = 1) \end{cases}$$

$$\tag{12.19}$$

The values for L_q, W_s, and W_q may then be found applying Little's law.

An alternative approach to obtaining performance measures for general queues is to approximate the general interarrival and service time distributions with *Erlang* or with *phase-type* distributions.[9] An Erlang random variable from an Erlang distribution of order k, denoted as E_k, is a sequence of k identically distributed exponential random variables; a phase-type distribution may be thought of as a generalization of the Erlang distribution. While simple formulae do not exist for the performance measures of a single isolated queue for all configurations of queues with phase-type or Erlang arrival and service processes, it is possible to set up systems of equations which may be solved to obtain these performance measures; thus exact solutions may be obtained numerically, if not analytically. Of course, the Erlang and phase-type models may still be approximations since the actual interarrival and service time distributions are being approximated with Erlang or phase-type distributions.

Networks of Queues

Most manufacturing and service facilities consist of more than a single queue; work orders in a factory may have to be routed through several stages of fabrication, subassembly, and assembly before being completed; patients in a hospital are received into the hospital, diagnosed, treated, and then discharged. Each workcenter or workstation in each of these facilities is a separate queue which may receive arrivals not only from outside the facility, but also from another workstation upstream in the manufacturing or service process. To properly evaluate such systems, it is necessary to consider entire systems or networks of interacting queues. A *queueing network* is comprised of several distinct queues, each of which is capable of sending or receiving entities to at least one other queue in the network. The individual queues may have any of the characteristics of an isolated queue—i.e., Poisson or general arrival and service processes, single or multiple servers, finite or infinite buffer capacity.

Several of the performance measures which were of interest for single isolated queues are also of concern in the analysis of queueing networks. For each queue in the network, one would like to know the percentage of idle time, the average queue size, the average waiting time, and so forth. However, networkwide performance measures are also of interest. In particular, one would like to know the throughput rate and the average time needed for an entity to pass through all stages of the network; this latter quantity is denoted the average *sojourn time* for the network.

There are essentially two different types of queueing networks which need to be examined: *open networks* and *closed networks*. In an open network of queues, entities continually enter and leave the system; as a result, the number of entities in the network is in constant flux. In closed networks, however, the number of entities in the system is fixed; entities simply pass from queue to queue in a never-ending cycle. The difference between these two types of networks is graphically represented in Fig. 12.7. In this section we shall primarily consider the simplest of all queueing networks: an open network of Poisson queues with infinite buffer capacities. This network is analytically tractable and will provide a basis for our discussion of the *approximations* for more general open networks. Finally, research for closed networks will be summarized.

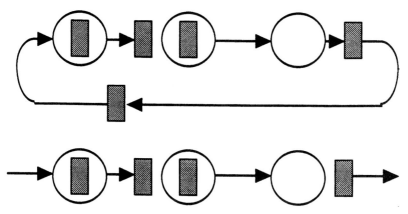

Figure 12.7 A closed network (*Top*) and an open network (*Bottom*). Each consists of three queues in series.

Open Poisson networks

Open Poisson networks were first analyzed by Jackson,[10] and hence they are also referred to as *Jackson networks*. In an open network of queues, the jobs arriving at a single queue may not only originate from outside the network but may be the output from another queue. Consider the three simple open networks of single-server queues presented in Fig. 12.8. The first network is a network of three queues *in series*; entities arrive from outside the network only to the first queue, which serves the entity before passing it on to the second queue; after being served by the second queue, the entity proceeds to the third and final queue, where it is served and then leaves the network. Thus the first queue is the only queue to receive *external arrivals,* i.e., arrivals from outside the network. In the second network, arrivals from outside the network come to queues 1 and 2, which serve the entities and send them to the third queue for final service before the entities leave the network; thus the output from the first two queues is *merged* into the input for the third. In the third network, entities arrive at the first queue and go either to the second or to the third queue, whereupon they are served before departing the network; the output from the first queue is *split* between the latter two queues. The service rates μ and the external arrival rates λ for each of the queues in each network are given in Fig. 12.9.

Several points may be drawn from these examples. First, not all queues in the network may receive external arrivals. Second, a queue may receive arrivals from one or more of the other queues in the network. Third, the departures from one queue may become arrivals to one or more of the remaining queues in the network. Thus in addition

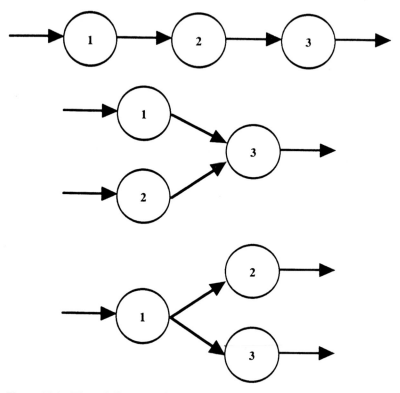

Figure 12.8   Three different configurations for a three-station open queueing network.

| Queueing Network | Queue | | | | | |
| | 1 | | 2 | | 3 | |
	λ	μ	λ	μ	λ	μ
1	2	4	0	3	0	5
2	2	4	2	3	0	5
3	2	4	0	3	0	5

Figure 12.9   External arrival rates and service rates at each queue for three open queueing networks.

to defining the standard parameters for each individual queue, one must define the *routing probabilities* which indicate where the departures from each queue are likely to go. In the first network, the probability that the output from the first queue will proceed to the second is one, while the probability that it will proceed immediately to the third queue or leave the network is zero. In the third network, however, the output from the first queue may go either to the second or to the third queue; the planner must define what proportion of entities served by the first queue will go to each of the latter two queues. For this example, the output of the first queue is evenly split between the second and third queues; 50 percent of the departures from the first queue proceed to the second queue, while the remaining 50 percent go to the third queue.

While the queueing network models discussed here assume that the service time distributions and the routing probabilities remain constant across all entities, this data can be seen as representing averages across all different *types* of entities. For example, the third network in Fig. 12.3 could represent a manufacturing facility which produces two products. Product 1 is processed by the first queue before being sent to the second queue, while Product 2 is also processed initially at the first queue but is then routed to the third queue. If the company produced equal amounts of Product 1 as of Product 2, then the routing probabilities for the first queue would indicate that one-half of all entities leaving the first queue proceed to the second, while one-half of all entities leaving the first queue go on to the third queue. Similarly, if Product 1 took 20 minutes per batch for processing at the first queue and Product 2 took 10 minutes per batch at the same queue, then the *aggregate* service time at the first queue would be $(\frac{1}{2}) \times (20 \text{ minutes}) + (\frac{1}{2}) \times (10 \text{ minutes}) = 15$ minutes or one-quarter hour per batch, for an aggregate service rate of four batches per hour. Thus facilities with multiple products or different types of customers may be readily handled by considering the relative proportions of each product or customer type at each queue.

For each queue which receives its arrivals from the output of other queues, an *effective* arrival rate may be determined by merging the arrival rates from the feeder queues. Of course, since all the queues have infinite buffer capacity, the total output rate from any queue is simply the total arrival rate to that queue. In the first network, the effective arrival rate to the second queue is simply the output rate of the first queue, which must be the external arrival rate to the first queue. In the second network, the effective arrival rate to the third queue is the sum of the output rates from the first two queues, which is the sum of their respective external arrival rates. Finally, the arrival rate for the second queue in the third network is the *split* portion of the output process from the first queue which it receives; this is the arrival rate to the first queue multiplied by the proportion of departures from the first queue which are routed to the second queue. These effective arrival rates, along with those for the remaining queues in the three networks, are presented in Fig. 12.10.

Queueing	Queue		
Network	1	2	3
1	2	2	2
2	2	2	4
3	2	1	1

Figure 12.10   Effective arrival rates at each queue in three open queueing networks.

Once the effective arrival rates for each queue have been determined, performance measures for each queue in the network may be obtained simply by applying the appropriate formula for the corresponding M/M/M queue to the effective arrival rate and the service rate for each queue in the network. This rather surprisingly simple result holds even for networks with feedback loops; each queue in the network behaves *as if it were* an isolated queue with Poisson input. Thus once the effective arrival rates have been determined, the network can be effectively "decomposed" into individual queues for analysis. Figure 12.11 shows the average number of entities waiting and in service at each queue for the three networks presented in Fig. 12.8. Note that since there is no blocking, the total throughput for each network is simply the total arrival rate to each network. In addition, for networks of single-server queues with no feedback loops, the average sojourn time may be derived from the average combined waiting and service times of the individual queues.

General and finite-buffered open networks

If the Poisson assumption is violated and general arrival and service processes are permitted, Jackson's results do not hold: a network of general servers with general external arrival processes cannot be decomposed into individual G/G/1 queues for exact analysis. The same holds true for networks of queues with finite buffer capacities; even Poisson queues with finite buffer capacities cannot be analyzed

Queueing	Queue		
Network	1	2	3
1	1.0000	2.0000	0.6667
2	1.0000	2.0000	4.000
3	1.0000	0.5000	0.2500

Figure 12.11   Number of entities waiting and in service at each queue in three open queueing networks.

in isolation from other queues in the network. In the absence of any analytical solutions to these important queueing networks, researchers have focused on developing good approximations for such systems. Of course, many of these models are handicapped by their reliance on approximations for the analysis of the decomposed isolated queues, since exact formulae for queueing systems with general arrival and service processes do not exist.

Kuehn[11] first proposed an approximate Jackson-like decomposition of open queueing networks with general arrival and service processes and infinite buffer capacity; Kuehn had to develop approximations for the SCVs of the departure processes from the individual queues, as these served as the basis for the SCVs of the arrival processes to the downstream queues in the network. Karmarkar[12] and Zipkin[13] took a simpler approach, decomposing a network of general queues into a set of isolated M/G/1 queues; this enabled them to use the Pollaczek-Khintchine equation in lieu of an approximation (e.g., Kraemer and Langenbach-Belz) in calculating queue performance measures.

Whitt[14,15] developed and tested a decomposition approximation termed the *queueing network analyzer* (QNA) for networks of multiple-server queues with general external arrival processes and general service processes. QNA decomposes a general open network into a set of GI/G/M queues, whereupon Whitt uses a variant of the Kraemer and Langenbach-Belz approximation to determine queue performance measures. Several simulation experiments were run to serve as benchmarks for comparing QNA against other queueing network models; QNA was shown to provide generally superior estimates of queue waiting time and system size when the arrival and service processes were not Poisson.

None of the queueing networks discussed above have considered the impact of finite buffer capacity in front of each queue. Capacity limitations on each queue may cause a single bottleneck server to effectively shut down an entire queueing network by blocking all upstream servers. Although decomposition may seem a promising method for modeling networks of more capacitated queues, this is difficult considering the extreme *interdependence* between the nodes. Hillier and Boling[16] proposed a method for decomposing an open network of Poisson queues in series with finite-capacity buffers into a set of isolated M/M/1/K queues. No simple formulae for the performance measures were found; numerical methods were required to determine the effective arrival and service rates. Much work has been conducted on developing approximations to open networks of finite-buffered queues in series with Poisson arrival and service processes: Altiok,[17] Perros and Altiok,[18] Brandwajn and

Jow,[19] and others have developed models. Other researchers have developed approximations for more general configurations of finite-buffered queues with Poisson arrival and service processes. Takahashi et al.,[20] Altiok and Perros,[21,22] Perros and Snyder,[23] Kerbache and Smith,[24] Smith and Daskalaki,[25] and Jun and Perros[26] have developed approximations for more general configurations, which are of course capable of approximating serial configurations. Recently, there has been an emphasis on developing queueing networks of queues with *general* arrival and service processes finite buffers. Altiok,[27] Gun and Makowski,[28] and Jun and Perros[29] developed methods for finite-buffered queues in series, while Kerbache and Smith[30] developed an approximation for more general configurations of finite-buffered queues.

Closed queueing networks

One of the early applications which motivated the development of closed queueing networks was the analysis of mining operations.[31] In the mining operations under analysis, a fixed number of surfaces were continuously mined by a fixed number of different machines. Each surface had to be processed by the different machines in a fixed sequence, which was repeated as the mining progressed. Since the number of surfaces was fixed, the queueing network developed to model the system was closed; the same entities, or surfaces, progressed though the machines, or queues, in a fixed repeating sequence. Thus closed networks were developed to model manufacturing systems which were effectively closed.

Gordon and Newell[32,33] developed results for closed networks of exponential queues which paralleled to some extent the development of open networks by Jackson. Unfortunately, Buzacott and Yao noted that the performance of closed queueing networks in modeling queueing systems deteriorates as the actual service time distributions become less similar to the exponential distribution; this result was also noted by Co and Wysk.[34] Furthermore, the analysis of closed queueing networks with general servers is difficult, leading Whitt to suggest that the preferred way to model closed networks is to approximate them with equivalent open models.[35]

Queueing Network Applications

Both open and closed networks have been used to model production systems. As noted by Buzacott and Yao,[36] closed queueing networks of queues with exponential service time distributions have been particularly popular in the modeling of flexible manufacturing systems.

A *flexible manufacturing system* (FMS), also referred to as a *flexible machining system,* is a set of machine tools dedicated to the fabrication of a family of parts which are similar in their manufacturing requirements. The machine tools are linked by an automated materials-handling system, which transfers parts to and from machines without human intervention. Thus the primary argument for using closed networks to model these systems is that the fixed number of entities in the network accurately represents the fixed number of pallets which are used to transport parts in an FMS. CAN-Q, a software package developed by Solberg[37] to facilitate the analysis of closed queueing networks, has subsequently been used to model flexible manufacturing systems.[38]

The QNA, developed by Whitt to model open queueing networks of G/G/M queues, has been used as an aid in factory layout at AT&T. McCallum[39] comments that QNA, which was discussed above (under "General and Finite-Buffered Open Networks"), was used to analyze manufacturing lines at several AT&T facilities. The ability of QNA to handle general arrival and service processes was considered particularly important at AT&T, since the arrival of work to some of the production lines was highly variable while the service times of some of the automated equipment were nearly deterministic. McCallum also noted that enhanced versions of QNA, which would enable the software to model partial yields and rework, were being developed.

Manuplan, a commercially available queueing network software package offered by System Dynamics Incorporated, was used by IBM in the design of a new printed-circuit-board factory;[40] decisions on equipment capacities, material-handling systems, material routings, and workcenter layouts were all made in conjunction with Manuplan. Manuplan has been specifically designed for ease of use in modeling manufacturing systems; although the IBM team used a version which ran on a mainframe, a microcomputer version with a spreadsheet interface is also available. In the microcomputer version, five different input screens are used to obtain data on the system to be modeled: data on systemwide constraints, equipment capabilities, parts requirements, possible routings, and machine operations are gathered at this stage. Manuplan then provides summary information on the mean flow time for each part, average work-in-process inventory levels, and equipment utilization. The group at IBM was particularly satisfied with Manuplan's speed; an alternative layout could be evaluated in under 60 seconds, compared with a 45-minute simulation run. After the range of layout options was substantially narrowed down, however, the detailed simulation model was used to obtain the final layout.

Conclusion

Queueing models are a viable alternative to simulation in the early stages of facilities planning. The models provide exact or approximate formulae for several performance measures of vital interest to the facilities planner: throughput, average waiting time, average inventory levels, and server idle time. Simple systems of isolated queues may be readily analyzed using the equations presented in this chapter; larger systems of several interacting queues, such as would be found in a factory or service facility comprised of several workstations, may be analyzed using one of the queueing network models developed by researchers. These models have been shown to provide the rapid feedback which is vital in the evaluation of alternative layouts; after clearly inferior layouts have been identified, a more detailed performance analysis using simulation may then be undertaken.

References

1. Pegden, Dennis C., *Introduction to SIMAN,* Systems Modeling Corporation, State College, PA, 1986.
2. Fredericks, Albert A., "Performance Analysis Modeling for Manufacturing Systems," *AT&T Technical Journal,* 65(4):25–34, 1986.
3. Gross, Donald, and Carl Harris, *Fundamentals of Queueing Theory,* Wiley, New York, 1985.
4. Kleinrock, Leonard, *Queueing Systems,* 2 vols., Wiley, New York, 1976.
5. Kraemer, W., and M. Langenbach-Belz, "Approximate Formulae for General Single Server Systems with Single and Batch Arrivals," *Proceedings 8th International Teletraffic Congress,* pp. 398–402, 1976.
6. Shanthikumar, J. G., and J. A. Buzacott, "On the Approximations to the Single Server Queue," *International Journal of Production Research,* 18(6):761–773, 1980.
7. Springer, Mark C., and Paul K. Makens, "Queueing Models for Performance Analysis: Selection of Single Station Models," *European Journal of Operational Research,* forthcoming.
8. Gelenbe, Erol, "On Approximate Computer System Models," *Journal of the Association for Computing Machinery,* 22(2):261–269, 1975.
9. Altiok, Tayfur, "On the Phase-Type Approximations of General Distributions," *IIE Transactions,* 17(2):110–116, 1985.
10. Jackson, James R., "Jobshop-Like Queueing Systems," *Management Science,* 10(1):131–142, 1963.
11. Kuehn, Paul J., "Approximate Analysis of General Queuing Networks by Decomposition," *IEEE Transactions on Communications,* C-27(1):113–126, 1979.
12. Karmarkar, Uday S., Sham Kekre, and Sunder Kekre, "Lotsizing in Multi-Item Multi-Machine Job Shops," *IIE Transactions,* 17(3):290–298, 1985.
13. Zipkin, Paul H., "Models for Design and Control of Stochastic, Multi-Item Batch Production Systems," *Operations Research,* 34(1):91–104, 1986.
14. Whitt, Ward, "Performance of the Queueing Network Analyzer," *The Bell System Technical Journal,* 62:9, 1983.
15. Whitt, Ward, "The Queueing Network Analyzer," *The Bell System Technical Journal,* 62(9):2779–2815, 1983.
16. Hillier, Frederick S., and Ronald W. Boling, "Finite Queues in Series with Exponential Service Times—A Numerical Approach," *Operations Research,* 15:286–303, 1967.
17. Altiok, Tayfur, "Approximate Analysis of Exponential Tandem Queues with Blocking," *European Journal of Operations Research,* 11:390–398, 1982.

18. Perros, H. G., and Tayfur Altiok, "Approximate Analysis of Open Networks of Queues with Blocking: Tandem Configurations," *IEEE Transactions on Software Engineering,* SE-12(3):450–461, 1986.
19. Brandwajn, Alexandre, and Yung-Li Lily Jow, "An Approximation Method for Tandem Queues with Blocking," *Operations Research,* 36(1):73–83, 1988.
20. Takahashi, Y., H. Miyahara, and T. Hasegawa, "An Approximation for Open Restricted Queueing Networks," *Operations Research,* 28:3, 1980.
21. Altiok, Tayfur, and H. G. Perros, "Open Networks of Queues with Blocking: Split and Merge Configurations," *IIE Transactions,* 18(3):251–261, 1986.
22. Altiok, Tayfur, and H. G. Perros, "Approximate Analysis of Arbitrary Configurations of Open Queueing Networks with Blocking," *Annals of Operations Research,* 9:481–509, 1987.
23. Perros, H. G., and P. M. Snyder, "A Computationally Efficient Approximation Algorithm for Analyzing Open Queueing Networks with Blocking," North Carolina State Working Paper CCSP-TR-86/18.
24. Kerbache, Laoucine, and J. MacGregor Smith, "Asymptotic Behavior of the Expansion Method for Open Finite Queueing Networks," *Computers in Operations Research,* 15(2):157–169, 1988.
25. Smith, J MacGregor, and Sophia Daskalaki, "Buffer Space Allocation in Automated Assembly Lines," *Operations Research,* 36(2):343–358, 1988.
26. Jun, K. P., and H. G. Perros, "Approximate Analysis of Arbitrary Configurations of Queueing Networks with Blocking and Deadlock," *Queueing Networks with Blocking,* H. G. Perros and T. Altiok, eds., Elsevier, Amsterdam, 1989, pp. 259–280.
27. Altiok, Tayfur, "Approximate Analysis of Queues in Series with Phase-Type Distributions and Blocking," *Operations Research,* 37(4):601–610, 1989.
28. Gun, Levent, and Armand M. Makowski, "An Approximation Method for General Tandem Queueing Systems Subject to Blocking," *Queueing Networks with Blocking,* H. G. Perros and T. Altiok, eds., Elsevier, Amsterdam, 1989, pp. 147–171.
29. Jun, K. P., and H. G. Perros, "An Approximate Analysis of Open Queueing Networks with Blocking and General Service Times," *European Journal of Operational Research,* 46:123–135, 1990.
30. Kerbache, Laoucine, and J. MacGregor Smith, "The Generalized Expansion Method for Open Finite Queueing Networks," *European Journal of Operations Research,* 32:448–461, 1987.
31. Koenigsberg, Ernest, "Twenty-Five Years of Cyclic Queues and Closed Queue Networks: A Review," *Journal of the Operational Research Society,* 33:605–619, 1982.
32. Gordon, William J., and Gordon F. Newell, "Closed Queueing Systems with Exponential Servers," *Operations Research,* 15:254–265, 1967.
33. Gordon, William J., and Gordon F. Newell, "Cyclic Queueing Systems with Restricted Length Queues," *Operations Research,* 15:266–277, 1967.
34. Co, Henry C., and Richard A. Wysk, "The Robustness of CAN-Q in Modeling Automated Manufacturing Systems," *International Journal of Production Research,* 24(6):1485–1503, 1986.
35. Whitt, W., "Open and Closed Models for Networks of Queues," *AT&T Bell Laboratories Technical Journal,* 63(9):1911–1979, 1984.
36. Buzacott, J. A., and David D. Yao, "Flexible Manufacturing Systems: A Review of Analytical Models," *Management Science,* 32(7):890–905, 1986.
37. Solberg, J. J., "A Mathematical Model of Computerized Manufacturing Systems," paper presented at the 4th International Conference on Production Research, Tokyo, August 1977.
38. Stecke, Kathryn E., and James J. Solberg, "The Optimality of Unbalancing Both Workloads and Machine Group Sizes in Closed Queueing Networks of Multiserver Queues," *Operations Research,* 33(4):882–910, 1985.
39. McCallum, Charles J., Jr., "Operations Research in Manufacturing," *AT&T Technical Journal,* 65(4):4–16, 1986.
40. Brown, Evelyn, "IBM Combines Rapid Modelling Technique and Simulation to Design PCB Factory-of-the-Future," *Industrial Engineering,* 20(6):23–26, 1988.

13

Computer Modeling Techniques for Facility Design and Evaluation

Edward F. Watson
Consultant, Systems Modeling Corporation
Sewickley, Pennsylvania

Randall P. Sadowski
Vice President, Systems Modeling Corporation
Sewickley, Pennsylvania

The design of a new system or modification of an existing system is often based on insight and experience. If the design is based on proven concepts and existing similar systems, this insight and experience is often sufficient to develop a workable design. In the past, many of these systems were based on mass production techniques that utilized proven concepts such as paced assembly lines and transfer lines. Today's systems often encompass new production concepts such as *just-in-time* (JIT), *flexible manufacturing,* and *agile manufacturing* (Nagle and Dove, 1991). In addition, automated material-handling systems with complex automatic control logic are often incorporated for the final system design. The concept is to design compact and flexible systems that carry minimal inventory and allow a fast reaction to customer orders as well as the introduction of new products.

These new systems tend to be complicated by both the processes used and the sophistication of the manufacturing equipment and material-handling devices. The integration of these components into the desired system can become a very difficult task. The system components tend to interact to produce undesirable results. The design process should yield a system which (1) has the right mix of system components, (2) is configured in such a way as to provide the desired performance, and (3) is cost effective over the life cycle of the system.

The design of a new system often occurs in three phases: conceptualization, configuration, and final design. The conceptualization phase involves the determination of the product mix to be required, processes to be used, equipment options to be considered, and performance measures to be achieved. In the configuration phase numerous equipment layouts are developed along with potential material handling in an attempt to determine which configurations are feasible. The final design phase takes a limited set of feasible configurations and incorporates control logic to determine which system design will meet the stated objectives.

As one progresses through the various phases, the number of potential options can become very large and extremely time consuming to consider without the aid of analysis tools. The computer has allowed the development of many modeling techniques that allow the system designer to consider a large number of options quickly. This chapter will present and discuss the basic types of computer modeling techniques available to the system designer. The types of modeling options and approaches will be presented first, followed by a more detailed discussion of these approaches, including examples.

Types of Models

Models, in general, are simplified representations of the real world and can take many forms: physical, mathematical, symbolic, and logical. Two popular approaches for analyzing manufacturing systems are *analytical modeling* and *simulation modeling*. Analytical modeling is a mathematical technique. Simulation modeling contains both logical and symbolic representations, but may contain mathematical and statistical submodels as well. Computer programs are readily available for both approaches.

Models of manufacturing systems can take various forms. Which form they take may dictate what modeling approach to use. In general, models can be either static, where time plays no role, or dynamic, where the system state is a function of time. A stochastic model is one in which statistical distributions are used to represent activities in the real system that are unpredictable. These models are also referred to as *nondeterministic,* while *deterministic* models contain no stochastic component.

Static/deterministic models are the simplest form of analytical models. They are sometimes referred to as *workcenter loading models* or *capacity models*. These models consist of "back of the envelope" calculations in that they require the calculation of workload by summing the amount of work allotted to each resource and comparing that with the available time of the resource. These models simply ensure that the average processing rate exceeds the average demand for processing. This type of calculation is used as a rough evaluation to size the system, e.g., choose the number of machines or determine the main bottlenecks. These models cannot incorporate the dynamics, interactions, and uncertainties in a system, and therefore they cannot estimate such things as flow times for parts moving through the system or time spent in a queue.

Dynamic models based on queueing theory are capable of explicitly modeling some of the variability encountered in day-to-day system operation as well as some system dynamics. Queueing models provide aggregate indicators of the dynamic system performance and are considered to fit into the dynamic (steady-state)/stochastic category.

The main advantage of queueing models over capacity models is that they can model such things as the time a part spends being processed as well as the time parts wait to be processed as they move through a facility. In practice, the time between successive arrivals of parts to a system and the service time for parts varies from part to part. This dynamic imbalance of flow rates can produce effects not predicted by the static/deterministic model, which simply ensures that the average processing rate exceeded the average demand for processing (Suri, 1985).

Simulation models provide a powerful and flexible means of representing dynamic systems. Simulation models can be classified as dynamic/stochastic models as well as dynamic/deterministic models. Simulation can be used to design new systems, improve existing systems, and provide evaluation capabilities for shop floor control. In general, simulation models are used over queueing models when more detail is required, when certain model peculiarities (e.g., complex control strategies, special configuration requirements, unique interactions between parts and material-handling system) exist, or when animation at any particular level of detail is required. Simulation is the only tool that allows systems interactions to be analyzed completely.

Modeling Approaches

Static modeling requires simple computations that provide rough-cut work allocation and equipment utilizations. They are often used at the start of the design process to provide gross estimates of equipment quantities. Since the calculations are relatively simple, they are typically performed with a hand calculator or a computer spreadsheet. Although these models provide rough estimates of equipment requirements, they do not consider the system variability that often exists due to the queueing characteristics most manufacturing systems exhibit.

In a simple *queueing system,* entities arrive at the system from time to time, join a waiting line or queue, are eventually served, and then leave the system. Depending on the type of system being modeled (production, communications, computer, etc.), an entity can represent a work order, data packet, information bit, and so on. Queueing systems are found wherever there is congestion or competition for scarce resources.

Queueing models provide an analyst with a powerful tool for designing and evaluating the performance of queueing systems. Typically, the analyst is interested in performance measures such as server utilization, time that a part spent in the system, and time that a part spent waiting in a queue. Typical input parameters to such models include the arrival rate, the service rate, and the configuration of the system. Both simulation models and analytical models can be used to predict performance measures for queueing systems.

In simple systems, performance measures can be determined mathematically. More complex models may also be solved using queueing theory by making simplifying assumptions. This mathematical approach provides the user with rough estimates of system performance with great savings of computer time and memory and human resources. The mathematical approach is also useful for determining relationships between certain performance measures as well as understanding the dynamic behavior of the queueing system.

In order to build realistic models of complex systems, simulation is generally required. Simulation models allow the user to incorporate any level of detail required for a thorough analysis. The trade-off is that a significant amount of time and resources can be expended to build and validate a simulation model. Simulation is often used to follow up with more detail than was provided by the rough estimates that were determined with mathematical modeling based on queueing theory.

Exact mathematical models of queueing systems provide the average value for a model parameter. In contrast, simulation models provide a statistical estimate of a model's performance measure, which includes the average value. It is therefore imperative that a thorough statistical analysis be performed on simulation model output to ensure that correct inferences can be drawn from the analysis. Simulation models are generally more accurate than mathematical models.

Traditionally, three approaches to simulation are identified: the *programming language approach,* the *simulation language approach,* and the *simulator package approach.*

Simulation with a programming language provides the most flexibility but is the most tedious and time consuming. The advantages of programming languages include small initial cost, ease of interfacing, and wide portability. The disadvantages are the lack of a standard set of modeling constructs and other common components of a simulation language such as statistics collection, file management, user interface, animation, and scheduling of events. This approach requires a high level of programming expertise and an extremely large time investment, and the resulting simulation is difficult to maintain or enhance by a second party.

In the simulation language approach, a discrete-event simulation language is used to build a simulation model of a particular system. This approach offers a standard set of model-building constructs, standard input-output capabilities, standard techniques to collect statistics, powerful statistical analysis tools, and powerful graphical animation capabilities. It is a very popular, if not the most popular, approach to modeling manufacturing systems. This approach provides flexibility in modeling, combined with a tool that is easier to learn and use than a programming language.

The simulator approach involves selecting a specific package and adapting the high-level constructs to fit the particular system being modeled. This approach is usually the easiest to learn and quickest to implement of simulation approaches. The downside is that it may not be able to model specific characteristics of the given system or allow much freedom to explore alternative control strategies. Thus, this approach provides an easy-to-use tool that often lacks the flexibility to model most complex systems accurately.

In general, analytical modeling and manufacturing simulators are used to identify and narrow down the feasible alternatives to one or two potential systems. Detailed analysis using a simulation language can then be performed to determine accurate performance measurements and operating characteristics.

To summarize the presented modeling approaches: Static models are easy to create and are used to obtain rough estimates of equipment requirements quickly. These models are most often used during the initial stages of system design to eliminate undesirable designs. Analytical models are mathematical models based on queueing theory that perform aggregate dynamic analysis of discrete manufacturing environments. They are used to predict system performance over a long-term time horizon, and hence they are not suitable for studying the impact of short-term decisions such as shop floor control applications. In addition, they cannot be used to evaluate extreme operating conditions. Analytical models give results for steady-state conditions only and are not suitable for estimating the transient behavior that occurs during system start-ups (e.g., recovery from machine breakdowns). Since the models are easy to use and require little time to run, they may be used whenever it is necessary to explore a wide range of alternatives quickly. This allows the users to use their imagination to investigate many alternatives in a short period of time.

Simulation models are models based on simulation language constructs that perform detailed analysis of discrete, continuous, and hybrid manufacturing environments. Simulation models are used for more detailed analysis of systems, where analytical models cannot consider certain characteristics of process flow, control logic, and so forth. Details of processing, material handling, storage requirements, production planning, and operating procedures are easily modeled using simulation languages. When standard constructs are not available to model a situation, programming language (e.g., FORTRAN, C) subroutines may be included to perform the necessary function.

Figure 13.1 summarizes some of the more important characteristics of analytical and simulation models.

The following sections discuss basic queueing concepts and present several computer analysis packages based on analytical models.

Analytical Models

Analytical models perform aggregate dynamic analysis of discrete manufacturing environments using mathematical models based on queueing theory.

Basic queueing terminology

Queueing theory refers to the mathematical and physical investigation of a class of problems characterized by several attributes. In the context of manufacturing systems, these attributes may be characterized as follows (Apple, 1972):

1. There is an input of units entering the system.

2. The units moving through the system are discrete.

3. The units requiring service are ordered in some fashion and receive service in that order.

4. A mechanism exists that governs the time at which a unit receiving service has its service terminated.

5. At least one of the two mechanisms, arrival or service, is not completely determined but can be considered a probabilistic mechanism of some sort.

When the arrival or service time is random, then the queueing model is classified as *probabilistic,* or *stochastic.* The variability

Modeling approach	Analytical models based on queueing theory	Simulation models
Applications	High-level models are used to evaluate particular manufacturing system designs and aggregate production plans.	Virtually any system can be modeled at any level of detail. Aggregate models are generally used to address high-level throughput and capacity issues.
	Applications are restricted to systems which can be modeled as simple queueing networks. Queueing network models make specific assumptions to keep the problem mathematically tractable.	Detailed models are used to address capacity and manpower planning, operational issues and specific sequencing, dispatching, and control logic issues. Typical models are dynamic/stochastic, but shop-floor control applications generally use dynamic/deterministic models.
	Queueing models are generally classified as dynamic (steady-state)/stochastic.	
Solution methodology	The following steps may be anticipated: • Obtain a queueing package. Learn the terminology and assumptions and general queueing theory (a technical background is generally assumed). • Collect the required input data from the shop floor and enter it into the model. • Execute the software. As it is based on queueing theory, the solution is generally obtained by solving a set of equations. • Obtain results. Exact values of specific output parameters, given the specific input parameters, are produced (specific requirements and assumptions vary across packages).	The following steps may be anticipated: • Obtain a language and model building training (a technical background and prior knowledge of simulation modeling is generally assumed). • Scope out project requirements (input, output, level of detail) and build and validate the simulation model. • Obtain the user-defined data from the shop floor and enter it into the model. • Collect observations of system behavior by simulating the system over time. • Conduct statistical analysis on the output to predict (or anticipate) the typical, worst case, and best case scenarios.
Typical user background	General knowledge of queueing theory Good knowledge of system being studied Average computer skills	Good knowledge of simulation modeling and theory, industrial statistics (to interpret output), and good computer skills Very good knowledge of simulation modeling language Good to very good knowledge of the system *Note*: Projects are often conducted by teams of individuals, each having a special area of expertise.

Figure 13.1   Summary of analytical and simulation model characteristics.

introduced by this stochastic process produces nonintuitive results; e.g., when the interarrival and/or service times are random, there may be a queue formed even if the mean service time is much less than the mean interarrival time. This will not occur when both times

Modeling approach	Analytical models based on queueing theory	Simulation models
Typical input	The mean values for specific parameters are required (each package has different requirements). Typical input parameters include: • Number of workcenters • Mean arrival rate • Mean service rate (setup, load, run) • Part routings • Failure data (MTBF, MTTR)	Strictly depends upon the user. Since the model is typically built from scratch, the modeler determines the level of detail, input parameters, and output performance measures. Statistical distributions are used to characterize random processes. Mean values are used where it is felt that the uncertainty is either insignificant or nonexistent.
Typical output	Queuing models yield steady-state predictions to *only* a limited set of questions.	Average, minimum, and maximum values of *any* output parameter defined by the user/modeler (also includes some measure of variation)
Facts	• Packages are readily available. • Packages are easy to learn and answers to fixed questions can be generated quickly. • Models are confined to the aggregate level and model assumptions are restricted to that dictated by the theoretical approach. • Queueing models cannot be used to study transient system behavior or the effects of extreme variation. • Packages generally do not provide good animation capability relative to simulation animation • Packages cannot provide a complete picture of system behavior over time as can a simulation model.	• Simulation languages are readily available. • Models can be constructed at any level of detail. All input parameters, output parameters, and modeling assumptions are user-defined. • Sophisticated graphical animations can be built at any level of detail and utilized for purposes of communication, presentation, validation, and training. Animations are particularly valuable to sell ideas and concepts for new systems or modifications to existing systems. • Simulation modeling provides the user with unlimited modeling capability. As the complexity of the system being studied increases, so do the modeling and analysis requirements.
When to use	Generally used to perform a rapid evaluation of the steady-state performance of many alternative systems (at a very aggregate level)	Generally applied whenever a substantial amount of information regarding system design, planning, and operation is required.

Figure 13.1 (*Continued*) Summary of analytical and simulation model characteristics.

are deterministic. This stochastic behavior causes randomly fluctuating waiting lines. For this reason, most applications of queueing theory are concerned with averages—e.g., average service utilization, average queue length.

Queueing analysis consists of a set of probabilistic models based on queueing theory that describes how a system will operate assuming certain arrival and service conditions. The key elements of queueing systems are parts (often referred to as *entities*) and servers. *Parts* generally refers to anything that requires service at some facility. For instance, parts could represent a job to be performed in a *flexible manufacturing system* (FMS), laundry at a laundromat, a tool at a service facility, or a patient in a hospital. A *server,* on the other hand, refers to anything that provides the requested service. For instance, a server might be a lathe, a washing machine, a repair technician, or a doctor.

Queueing systems can generally be characterized by the following parameters: arrival pattern, service pattern, system capacity, queue (service) discipline, and the number of service facilities.

The *arrival pattern* of parts is generally specified by the interarrival time, or the time between successive arrivals to the system. It is deterministic if the parts arrive at equally spaced intervals; it is probabilistic, or stochastic, if the time between part arrivals varies according to a probability distribution.

Arriving parts can be said to balk, renege, or jockey. *Balking* is considered to have occurred when an arriving part declines to enter the queue because the queue is too long. A part is said to *renege* when the part enters a queue but eventually leaves the queue because the wait is too long. When there are multiple facilities, a part may *jockey* between queues in the hope of reducing the wait time.

The *service pattern* is generally specified by the time required for a server to serve a part. The service pattern can be deterministic or probabilistic.

The *system capacity* is simply the maximum number of parts allowed in the system. The system is defined by both the queue and the server.

The *service discipline* refers to the order in which arriving orders are processed. This can also be referred to as *queue discipline* if one assumes that parts are served in the order in which they wait in the queue. The most common discipline is "first-in–first-out" (FIFO), to indicate processing in the order of arrival.

The *number of service facilities* refers to the number of servers, and, inherently, the configuration of servers. There can be a single server or multiple servers either in parallel or in series.

Queueing models offer a number of *performance measures.*

Job flow time: The average total time spent by a job between arrival and departure

Utilization of the servers: The average amount of time the server is busy

Average number of jobs waiting for service: The average number of jobs in the queue

Average total number of jobs in the facility: The total number of jobs in the queues and servers.

Average waiting time: The time spent by a job waiting for a server

These measures are not inclusive but they do provide a good idea of the value that queueing models have for manufacturing systems analysis.

A queueing system can be said to be in one of two states—either transient or steady state. When a queueing system is just beginning its operation, the state of the system is greatly affected by the initial state of the system and by the elapsed time since the start of operation. These conditions are known as *transient conditions*. After a certain time period, the state of the system will be independent of its initial state and the elapsed time and is said to have reached *steady-state conditions*. Hence, steady state is reached when the system's performance measures are equal to their mathematical limits as the elapsed operation time goes to infinity.

A common mistake is to interpret *steady state* to mean that the performance measures being studied become constant after some certain elapsed time. Actually, *steady state* means that the distribution of the values of the performance measure becomes invariant (Law and Kelton, 1991). Queueing theory applications have focused on steady-state conditions, since transient conditions are very difficult to study analytically.

A network of queues

Manufacturing system performance is based on the results of the processing of various parts through the servers of the system. In manufacturing terminology, *servers* denotes machines or more generally resources. Hence, as the parts work their way through the system toward completion, they may visit a series of resources. When a particular resource is busy, the part must wait in a queue. When one or more queues are encountered during a part's processing through the system, the part is said to have encountered a *network of queues*.

Perhaps the most intuitive way of dealing with a network of queues is through some decomposition strategy, whereby each queue may be analyzed separately. Unfortunately, this does not always work. System interactions are often overlooked, and certain queueing networks are not easily decomposed. Interactions between queues are often the most critical part of the analysis.

The decomposition strategy has been effective in dealing with a particular class of queueing networks. Assuming that the required conditions are satisfied, each queue and its associated resource may be modeled as if it were an independent Markovian queue (e.g., the probability of an arrival to or departure from the queue in some future time interval is unaffected by the past history of the queue). Such a process is relatively easy to analyze and serves as a good approximation to many real-life queueing processes.

The necessary conditions to assume the Markovian principle are (Ravindran et al., 1987):

1. All external arrivals to the network occur in independent Poisson streams. There may be several different streams entering at different points.
2. All service times are negative exponentially distributed with rates that depend at most upon the local queue.
3. All queues have unlimited capacity with no blocking or balking.
4. Any branching of the part streams is probabilistic and independent of everything except the position in the network.

These conditions define what is known as an *open network having a product-form solution*. The term *open* implies that the network can potentially have unlimited numbers of parts at any point in time. In an open queueing network, a part that completes service at a service center immediately enters another service center, reenters the same service center, or departs from the network. *Product-form* refers to the idea that the network factors into independent subsystems.

A *closed network* is one with a fixed number of parts. A part that completes service at a service center immediately enters another service center or reenters the same service center. Though the closed-network class is more difficult to model, it has the advantage of representing some dependency among the subsystems. Some open networks may be modeled as closed networks to take advantage of this. Since closed queueing networks are used to represent systems in which a fixed number of parts contend for resources, a dummy resource would be needed (to act as a generator of arrivals) to convert a superficially open network into a closed one. The product form of a closed network satisfies all the conditions listed above except the first (Hillier and Lieberman, 1990).

Queueing models may be evaluated based on the form the solution takes. There are three general solution types (Buzacott, 1985): closed-form algebraic solutions, computational algorithms, and approximate solutions.

Closed-form algebraic solutions imply that formulae for the performance measures can be found. This approach is based on the queueing network model given by Jackson (1963). Jackson modeled the shop as a network of queues and established sufficient conditions for decomposing the network into a set of independent queues. The manufacturing system of M machines is modeled as a network of queues with the following assumptions (Buzacott, 1985):

1. Service times are exponentially distributed.

2. Job routing is described by a probability matrix.

3. Queue discipline is FIFO with no restrictions on queue lengths.

4. External arrivals have exponential interarrival times.

5. The first operation for a job is described by a probability vector.

This model is said to result in a *product-form solution*. This model may represent a job shop as an open queueing network. In addition, it may represent an FMS as a closed queueing network.

Computational algorithms can be used for dealing with complex product-form expressions as well as models that do not have product-form solutions. For instance, the *mean-value analysis* (MVA) technique is concerned with the efficient evaluation of complex product-form formulae. Most performance prediction models that do not have product-form solutions are based on a Markov chain model of the system (Buzacott, 1985). A *Markov chain* is a stochastic process in which the conditional probability of a future state is independent of the past states and depends only on the present state. A Markov chain takes on a finite, or countable, number of values or states.

Approximate methods are available when computational algorithms require excessive computer time or memory. The common approximation methods (aggregation, decomposition, iteration, and simplification) are used by relaxing one of the model assumptions so that a solution may be obtained. The general idea behind *aggregation* is to replace all parts or any part of a queueing network with a simpler system. *Decomposition* refers to decomposing a queueing network into a set of isolated queues, each with its arrival process independent of the behavior of other queues. *Iteration* is generally used along with decomposition since the arrival at one queue depends on the activities at other queues. *Simplification* refers to the process of simplifying an insoluble model until a solution is realized.

Consider a job shop in which jobs arrive according to a Poisson process and where jobs are dispatched to the job shop with unlimited capacity. Assuming exponentially distributed service times and FIFO queue discipline, the steady-state probability of the number of jobs at the machines has a product-form solution. Hence, each machine can be studied in isolation.

Computer solution methods for networks of queues

Numerous analytical models based on queueing networks have been developed to aid in the analysis of manufacturing systems (Snowdon and Ammons, 1988). These non-simulation-based computational tools are generally available in the form of computer programs. The following provides an overview of several of these packages.

Computer Analysis of Networks of Queues (CAN-Q). CAN-Q is a computer program, developed by James Solberg at Purdue University, that implements a mathematical model of work flow in production systems. CAN-Q models a closed queueing network with processed jobs representing parts and machines representing resources.

The basic assumptions of the model are as follows (Buzacott and Yao, 1986):

1. The total number of jobs in the system is fixed. Since each job requires a pallet, the total number of pallets available in the system is fixed, as in most FMSs.

2. All machines have FIFO queue discipline, all job classes must have the same service rate parameter, and service time distributions must be exponential.

3. There is no blocking and each queue has infinite capacity, large enough to hold all the pallets in the system. When processing is complete, parts will always have a space to exit to and will never block the entrance of another part to that machine.

4. Machines are either busy or idle. There is no machine downtime, repair time, retooling, etc.

It is pointed out that according to studies investigating model robustness (Suri, 1983), performance measures are insensitive to system parameter estimation errors such as service rates. Furthermore, if the service time distributions of FIFO queue disciplines are not exponential, then the model will not, in general, yield satisfactory performance evaluations.

Other characteristics of CAN-Q are discussed in a paper by Solberg (1977). For instance:

1. CAN-Q represents a connected system of interdependent components; hence, it does not rely upon an assumption that the components of the system operate independently. This provides more modeling flexibility than traditional network models.

2. CAN-Q has the ability to model some aspects of material handling such as the delay required to transport material between areas. This aspect is modeled by a special station called a *transport station*.

3. There is no restriction on routings; i.e., the choice of a route can be either fixed or random.

The way CAN-Q handles the release of work orders into the system allows the user to determine the best trade-off between short flow times and high utilizations. It is assumed that the work-in-process (WIP) will be constant by releasing a part into the system only when a part is released from the system. The number of items in the system is an important management function. Large values will provide enough in-process inventory to ensure that the busiest stations almost always have items waiting, which results in high utilization and long flow times (Solberg, 1980).

Mean-Value Analysis of Queues (MVAQ). MVAQ is a computer program based on the mean-value analysis of queues. It is generally used to predict the performance of FMS cells by computing throughputs, utilizations, and mean queue lengths in closed networks of queues. The name MVAQ was originally derived from the fact that it deals mainly with the mean value (first moment) of statistical distributions associated with the problem. MVAQ's contribution to FMS modeling is to extend the basic functionality of CAN-Q with additional unique features (Suri and Hildebrant, 1984), such as:

1. *Multiple part classes modeled.* A fixed number of pallets, upon which parts flow through an FMS, can be dedicated to individual parts, and performance measures are obtained on a part-by-part basis.

2. *Low computer memory required.* MVAQ uses an algorithm that enables implementation with very low memory requirements in comparison to other queueing network algorithms, while allowing multiple-resource stations to be analyzed.

3. *Accurate large-system performance.* MVAQ maintains efficiency of calculation and good numerical stability when the number of parts and machines in a system becomes very large.

On the other hand, the theory behind MVAQ makes certain assumptions that cannot always be satisfied by an FMS (Suri and Hildebrant, 1984). For instance, in an FMS, processing times are generally known, whereas MVAQ assumes exponentially distributed processing times. In addition, in an FMS, the routing sequence is generally predetermined, whereas the MVAQ model assumes that the routing of a part to the next station is strictly probabilistic.

One of the difficulties with the MVAQ analysis is that it assumes exponential service times; or, more exactly, it assumes that the service time's coefficient of variation is 1.0 (Sanders, 1986). This causes difficulties when analyzing assembly systems since some stations have zero jam rates implying zero coefficient of variation, and some have significant jam rates implying significant coefficient of variation. (Whenever a station receives a defective assembly, it is said that the station is *jammed.*) For this reason, it is necessary to account for varying station variances. Fortunately, a tool was developed to take

this into account. Whitt (1985) developed a second-order approximation for the analysis of station delays for these queues and then
developed an approximation analysis for networks of these queues.

MANUPLAN. MANUPLAN II is perhaps the most commercialized of
the analytical computerized modeling techniques geared toward
manufacturing systems analysis. MANUPLAN II is based on an
open-network queueing model with multiple classes of parts
(Network Dynamics Inc., 1988) and is solved using a node decomposition approach. Each node is analyzed with an approximation for the
mean waiting time based on the first two moments of the arrival and
service distribution. MANUPLAN estimates the dynamics of the
interactions between the resources and the parts in the system and,
hence, the time that a part spends waiting at each resource (Suri and
Diehl, 1985). The solution takes into account the interconnection of
the nodes and the impact of failures on the service time and departure distributions. This leads to predictions of the production rates,
the utilization of equipment, and the average number of parts waiting at each resource.

Inputs to MANUPLAN are the basic system design data: information on part routing, equipment capacities and reliabilities, and production requirements. Input variability can be modeled. The output
values from MANUPLAN II, such as flow time and WIP quantities,
are "steady-state average values." Outputs of MANUPLAN include
part flow times, WIP levels, equipment utilizations, and production
rates achieved (Suri and Diehl, 1985).

All queue disciplines are FIFO, implying that no priority rules are
established for parts. MANUPLAN does not model limited buffer space
(i.e., blocking is not possible), but does model certain dynamics intrin-
sic to the system. Such dynamics would include resource contention,
variability in arrival rates and service rates, reject probabilities, and
failure rates. More involved interactions such as from material-handling vehicle movement or merging conveyors cannot be modeled.

The node decomposition and approximation techniques used in
MANUPLAN II are based on the MVA (Suri and Hildebrant, 1984).
The basic concepts of the underlying theory in MANUPLAN II can
be found in the user's manual for Release 1.1 (System Dynamics
Inc., 1988).

The Queueing Network Analyzer (QNA). QNA (Whitt, 1983) is a performance analysis software tool developed by AT&T Bell
Laboratories that calculates approximate congestion measures for
queueing networks. A feature of QNA, distinguishing it from many
analysis packages, is that arrival processes do not have to be

Poisson and service times do not have to be exponential. This ability to treat non-Markovian networks provides QNA with flexibility not found in existing exact methods.

QNA analyzes open networks of multiple resources, each having the FIFO queue discipline with no queue capacity constraints. Each service station can have any number of identical independent machines (serving one part at a time). Multiple part classes are allowed, but parts cannot change classes. Service times may depend on the part class, but are otherwise independent of the network history and are mutually independent and identically distributed.

Two parameters are used to characterize the arrival process and service process, one to describe the rate and the other to describe variability. The nodes in the network are then partially characterized by the first two moments of the interarrival and service time distributions. Given the approximation that the nodes are stochastically independent, congestion measures for the network can then be determined.

Since the nodes are analyzed separately after the parameters for the internal flows are determined, the QNA approximation method may best be described as a parametric-decomposition method. In addition, since the aggregate network analysis requires that the nodes be treated as stochastically independent, this independence may be considered a generalization of the product-form solution which is valid for Markovian networks.

The output from QNA includes several different congestion measures for each node such as utilization, expected number of busy machines, and the mean and variance of the equilibrium delay. Measures for the aggregate network are also calculated.

The basic underlying assumptions of the general algorithm used by QNA are:

1. The network is open rather than closed. Parts come from outside, receive service at one or more nodes, and eventually leave the system.

2. There are no capacity constraints. There is no limit on the number of parts that can be in the entire network and each service facility has unlimited waiting space.

3. There can be any number of machines at each node. They are identical independent machines, each serving one part at a time.

4. Parts are selected for service at each facility according to the FIFO discipline.

5. There can be any number of part classes, but parts cannot change classes. Moreover, much of the analysis in QNA is done for the aggregate or typical part.

6. Parts can be created or combined at the nodes.

An illustrative example

A simple example is constructed to illustrate various approaches to analytical modeling. Three approaches are reviewed: *simple composite analysis* (COMQ), the CAN-Q approach, and the MANUPLAN approach. COMQ computes an equivalent weighted single arrival time and process time for each machine and then uses simple queueing equations for a single queue. The intent is not to present a detailed analytical comparison of each approach, but rather to try to characterize the requirements and expectations when using such approaches.

Example problem input. A simple, modified, flow shop with five part types and five machines is studied. Each part type has a unique routing that consists of two to five sequentially ordered machine visitations. Each job has a lot size of one. Setup times are assumed to be zero. The arrival, or system demand, data are shown in Figure 13.2.

Since the three approaches each require different demand input, this information has been expressed as part demand percent (CAN-Q), part interarrival time (COMQ), and part yearly demand (MANU-PLAN). All three arrival patterns are equivalent.

The routing and process time information for each part is shown in Fig. 13.3.

The input provided to COMQ consists of the part interarrival times, part routings, and associated processing times. COMQ assumes that parts arrive at the system according to the interarrival times and are processed. There is no limitation to the number of parts that can enter the system (i.e., an open system), and since the computations are very aggregate, the actual sequence of machine visitations is of no significance. The input process literally takes seconds.

Part type	Demand percent	Interarrival time	Yearly demand
1	43	20	5031
2	11	80	1287
3	14	60	1638
4	19	45	2223
5	12	70	1404

Figure 13.2 Example problem arrival data.

Part type	Operation number	Machine	Processing time
1	1	1	4
	2	2	5
	3	4	10
	4	5	11
2	1	2	6
	2	3	4
	3	5	9
3	1	1	10
	2	2	12
	3	3	3
	4	4	14
	5	5	1
4	1	1	2
	2	3	7
	3	4	4
5	1	2	5
	2	3	8

Figure 13.3   Example problem routing and process data.

The input session with CAN-Q is similar. Aside from the part routings and process times, the product mix (demand percent) and number of items in the system are required. The differences in input suggest the differing approaches to solving the problem. CAN-Q drives production based on the number of items allowed in the system (i.e., a closed system). In the example, 10 items are always in the system. Whenever an item exits the system, another item enters.

The MANUPLAN II approach requires the Lotus 1-2-3 spreadsheet package for execution. Data are input through a Lotus worksheet. The worksheet can be created either with the Lotus interface or through an ASCII text file conversion program. The MANUPLAN package offers a significant number of features relative to the other packages mentioned. These features are apparent during the input process.

In addition to the standard input requirements (i.e., routing, processing times, yearly demand), MANUPLAN allows an array of other options. The user must specify a maximum utilization limit for every workstation. MANUPLAN will not allow a value greater than 95 percent since, at these higher levels, more detailed aspects of the system dictate utilization and are not within the scope of accuracy for MANUPLAN. Variability in arrivals and equipment can be specified to represent late/early arrivals of raw/purchased parts and nondeterministic process times, respectively. Since MANUPLAN is not designed to model completely deterministic systems, the minimum variability is 10 percent. Equipment failure information (mean time between failures and mean time to repair), machine overtime, setup times, and various other factors are used to define the system under study.

Assuming that the MANUPLAN user uses Lotus and experiences no problems installing MANUPLAN, an additional one or two days may be spent reading the user's manual, running the tutorial problem, and familiarizing oneself with the functionality of the software. Each session following this warm-up period may take as little as 5 minutes or as much as 60 minutes, depending upon input requirements. It is assumed that data are available and that the user is capable of defining the system with the user interface provided by MANUPLAN.

Example problem output.   In order to summarize the output provided by each approach, a common production rate was assumed. The CAN-Q program was executed to generate a base case production rate. The interarrival times specified in COMQ were modified so that the production rates generated by CAN-Q could be achieved. Similarly, product demand and available time were adjusted in MANUPLAN. Changing the production rate in COMQ is simply a matter of altering the interarrival times. In MANUPLAN, the product demand and/or the available time must be altered to affect the production rate. The values shown in Fig. 13.2 reflect these changes.

CAN-Q approaches the problem by producing as much as possible, subject to the number of parts allowed in the system and the number of material-handling devices (e.g., transporters). Hence the production rate may be changed by increasing the number of parts allowed in the system or by increasing the number of transporters.

The utilizations calculated by each approach were almost identical, as planned by the input. They were approximately 0.41, 0.60, 0.37, 0.82, and 0.68 for Stations 1 through 5 respectively. Note that CAN-Q assumes a transporter that essentially becomes a sixth machine. For this example, it was assumed that there was a single transporter with a transport time of two minutes which resulted in a utilization of 0.82. Changing the number of available transporters or the transport time would alter the results from CAN-Q.

The job flow times calculated by each approach varied significantly in magnitude, as shown in Fig. 13.4. Each approach does rank the flow times similarly, as would be expected (i.e., job 3 has the largest flow time, job 5 has the smallest flow time).

Similar to job flow times, the average number of jobs in queue varies significantly, as shown in Fig. 13.5. In general, increasing the number of jobs in queue (WIP) causes an increase in the time each job spends in queue, which causes an increase in the job's flow time.

The COMQ package calculated an average of 9.5 jobs in the system. CAN-Q, by design, only allows 10 jobs in the system at a time.

	Station				
	1	**2**	**3**	**4**	**5**
COMQ	0.29	0.88	0.22	3.80	1.44
CAN-Q	0.25	0.70	0.19	2.01	1.08
MANUPLAN	0.16	0.44	0.12	1.95	0.73

Figure 13.5   Average number in queue.

	Part type				
	1	**2**	**3**	**4**	**5**
COMQ	103.6	49.9	116.9	62.3	25.7
CAN-Q	85.9	51.8	102.8	49.9	31.2
MANUPLAN	68.0	34.0	79.0	38.0	19.7

Figure 13.4   Part flow times.

MANUPLAN calculated an average of 6.28. As would be expected (based on average number of items in each workstation queue), the average station waiting times vary significantly across approaches. The resulting values are given in Fig. 13.6.

General observations.   Analytical models generally yield a worst-case analysis—that is, they tend to give results that are worse than what would happen in the real world. Part of this is due to the high degree of variability (i.e., exponential process times) that many of them assume. In practice, process times have very little, if any, variation and arrivals tend to occur at fixed periodic times, such as at the start of each week. Also, real systems tend to be controlled by humans who attempt to minimize bottlenecks and large queues. However, for rough-cut analysis where many options are to be considered, analytical models provide a fast and economical tool.

	Station				
	1	**2**	**3**	**4**	**5**
COMQ	3.2	9.4	3.3	42.8	18.2
CAN-Q	2.2	6.1	1.7	17.4	9.4
MANUPLAN	1.8	4.8	1.9	22.2	9.3

Figure 13.6   Average waiting time.

Finally, one should avoid making direct comparisons of the data shown above. Although COMQ gave results that were similar to CAN-Q and MANUPLAN, it is not to be assumed that this would happen under different circumstances, particularly as the studied system becomes more complicated.

Summary remarks

Analytical models have been found to be outstanding tools for providing insight into the study of manufacturing systems. They aid in the identification of key design parameters and their effects on the system. On the other hand, representing abstract models of the real world mathematically has proven to be a challenging task for researchers as well as systems analysts.

According to Buzacott and Yao (1986), analytical models are useful in a wide range of applications from establishing basic design concepts to detailed design to scheduling and control. As noted, the models can be used to provide insight into a number of key design issues (i.e., central versus local storage, machine grouping and loading, and work order release) and to determine how the overall production capacity of the system is affected by, for instance, the part mix, the machine capacities and capabilities, the maximum inventory levels allowed, and the type or number of material-handling facilities available. It is further noted that, in certain cases, stochastic models which use summary information have closely agreed with simulation models which used more detailed information.

At the detailed design stage, there are at least two good reasons to include analytical modeling in the decision-making process. First, when analytical solutions exist, they can be compared to the simulation results as part of the simulation model validation process. Secondly, analytical models (if applicable) can utilize simulation output to perform sensitivity analysis to determine the effects of changes in system parameters. Hence, when a large amount of simulation output analysis is required, analytical models may help expedite the analysis process.

Computer Simulation

Computer simulation (Pegden et al., 1990; Banks and Carson, 1984) involves creating a model of a real or proposed system for the purpose of evaluating the system's behavior for given sets of conditions. It allows the analyst to draw inferences about new systems without building them or make changes to existing systems without disturbing them. It is perhaps the only tool that will allow total system interactions to be analyzed.

Simulation (as is the case with queueing) is not an optimization tool, in that it will not automatically determine the best system design. Using a simulation to design or evaluate a system is more of a "what if" process. Once the simulation has been constructed, the analyst queries the model by asking "what if" questions. The model responds with the results in the form of system performance measures and/or a picture in the form of an animation. By interpreting these results and asking successive "what if" questions, the analyst can develop the system of choice.

The flexibility of simulation allows the model to be constructed at almost any level of detail and to include accurate system control logic. Because of this flexibility, part of the simulation process requires the analyst to determine what must be modeled and at what level of detail. The following sections describe the overall simulation process, illustrate the mechanics of simulation, outline the relevant components of a simulation language, and present a case study to illustrate the process.

The simulation process

Associated with a simulation study is a finite number of project activities typically addressed by the simulation analyst (Sadowski, 1989; Thesen and Travis, 1991). These activities include, but are not limited to, problem identification, definition of problem objectives, model formulation, data collection, model translation, verification and validation, experimentation and analysis, documentation, and selling the results. This section briefly discusses each of these activities.

What actually precedes a simulation study is the *problem identification*. This activity involves determining what exactly is the problem and what is the best method to address and resolve it. It may be the case that simulation is not an alternative. If the problem involves a complex system with dynamic (e.g., time-dependent) components that interact in an uncertain (e.g., random) manner, then simulation is often the only tool capable of performing a thorough study. An accurate definition of the problem can dictate the level of detail required in the model and may indicate specific areas where special care must be taken.

If simulation is chosen as the modeling tool, a clear statement or *definition of project objectives* must be set forth. A critical factor in project success is the ability to capture accurately the appropriate functionality of the real system. An accurate definition of the project objectives often dictates the level of detail required in the simulation model. It also aids in establishing the boundaries of the simulation model. While the model is being built and while the results are being analyzed, the analysts must always keep the project objectives clear in their minds.

The *model formulation* stage should be a mental activity that attempts to develop an overall strategy for the simulation study. It is during this time that alternative modeling approaches and data structures are considered. This effort is often initiated with the development of a functional specification for the simulation model. Developing a functional specification requires that the modeler completely define all inputs to and outputs from the simulation, describe each of the simulation components, define all control logic for the model, and present at least a short discussion of possible extensions under consideration for the future. The final functional specification is a document of agreement among all parties involved in the simulation study. Experience has shown that if a functional specification is developed, much of the model formulation is completed by the time the specification is accepted. The result of the model formulation stage is a blueprint of the subsequent model. The model construction phase then becomes an integration of the data structure and the modeling framework into a working simulation model.

During the model formulation process, a determination of what data are available and what data are required is made. The importance of this *data collection* phase cannot be overstated. The adage "garbage in, garbage out" is a constant reminder. The amount of data and type of data are dictated by the level of detail defined during the establishment of objectives and the formulation of the model. A very detailed model may require exact distributions to characterize some processes, whereas time estimates may be appropriate for a more aggregate approach.

Data collection is not a one-step process, but can continue through the simulation process. In some cases, the initial data may consist of reasonable estimates that are sufficient for the initial analysis. Once the model is available, sensitivity analysis can be performed to determine which data are critical. Once the critical data have been identified, a more thorough data collection can take place.

Model translation involves the actual construction of the model, following the details of the functional specification and the developed data structures, using the simulation language constructs. This effort generally requires a familiarity with the system modeled and with the software used to implement the model. A simulation language or package may be employed at this stage. Even though the concepts and details for the resulting model have already been defined, translating them into a usable computer simulation model can take a significant amount of time and effort.

Simultaneous to this process is *model verification* (Sargent, 1991)—that is, ensuring that the model behaves in the way it was intended. A large simulation model can become very complex and difficult to verify. Thus, verification should be an ongoing process where each model component is verified as it is completed. The use of animation is highly recommended during this phase as it provides a mechanism to view the complex interactions that often occur. These animations can be relatively simple as long as they illustrate the desired activities and interactions.

Model validation is the process of ensuring that the model behaves in the same way as the real system. Validation is accomplished with the completed model. However, if the modeled system can be easily decomposed into several smaller subsystems, these subsystems can be verified and validated separately prior to the complete system validation. If the system currently exists, then some kind of comparison can be made to ensure that the model represents the real world. If the system does not exist, but similar ones do, then the simulation results can be compared to the similar system and at least a partial validation can be performed. If there is no real system to compare with the simulation, then formal validation cannot be performed. If this is the case, then it is recommended that extra effort be devoted to the verification and that those who are familiar with the system design be closely involved with the simulation effort. A partial validation can be performed using individuals who are extremely familiar with the system concepts. As mentioned earlier, animation is an excellent tool for this process.

Experimentation with the simulation model involves the definition of input parameters and the generation of simulation output performance measures (Seila, 1991). In large-scale simulation studies, it may be necessary to employ experimental design concepts to ensure that the experimentation process is tractable. We normally think of experimentation as being separate from the analysis. However, in most cases experimentation and analysis are an iterative process where an experiment is conducted, the results are analyzed, and, based on these results, further experiments are defined.

The basic issues to address during the *analysis of the results* are of a statistical nature. It is imperative that the analyst have, at a minimum, a solid background in practical statistics. Actually, prior to the experimentation the analyst must determine whether the study system is terminating or nonterminating, the time the system is in a transient state, the number of observations that must be collected, and the types of statistical test that are to be performed.

The nature of the analysis should be determined during the problem definition activity. The type of analysis required should be consistent with the types of decisions to be made. If the model was designed to predict the performance of a system, then the statistical issues can be very important and the analysis will generally require much longer runs than if a comparative analysis is being performed.

At a minimum, the *model documentation* should describe the data structure, the key elements of the model, the general flow logic, and all variables, queues, and so forth. Sufficient documentation might consist simply of comments throughout the model. The depth of documentation depends on the user, the expected life of the model, and the complexity of the model. In addition, the simulation results should be well documented and easy to understand. Nothing is worse than losing credibility due to poor communications.

The final activity associated with a successful simulation project is that of *selling the results* (Sadowski, 1992). As with any project, a presentation or report is normally required. The analysts should report their findings at a high level, but be prepared to talk details. With simulation projects, it is important to "sell" throughout the entire project, not just at the final presentation. This starts with selling the idea of using simulation—as the benefits should outweigh the costs. However, the actual benefits are often very difficult to quantify, whereas the costs are very easy to identify. The validity of the project may hinge on doing a successful job of selling the merits of the simulation study to those involved in designing and operating the system since you will depend upon them for information, insight, and honest assessment of the planned or modified system. After developing the model, the analyst must establish the credibility of the model with the project team and the major players who understand the system under study by proving that the simulation does behave like the proposed system. A successful simulation project is the result of a unified team effort, not the efforts of a single individual.

The mechanics of simulation modeling

Simulation modeling involves the process of designing a model of a real or proposed system and conducting experiments with the model for the purpose of learning more about how the system works and how the system can be improved. Regardless of the software tool used, the underlying concepts of how a simulation works are very much the same. This section will introduce the reader to the mechanics of simulation modeling as well as to common terminology used and assumptions made.

The actual thought process used by a simulation modeler will be described with a trivialized example, the single-machine problem. At the conclusion of this example, the reader should begin to understand the basics of model development, how a simulation works, and the impact of uncertainty on a dynamic system.

The single-machine problem

Consider a single machine that processes two job types. Type 1 jobs arrive every ten minutes and require four minutes to process. Type 2 jobs begin arriving at Time 3 and continue to arrive every six minutes. Type 2 jobs require three minutes to process. The transfer time from the arrival point to the machine is one minute, and the transfer time from the machine to the exit point is one minute. It is assumed that material handlers are always available (i.e., they don't need to be modeled). Our objective is to evaluate how long it takes jobs to flow through the system and to define what size of buffer is required in front of the machine. Additional statistics on machine utilization and the number of jobs in the system will also be collected.

When studying such a system, the modelers often put themselves "in the shoes of the entity." An *entity* is defined as any person, object, or thing whose movement through the system causes changes in the state of the system. An entity may be either physical, such as parts that move through the system, or logical. A *logical entity* might control operator breaks. Such an entity would not leave the system, but at predetermined time intervals cause operators to stop working and go on break. Some time later, the same logical entity would make the operators available for work. This entity would continue this cycle for the duration of the simulation. In the single-machine problem, the job types or parts are referred to as *physical entities*.

The development of a simulation model requires that the modeler describe the system as a sequence of processes that the entities will encounter. For instance, the following sequence of processes may be used to describe the single-machine problem:

- Create the job.
- Transfer the part from the arrival area to the machine.
- Queue for the machine to become available.
- Process the part on the machine.
- Transfer the part to the exit area.
- Collect statistics.
- Dispose of the job.

The modeler will also specify how many jobs to create or how long to simulate the system so the computer model can be simulated for as long as required by the modeler. As each job moves through the system, it causes various *events* to occur. Whenever an event occurs, the system state changes. Statistics are collected to characterize the state of the system over time. For instance, in the above example, five events can be defined:

Event 1: Arrive

Event 2: Enter queue

Event 3: Start process

Event 4: End process

Event 5: Exit system

A part arrival causes the number of parts in the system to be increased by one and causes a delay for the transfer of the part from the arrival point to the machine. As the part arrives at the machine, it might enter the queue if the machine is busy, or if the machine is idle and available, it would cause the machine to start processing the part. At the end of a process, a delay is encountered for transfer of the part to the exit area, and if there is another part currently in the queue, it would start the process on that second part. Otherwise, the machine would be set to idle. A part exiting from the system will cause statistics to be collected and the number of parts in the system to be reduced by one. In all cases, the *state* of the system has changed.

Figure 13.7 shows the sequence of events that would occur during the first 20 time units of the simulation. It also shows how the number of parts in the system and queue changes after each event. The first part, Type 1, arrives at the system at Time 0 and causes the

Time	Event	Part type	Arrival number	Number in system	Number in queue
0.00	Arrival	1	1	1	0
1.00	Enter queue	1	1	1	1
	Start process	1	1	1	0
3.00	Arrival	2	2	2	0
4.00	Enter queue	2	2	2	1
5.00	End process	1	1	2	1
	Start process	2	2	2	0
6.00	Exit system	1	1	1	0
8.00	End process	2	2	1	0
9.00	Arrival	2	3	2	0
	Exit system	2	2	1	0
10.00	Enter queue	2	3	1	1
	Start process	2	3	1	0
	Arrival	1	4	2	0
11.00	Enter queue	1	4	2	1
13.00	End process	2	3	2	1
	Start process	1	4	2	0
14.00	Exit system	2	3	1	0
15.00	Arrival	2	5	2	0
16.00	Enter queue	2	5	2	1
17.00	End process	1	4	2	1
	Start process	2	5	2	0
18.00	Exit system	1	4	1	0
20.00	End process	2	5	1	0
	Arrival	1	6	2	0

Figure 13.7 Single-machine example event sequence.

number in system to increase to 1. The part is transferred and scheduled to arrive at the queue at Time 1. Upon arrival, the part passes through the empty queue and starts its process of 4 time units. Thus, it will end its process at Time 5. Before the first part has completed its process, the second part arrives at Time 3 and is scheduled to arrive at the queue at Time 4. Upon arrival it enters the queue and remains, as the machine is currently busy processing the first part. At Time 5 the first part completes its process and the second part is removed from the queue and its process is started. The first part is transferred to the exit and scheduled to depart at Time 6.

At each event in the simulation process, the next event and its time are determined. This process of determining and scheduling when the next event for each entity will occur is the basis for *next-event simulation*. If the arrival, transfer, or process times were stochastic, the actual time would be generated, and then the next event would be scheduled. This process is the fundamental basis for all discrete-event simulation packages. If a machine is busy when a part arrives at its queue, the part is placed in the machine's queue to await processing. It is important to understand that a simulation model keeps track of all the various activities and events that occur over time. If there were six machines in the system, all six machines would be operating simultaneously, and the simulation model would capture and schedule all the events that would occur.

Since all arrival, transfer, and process times are deterministic, the outcome of a simulation of this system is fairly easy to predict. The output generated for a run of 48,000 time units of this single-machine system is shown in Fig. 13.8. The average flow times are 6.67 and 6.00 for part types 1 and 2 respectively, with very little variation. Although the machine was busy 89.9 percent of the time, there was very little congestion as shown by the 0.233 average number of parts in the queue. In fact, there was never more than one part wait-

	Average	Minimum	Maximum	Observations
Flow time, type 1	6.67	6.00	8.00	4800
Flow time, type 2	6.00	5.00	8.00	7999

	Average	Minimum	Maximum
Number in queue	0.233	0	1
Machine busy	0.899	0	1
Number in system	1.666	0	2

Figure 13.8 Simulation results for single-machine example.

ing for the machine. Finally, there was an average of 1.666 parts in the system and never more than 2.

It is interesting to look at what happens if variability is incorporated into the between-arrivals, transfer, and machine process times. This scenario would more closely represent reality since machine shops rarely can predict exactly when the next order will arrive, or exactly how long it will take to process. The "worst-case scenario"— the one that usually causes the most congestion—is the one that would be modeled by characterizing these time delays as exponentially distributed random variables. Figure 13.9 shows the event

Time	Event	Part type	Arrival number	Number in system	Number in queue
0.00	Arrival	1	1	1	0
0.48	Enter queue	1	1	1	1
	Start process	1	1	1	0
2.65	End process	1	1	1	0
3.00	Arrival	2	2	2	0
3.52	Enter queue	2	2	2	1
	Start process	2	2	2	0
4.05	End process	2	2	2	0
4.20	Exit system	2	2	1	0
4.35	Exit system	1	1	0	0
6.25	Arrival	2	3	1	0
6.82	Arrival	2	4	2	0
7.04	Enter queue	2	4	2	1
	Start process	2	4	2	0
7.67	Enter queue	2	3	2	1
9.39	End process	2	4	2	1
	Start process	2	3	2	0
10.37	Exit system	2	4	1	0
13.68	Arrival	1	5	2	0
13.75	End process	2	3	2	0
14.45	Enter queue	1	5	2	1
	Start process	1	5	2	0
14.60	Exit system	2	3	1	0
15.13	End process	1	5	1	0
15.47	Exit system	1	5	0	0
15.96	Arrival	1	6	1	0
16.14	Arrival	2	7	2	0
16.60	Enter queue	2	7	2	1
	Start process	2	7	2	0
16.95	Arrival	1	8	3	0
16.99	Enter queue	1	6	3	1
17.57	Enter queue	1	8	3	2
17.85	Arrival	2	9	4	2
20.03	Enter queue	2	9	4	3

Figure 13.9 Event sequence with exponential times.

sequence for this scenario. The first indication of the induced variability is the fact that the first arrival, at Time 0.00, is processed before the second arrival, at Time 3.00, and yet the second part exits the system first (at Time 4.05 versus 4.20 for the first part). This is caused by the differences in the transfer times from the machine to the exit. The first part has a large transfer time, 1.70, whereas the second part has a small transfer time, 0.15. The effect of variability can be observed throughout the sequence. Most notable is that there are nine arrivals, there were six before, and three parts in the queue at the end of the sequence, zero before.

From looking at the first example sequence, Fig. 13.7, one might conclude that it would be fairly easy to perform a simulation experiment by hand or to write your own software. The value of simulation software starts to become apparent when looking at the second sequence, Fig. 13.9. If one takes the next logical step, adding more parts, machines, and logic, then the value of available software is realized.

The output generated from this revised simulation model is shown in Fig. 13.10. As expected, the results are very different. The flow times for the parts are approximately 5 times greater than for the deterministic simulation, with the maximum flow times being more than 20 times larger. The machine utilization times are essentially the same. However, the number in queue and number in system are far greater than before, the maximum number in system being 57, versus 2 for the first run.

Although the differences between the two simulation results are drastic, the real world rarely has this high a variability. Thus the results obtained from a real system, as described above, with a limited amount of variability would fall between these two extremes. The value of simulation is that one can identify the degree of variability and use those estimates to generate accurate predictions of system behavior.

To illustrate how one might develop a simulation for these examples using available software, a flowchart of this model using the

	Average	Minimum	Maximum	Observations
Flow time, type 1	31.10	0.29	165.92	4790
Flow time, type 2	30.54	0.13	169.41	7947

	Average	Minimum		Maximum
Number in queue	6.742	0		55
Machine busy	0.897	0		1
Number in system	8.172	0		57

Figure 13.10 Simulation results with exponential times.

SIMAN simulation language is shown in Fig. 13.11. Although this is a very simple example, it indicates the type of thought processes and the results that a simulation analyst might expect. The SIMAN example also shows the basic types of modeling constructs that are available in a simulation language.

Key components of a simulation

The single-machine problem discussed above was modeled according to a process orientation. In this manner, the model consists of a description of the processes through which the entities move as they progress through the system. Aside from an entity as mentioned above, the key components of a simulation include *variables, resources, control logic, and statistics.*

In terms of the single-machine problem, entities were the parts that flowed through the system. Since there were two part types, it was necessary to distinguish between them. This would normally be accomplished with a local variable called an *attribute.* An attribute is a value that is carried with, and is unique to, each entity. It might record the part type, the time the entity entered the system, the next workstation, the processing time, and so on. These attributes can also have different interpretations for different types of entities. The analyst must define the number of attributes required for each entity to maintain.

The simulation model will also have *global* variables that are available to all entities. Examples would be number of parts in the system, number of parts at a workcenter, number in a specific queue, or status of a machine. Some of these global variables are automatically included and updated by the simulation—e.g., number in queue and resource status. Others are defined and updated by the analyst—e.g., number in system and number at a workcenter.

As the entities or parts move through the system, they typically require the use of resources. The resources that are of primary importance in a simulation model are the ones that are constrained. In fact, the predominant use of simulation modeling is to address resource contention issues. Typically we think of resources as machines, fixtures, and tools. However, material-handling devices, workers, and even space can be considered and modeled as a scarce resource. In many situations, an entity may require several resources (i.e., material handler, machine, fixture, and operator) before it can continue on its process plan.

The control logic is the main means of describing how all the components of a system interact. Incorporating accurate control logic is probably the most difficult part of creating a good simulation model. Control logic can take the form of order release, local dispatching, operator assignments, material-handling control, and

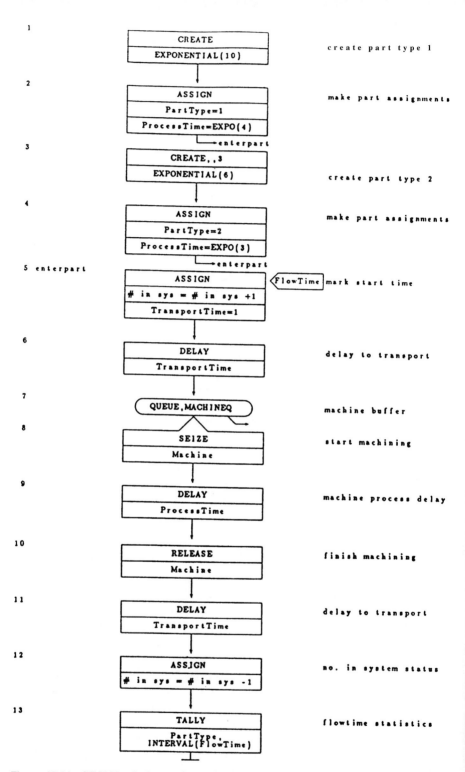

Figure 13.11 SIMAN code for single-machine problem.

so forth. The control logic for an automated system is fairly easy to incorporate because the decisions are made by some type of computer; thus, they are easy to replicate in a simulation. Control logic or decisions that are made by human operators are much harder to include as it is frequently very difficult to quantify the decision-making process.

Finally, the statistics collected during a simulation are the predominant means of characterizing the expected performance of the model. They provide important information to the decision-making process. Statistics can take many forms: counters, observational statistics, time-persistent statistics, and frequencies. *Counters* are single integers that can be incremented and decremented during the simulation with only the final value being reported. *Observational statistics* are typically composites of several occurrences during the course of the simulation (e.g., cycle times, due date performance, waiting time). These are normally reported with an average, minimum, maximum, number of observations, and a measure of the variance. Time-persistent statistics are time-weighted values. These usually report utilizations, number in queue, amount of time failed, etc. Their output normally includes an average, minimum, maximum, and sometimes a measure of variation and the final value. *Frequency statistics* are time-persistent statistics that report frequencies of different states that occur (e.g., busy, idle, setup, fail, and load/unload states for a machine).

Simulation tools today generally provide a minimal set of features that are fundamental to the modeling process. A time-advance mechanism functions in two ways: first, it acts as a "simulation" clock to advance time based on the occurrence of events; second, it schedules events and executes events when they are scheduled to occur. An internal structure is required to update continuously the data used to describe the state of the system, the data associated with entity movement through the system (i.e., number of entities in a queue), and the input and output data. Experimental conditions must be maintained and accessed by the user to describe the initial conditions of the simulation, the run length, the resource capacities, and so forth. In addition, a mechanism to collect and report statistics automatically is required.

As discussed in the single-machine example, since random variables are typically required to characterize various process delays, a means of generating random variants is also required. Hence, a random-number generator and random-variable generator are also standard components. Report generation, error checking, user-interaction facilities, and animation are all key to a robust modeling strategy.

Case study

A new simulation user usually starts by purchasing a simulation software package and training in that software. These software packages are well documented and provide a structured way to construct a simulation model. However, deciding how to tackle a simulation study, choosing what to include in the model, and understanding how long it will take is something that can only be done by someone with modeling experience. Since it is not possible to provide such expertise here, a case study of a real-world simulation experience is presented to give some idea of the application of the simulation process and the time required to complete such a project.

The following case study is on an actual project, but the company name and product are withheld to maintain confidentiality. The production methods and terminology discussed have been slightly altered to protect the proprietary nature of their technology. The modeling and analysis methods discussed closely resemble the sequence of events and discoveries realized during the study of this system.

Problem discussion and definition. Prior to this study, the company's manufacturing facilities were based on a philosophy that was quickly becoming outdated. The concern of upper management was that in order to maintain market share and to keep the upper hand on innovation, the company needed to make a smooth transition into new product lines and also to smoothly accomplish proposed changes to the technological and philosophical aspects of manufacturing. One of the major problems was the need to change from the current "manufacture-to-inventory" to a "produce-to-order" concept.

Their strategy included opening a new plant to alleviate the overflow at an existing domestic plant. However, the main emphasis was to be on incorporating a whole new product line family. The company had the fundamental manufacturing know-how, but it required the ability to predict manufacturing requirements and potential problem areas at the new facility. The objective of this new facility was to produce products with a much shorter cycle time, thereby lowering inventories and improving customer satisfaction. In addition, the facility was to be designed in such a way that it would facilitate new product introductions with minimal interruptions.

Facility characteristics. The new facility under study depicts discrete-parts manufacturing in a make-to-order (MTO) environment. The manufacturing floor contains five primary departments: raw material and purchased parts receiving, machining, and general processing;

in-process storage and assembly; and final storage and shipping. The machining and general processing areas are divided into "workcenters"; each workcenter may be populated with either sophisticated computer-controlled machines, manually controlled machines, or a combination of both. Workcenters, in general, perform different functions and contain multiple machines.

Orders flow from one area to the next via fork trucks and manual hand trucks. The in-process storage area consists of a mini-AS/RS (automated storage and retrieval system) for medium to small parts and a manual rack storage system for large and bulky parts.

The machines available in the workcenter offer a trade-off to the designer of the system: the more sophisticated machines, although more expensive, have a faster production rate, require less operator intervention, and need less setup. The basic machines are cheaper, manually intensive, and require more setup time.

Two different configurations were to be evaluated in the final assembly area. First is the traditional paced assembly line approach. Each assembly line generally accommodates one of the many part family product lines. Alternatively, a new team assembly approach is advocated to allow employees to have more responsibility and ownership. With the team assembly approach, all the subassembly and final assembly for a particular unit are done by the same three operators in the same general assembly area.

Product characteristics. The end-items produced at the new facility generally have a multilevel bill of materials (BOM): machining-subassembly-assembly. The products are produced from the same part family, and, therefore, have similar process requirements. There is a wide variety of product variations, each variant requiring a different sequence of processes and process requirements. An erratic demand pattern is expected due to the nature of the product. Marketing provides a five-year, long-term forecast that is generally not acknowledged with any seriousness.

The master production schedule (MPS) extends for twelve months and is the primary component of the final assembly schedule (FAS) generation. The FAS uses a pull concept. The manufacturing lead time is expected to be reduced from an historical three months to an aggressive four to six weeks. From the final assembly schedule, the production planners generate a production schedule (PS) that drives the machining and basic processing department in a push fashion.

An average part order visits five workcenters before being sent to in-process storage to await final assembly. The range of workcenter visitations is between 3 and 15.

Objectives.   The objective of the new facility design is twofold. First, to reduce the manufacturing lead time, thus providing a fast reaction time to customer orders, minimal levels of WIP inventories, and little or no final product inventory. Second, the design must be cost effective. Given that no customer orders are to be turned down and full customer satisfaction achieved, the costs associated with capital equipment purchases, floor space requirements, and inventory levels must be minimized.

The types of questions that need to be addressed are these: What buffer sizes are expected in both machining and assembly areas? How many material handlers are required for the whole system? What size AS/RS and rack system are required to handle anticipated demand? What lead time can be expected for the proposed systems? In addition, the feasibility of the planned approach needs to be proven.

The design team identified a number of areas where modeling and analysis could have the greatest impact:

- Determine machine capacity requirements in machining and assembly areas.

- Determine best layout configuration (paced versus team assembly) and capacity requirements in assembly area.

- Determine storage requirements for in-process parts.

- Determine the material-handling requirements.

- Assess the impact of potential productivity gains versus cost effectiveness for machine center alternatives (manual versus computer-controlled machining).

- Develop and test scheduling logic and operational policy.

- Determine the best overall systems integration.

Initial analysis.   Prior to developing the initial design, a project team was created that consisted of eight diverse individuals, with a wide variety of skills and specialties. The team ranged from a systems control individual to the production supervisors who were to assume responsibility for the facility when completed.

The first step in the overall design was to develop an aggregate facility layout (which located the primary departments). A mixture of experience and layout optimization techniques (to reduce the overall distance traveled for anticipated part mix and production volumes) were utilized to find a balance between qualitative and quantitative factors.

Static analysis using a standard spreadsheet package was employed to determine approximate capacity requirements (e.g., number of workcenters of each type) on the manufacturing floor. This preliminary study established system functionality and required significant data collection efforts to determine process times, failure rates, reject rates, equipment specifications, and operator standards.

Analytical models (i.e., queueing models) were employed to evaluate machine requirements at each workcenter. A number of steady-state performance measures were calculated, such as WIP estimates, machining requirements at each workcenter, system throughput times, workcenter buffer sizes, and material-handling estimates. Various part mix strategies were evaluated to ensure that adequate capacity existed to maintain the required level of throughput and lead time.

This initial analysis resulted in several alternate designs, each of which appeared to provide the desired performance. A rough-cut economic analysis was then conducted to determine the feasibility of continuing the project. With a positive economic picture, the next step was to develop the detailed final design, including equipment selection. The primary tool chosen to aid in this process was simulation.

Simulation analysis. The simulation portion of the project was launched with a two-day meeting with the original project team assigned to develop a first draft of a complete functional specification for the simulation activity. The first half day was spent with the entire project team developing overall project goals and objectives. Part of this process included documenting all of the questions and issues that each team member wanted to test with the final simulation model. The remaining time was spent with the simulation analysts meeting individually with each team member to obtain system information specific to that member's expertise.

A formal document was created and each team member was requested to review the entire document for accuracy and completeness. Based on this review process, the document was revised several times until a complete agreement was reached by the entire team. Each team member then signed that document, signifying their acceptance.

The acceptance of the functional specification allowed the actual simulation model formulation and translation to begin. During this time, the project team members were periodically required to provide additional data and control logic that had not been detailed in the functional specification. In some cases, the best approach to defining system logic was not clear (e.g., determining in what sequence to gain control of parallel resources). When this occurred, the simulation was written to incorporate alternative control strategies. Each alternative could then be evaluated with the final system design to determine the best strategy.

Although several alternative designs were evaluated, a single simulation model was developed. This model was intentionally designed so it could easily be configured to include the various aspects of all the designs under consideration. Later, during the analysis, this provided the capability to consider combinations of options that had not been included in the initial system designs. In fact, the final system design was considerably different than any of the initially proposed designs.

Upon completion of the model, the project team was reassembled for a one-day meeting to verify and validate the model. Since the proposed system was only a concept, an animation was critical as the primary vehicle for communication with the team members. One of the initial designs was selected and subjected to intense scrutiny. The results of this activity required that several minor changes be made in the simulation model. Once the validity of the model was established with the project team, the analysis was scheduled to begin.

The actual analysis was conducted over a period of several weeks. Members of the project team submitted questions and design configurations to the simulation analyst; the appropriate simulation runs were then conducted. The results were given to the team members and additional runs were made. This interactive and iterative process prompted further questions and analysis. Thus, the final system design was the result of an evolutionary process with the team members providing the system expertise to determine the best system and control options.

Simulation was first employed to evaluate all the options in the primary machining area. The primary consideration was optimal use of labor. The computer-controlled machines were evaluated against and in combination with the manually controlled machines. Since computer-controlled machines required fewer operators but were more expensive, a balance had to be established for the best configuration of machines and operators.

Although the analytical models could estimate certain system performance parameters fairly well, the simulation effort provided important insight into potential operational problems not previously identified. Because of the unstable product demand, the simulation model recognized that the primary bottleneck actually involved two workcenters in an alternating fashion. For a period of time, the first workcenter was the bottleneck; but as the product mix changed, the second workcenter became the bottleneck. Both workcenters performed similar operations, but their machines were not interchangeable. To resolve the problem, the manufacturing engineers added an additional workcenter that could be used interchangeably to handle any overload at those two bottleneck workcenters. Since a queueing model can only provide averages, it did not anticipate the problem of shifting bottlenecks.

The simulation modeling approach also allowed the designers to evaluate accurately the trade-offs between the various workcenter configurations. A balance between the manually controlled machines and the computer-controlled machines was, in many cases, found to be a hybrid mix of the two types of machines. For instance, instead of having two computer-controlled machines, a better alternative was to have one computer-controlled machine with two manually controlled machines. The advantage to having more machines was in increased flexibility. One operational assumption was that if two identical machines were both available, the next part requiring process would choose the faster machine.

Having configured the machining area, the simulation was utilized to evaluate the integration of the manufacturing, storage, and assembly areas. Specifically, the two approaches to final assembly were first compared without limits on the intermediate storage area to determine which configuration provided the best performance and minimized labor costs. After a detailed study, it was found that the team approach to assembly, despite its increased floor space requirements, provided the minimum-cost solution as well as the best workplace environment. With both the machining and assembly areas configured, the simulation was exercised to determine the size requirements of the mini-AS/RS and the rack system.

Finally, simulation modeling was employed to test the sensitivity of the system to variations in product mix as well as future products and processes. Certain product lines were anticipated that required new technology. The new technology was never given the opportunity to be "tried and tested"; uncertainty surrounding processing times was apparent. The design team isolated the areas of uncertainty and used sensitivity analysis to determine the impact of varying process times. The assembly area was found to be sensitive to the upper extremes of process times anticipated. Graphical displays of queueing and resource utilizations showed that the worst-case scenarios occurred at isolated points in time. Instead of changing the design, it was decided that overtime would be used to handle these occasional peaks of excess requirements.

Because the system design was based upon several new concepts and anticipated product lines, much of the data used for the analysis was derived from the best estimates available at that time. Thus, before the final design was accepted, a complete review of the data accuracy was conducted. The simulation was then used to test the sensitivity of the system performance to all data suspected of being inaccurate. In several cases, the system was found to be very sensitive to certain data types. In these cases, additional effort was made to secure better data estimates. These new estimates were then entered in the model and it was determined that the system would still perform as expected.

Summary remarks. The final design resulting from the simulation project was found to be substantially different than what would have been implemented otherwise. The simulation effort uncovered and helped resolve several problem areas. In addition, it was found that the mini-AS/RS and manual rack storage areas could be smaller than originally planned. Some of these issues might have been found and resolved during the normal planning process, but the entire project team agreed that the simulation effort more than paid for itself. No attempt was made to attach a dollar value to the economic benefit of simulation, but it was agreed that it had been considerable.

Currently, the facility has been built and is running as expected, despite the fact that most initial estimates of process times and failure rates were somewhat different than those of the actual operating system. The simulation model was used to evaluate and resolve minor problems during installation and start-up and is continually being used to evaluate new product mix strategies and new product introductions.

The entire simulation process from start to acceptance of final design covered a period of approximately four months, although not all this time was devoted to model construction. The functional specification required over two weeks, although more than half of this time was spent obtaining information and reviewing the document. The modeling phase accounted for approximately eight weeks from start to finish. As with the functional specification, some of this time was spent obtaining additional information and data required for the model. The analysis covered a four-week period, but only three person-weeks were expended. The total simulation effort was estimated to have required approximately five person-months to complete. This estimate included the model formulation and conceptualization, development of the functional specification, model development, statistical analysis, model revisions, and animation. However, the perceived savings in the design and later operation costs more than offset this investment. No attempt was made to quantify the additional cost that would have occurred during start-up due to issues uncovered and resolved with the simulation.

It should be noted that two other benefits of doing a thorough simulation study are that, first, it forces organization and communication, and, second, it is the only known means of studying the behavior of such complex systems. During this study, the graphical animations developed were the primary means of communicating the part flow, control logic, and system behavior. These animations inspired all team members to become involved, feel ownership, and contribute accurate information in a timely manner. In the process of gathering information, the engineering staff was forced to think ahead and understand how their area tied into, and affected, all other areas.

Such involvement occurred across all levels of the organization: shop floor, engineering, quality, and management.

The Future of Computer Modeling

Computer modeling of manufacturing facilities started in the 1950s and 1960s. At that time, it was only used by a few companies as it was very time consuming and expensive, partially due to the cost of computer time on the mainframe computer. Several factors have led to an explosion of computer modeling during the 1980s. The arrival of the personal computer, animation, and user-friendly commercial software were the three primary factors contributing to this explosion. Today there are numerous packages available, from simple static analysis to complex simulation languages. However, there are still several problems that must be overcome before computer modeling becomes an industry standard.

The primary obstacle is that there is no single package currently available that will provide the full functionality required (static and queueing analysis, stochastic analysis—from aggregate models to more detailed models, operational analysis, and real-time scheduling). If one examines the potential life cycle of a manufacturing system and the computer modeling tools required, the list is quite varied. At the beginning, one needs a tool that will allow what is commonly called *rapid prototyping*. This could be strictly a static tool or it could incorporate queueing models. This tool would be used to examine initial concepts and different types of systems. It would require limited amounts of data as inputs and only provide rough estimates of the system performance. As the initial designs are developed, an easy-to-use simulation is required. As with the rapid prototyping tool, it would require only a limited amount of data, but provide a stochastic analysis tool.

As the design process continues, the level of detail in the simulation model is increased and a final design is created. The final simulation may be quite complex when compared to the first set of models. Once the system is in operation, it is desirable to have another tool that would serve as an on-line decision support tool to assist in scheduling work through the system. This tool would be able to predict due date performance and could be used to play "what if" games to evaluate alternative operating strategies. If modifications are to be made to the system, each model should be able to easily incorporate each change and allow the user to evaluate various alternatives. These revised models could then be used to determine the best strategy for the proposed changes.

Not only would it be desirable to have all these capabilities in one software package, it should allow for easy and friendly use by nonmodeling and nonsimulation experts. In other words, it should provide ease of use without restricted flexibility. It should also provide an extendable modeling system and support application-based terminology. Thus, if the user is modeling a machine, the first model may be very simplistic. As the modeling requirements become more detailed, the same machine might require setup, maintenance, failures, operators, etc. The simulation should be able to include these additional features when requested, without additional coding. In other words, a simple "yes" to failures, along with the failure and repair times, should be the only entry required by the user. Current simulation software provides most of these capabilities (Pegden and Davis, 1992).

Incorporating analytical modeling and simulation into a single package has already been proven feasible (Deuermeyer et al., 1991). A single set of inputs can be used for the queueing portion or for a simple simulation. However, once the logic within the simulation becomes complex, it is very difficult if not impossible to capture this in an analytical model. If the use of the analytical model is restricted to simple constructs, then these capabilities can be packaged together. Simulation has already proved to be a valid tool for scheduling and control, although it needs to be applied with slight differences. For example, the system must be preloaded to represent the current status, and the incoming parts must be taken from the actual release schedule. The output from a simulation scheduler is also different, but all the information required is already available in most simulation packages.

The manufacturing industry has seen significant advances in computer modeling techniques. These techniques have been greatly aided by the advances in computer software and hardware technology. It is expected that continuous improvements in functionality and user friendliness will be made in the future. It is anticipated that these enhancements will provide the user with an integrated set of powerful modeling and analysis techniques. Hopefully, this will allow these robust techniques to become an industry standard.

Further Reading

General modeling

Brown, E.: "IBM Combines Rapid Modeling Technique and Simulation to Design PCB Factory-of-the-Future," *Industrial Engineering,* vol. 20, no. 6., June 1988.

Hillier, F. S., and G. J. Lieberman: *Introduction to Operations Research,* 5th ed., McGraw-Hill, New York, 1990.

Leung, Y., and R. Suri: "Performance Evaluation of Discrete Manufacturing Systems," *IEEE Control Systems Magazine,* June 1990.

Ravindran, A., D. T. Phillips, and J. J. Solberg: *Operations Research Principles and Practice,* Wiley, New York, 1987, pp. 77–86.

Analytical modeling

Buzacott, J. A.: "Modelling Manufacturing Systems," *Robotics and Computer-Integrated Manufacturing,* vol. 2, no. 1, pp. 25–32, 1985.

Buzacott, J. A., and D. D. Yao: "Flexible Manufacturing Systems: A Review of Analytical Models," *Management Science,* vol. 32, no. 7, 1986.

Douglas, D. E.: "Queueing Analysis Techniques Measure Random Wait Time," *Industrial Engineering,* February 1986, pp. 26–29.

Jackson, J. R.: "Jobshop-like Queueing Systems," *Management Science,* vol. 10, no. 1, October 1963, pp. 131–142.

Reiser, M., and S. S. Lavenberg: "Mean-Value Analysis of Closed Multichain Queueing Networks," *Journal of the Association for Computing Machinery,* vol. 27, no. 2, pp. 313–322, April 1980.

Seila, A. F.: "Output Analysis for Simulation," *Proceedings of the 1991 Winter Simulation Conference,* B. L. Nelson, W. D. Kelton, and G. M. Clark (eds.), Institute of Electrical and Electronics Engineers, Piscataway, New Jersey 1991.

Solberg, J. J.: "A Mathematical Model of Computerized Manufacturing Systems," *Proceedings of the 4th International Conference on Production Research,* Taylor & Francis Ltd., London, 1977.

Solberg, J. J.: "The Optimal Planning of Computerized Manufacturing Systems," CAN-Q User's Guide, NSF Grant no. APR74 15256, Report no. 9, July 1980.

Snowdon, J. L., and J. C. Ammons: "A Survey of Queueing Network Packages for the Analysis of Manufacturing Systems," *Manufacturing Review,* vol. 1, no. 1, pp. 14–25, March 1988.

Suri, R.: "RMT Puts Manufacturing at the Helm," *Manufacturing Engineering,* February 1988, pp. 41–44.

Suri, R.: "Quantitative Techniques for Robotic Systems," in *Handbook of Industrial Robotics* S. Nof. (ed.), New York, 1985, pp. 605–638.

Suri, R.: "Robustness of Queueing Network Formulae," *Journal of the Association for Computing Machinery,* vol. 30, pp. 564–594, 1983.

Suri, R., and R. R. Hildebrant: "Modelling Flexible Manufacturing System Using Mean-Value Analysis," *SME Journal of Manufacturing Systems,* vol. 3, no. 1, pp. 26–38, 1984.

System Dynamics Inc.: MANUPLAN II User's Manual, 2d ed., Network Dynamics Incorporated, Burlington, MA, 1988.

Whitt, W.: "The Queueing Network Analyzer," *The Bell System Technical Journal,* vol. 62, no. 9, pp. 2779–2815, November 1983.

Simulation modeling

Balci, O.: "Guidelines for Successful Simulation Studies," *Proceedings of the 1990 Winter Simulation Conference,* O. Balci, R. P. Sadowski, and R. E. Nance (eds.), New Orleans, 1990.

Banks, J., and J. S. Carson, III: *Discrete-Event System Simulation,* Prentice-Hall International Series in Industrial and Systems Engineering, Prentice-Hall, Englewood Cliffs, NJ, 1984.

Law, A. M., and W. D. Kelton: *Simulation Modeling and Analysis,* 2d. ed., McGraw-Hill, New York, 1991.

Pegden, C. D., and D. A. Davis: "ARENA: A SIMAN/Cinema-Based Hierarchical Modeling System," *Proceedings of the 1992 Winter Simulation Conference,* December 1992.

Pegden, C. D., R. E. Shannon, and R. P. Sadowski: *Introduction to Simulation Using SIMAN,* McGraw-Hill, New York, 1990.

Sadowski, R. P.: "Selling Simulation and Simulation Results," *Proceedings of the 1992 Winter Simulation Conference,* December 1992.

Sadowski, R. P.: "Avoiding the Problems and Pitfalls in Simulation," *Proceedings of the 1991 Winter Simulation Conference,* B. L. Nelson, W. D. Kelton, and G. M. Clark (eds.), Phoenix, AZ, December 1991.

Sargent, R. G.: "Simulation Model Verification and Validation," *Proceedings of the 1991 Winter Simulation Conference,* B. L. Nelson, W. D. Kelton, and G. M. Clark (eds.), Phoenix, AZ, December 1991.

Solomon, S. L.: *Simulation of Waiting-Line Systems,* Prentice-Hall, Englewood Cliffs, NJ, 1983.

Thesen, A., and L. E. Travis: "Introduction to Simulation," *Proceedings of the 1991 Winter Simulation Conference,* B. L. Nelson, W. D. Kelton, and G. M. Clark (eds.), Phoenix, AZ, December 1991.

Proceedings of the 1990 Winter Simulation Conference, O. Balci, R. P. Sadowski, and R. E. Nance (eds.), New Orleans, December 1990.

Proceedings of the 1991 Winter Simulation Conference, B. L. Nelson, W. D. Kelton, and G. M. Clark (eds.), Phoenix, AZ, December 1991.

Combined modeling

Deuermeyer, B. L., G. L. Curry, and R. M. Feldman: "An Approach to Automatic Modeling of Semiconductor Fabrication Facilities," Working Paper INEN/OR/WP/06/8-91, Texas A&M University, 1991.

Haider, S. W.: "Experiences with Analytic and Simulation Modeling for a Factory of the Future Project at IBM," *Proceedings of the 1986 Winter Simulation Conference,* J. Wilson, J. Henrikson, and S. Roberts (eds.), December 1986.

Sanders, J. L.: "Issues in the Design and Modeling of Automatic Assembly Systems," *Proceedings of the 1986 Winter Simulation Conference,* J. Wilson, J. Henrikson, and S. Roberts (eds.), December 1986.

Shimizu, M., and D. V. Zoest: "Analysis of a Factory for the Future Using an Integrated Set of Software for Manufacturing Systems Modeling," *Proceedings of the 1988 Winter Simulation Conference,* M. Abrams, P. Haigh, and J. Comfort (eds.), December 1988.

Suri, J., and G. W. Diehl: "Manuplan: A Precursor to Simulation for Complex Manufacturing Systems," *Proceedings of the 1985 Winter Simulation Conference,* 1985.

Suri, R., and M. Tomsicek: "Rapid Modeling Tools for Manufacturing Simulation and Analysis," *Proceedings of the 1988 Winter Simulation Conference,* M. Abrams, P. Haigh, and J. Comfort (eds.), 1988.

General manufacturing

Apple, J. A.: *Material Handling Systems Design,* Wiley, New York, 1972.

Material Handling Handbook, 2d ed. (ed. in chief, R. A. Kulwiec), sponsored by the American Society of Mechanical Engineers and the International Material Management Society, Wiley, New York, 1985.

Nagle, R. N., and R. Dove: *21st Century Manufacturing Enterprise Strategy: An Industry-Led View,* vol. 1, Iacocca Institute, Lehigh University, November 1991.

Sadowski, R. P.: "Are Factory Performance Measures the Cause of Poor Productivity," *CIM Review: The Journal of Computer-Integrated Manufacturing Management,* vol. 1, no. 1, Fall 1984.

Planning for Integrated Systems across a Facility

Carter C. Utzig
Senior Manager
KPMG Peat Marwick
Charlotte, North Carolina

The design of new facilities takes on a number of new facets in today's highly competitive environment. Today, the design team must blend the aspects of human behavior, organizational design, compensation, procedures, information technology, and myriad other concepts such as total quality management (TQM) and just-in-time (JIT) to help assure the short- and long-term benefit of the facility to the competitive position of the company.

In the past, information technology planning was usually separated from the design of the facility or split amongst different organizations (i.e., engineering and information systems). This approach almost always resulted in wasted resources (human or monetary) because of the redundant efforts and the arguing associated with determining who was responsible for what. Companies with facilities planned this "old fashioned" way are now finding out that they have numerous systems which either cannot be technically integrated in a cost-effective manner, or have institutionalized "non-best practices." More importantly, as marketplace shifts occur or the organization running the new facility grows more sophisticated, companies quickly become aware that the current systems are not capable of meeting their needs to run the business and that the facility is losing its benefit to the company.

Little if any recognition during the planning process is given to the relationship between information technology and other initiatives companies might launch to sustain their competitive advantage. Imagine the impact on the momentum of an employee involvement program if little or no information is available for the teams to evaluate and resolve problems. Could one ever implement ISO9000 without some automation of the documentation of procedures? Consider a design-of-experiments team working on a product attribute problem critical to maintaining a key customer's business. The team isn't likely to be successful if they must manually capture, enter into a spreadsheet, and then analyze all their data. It is likely these non-systems initiatives would slowly die, wasting the resources invested, negatively influencing the facility's culture, and, most importantly, losing revenue.

Because of the dynamic nature of the information technology industry, this chapter will not discuss the current state of information technology or its trends. This chapter will address key issues which the design team should give consideration to during the development of an information technology plan for a new facility. A planning approach is also outlined in this chapter which focuses on the nontypical issues which a cross-functional design team should address when determining the role of information technology in a new facility.

Key Planning Considerations

To effectively plan for information technology, the design team should understand the role information technology will play in the new facility. Information technology is only one of a group of intertwined components that the design team must plan for. The following list includes the key components the team must consider:

Strategy: This component describes the intent of the plant. It outlines how the facility will support the firm's market and product strategies by defining key operating measurements.

Organization: This component describes the structure, role, and responsibilities for the organization.

Procedural process: This component documents the flow of activities within the business processes.

Policies: This component contains a number of policies, such as lot sizing, safety stock, and service levels.

Technology: This component includes the information systems and process technology utilized by the activities in the various business processes.

Culture: This component defines the cultural attributes that the new organization should aspire to.

Measurement: This component defines how performance should be assessed for each of the business processes. It also defines the basis for compensation.

It is important that the team consider each of these components concurrently because they are interdependent. The team might identify new technology that provides a quantum change in the performance of a key business process. This quantum leap would constitute a new competitive advantage and would therefore require the company to rethink its market and product strategies. Other examples which reinforce the need to understand the components and their interdependency are:

- Procedures may specify that the receiving clerk checks receipts and then enters all receipt information on-line. This could represent an opportunity for the company to immediately transfer funds electronically to that vendor's bank account. This implies a number of information technology and organizational issues.

- If the strategy is to produce specialty products, this implies business processes which have few non-value-added activities, process technology which supports small-lot production, and information technology to quickly translate customer needs into specialized products.

The design team must also take into consideration the current and anticipated infrastructure supporting the location where the facility will be located. The most important infrastructure question revolves around the educational level and training of the workforce which will initially work in and support the facility. This issue has two aspects: knowledge of information technology and understanding of business processes. An example of a situation where considering the workforce's knowledge of information technology is essential would be trying to install a new facility in what was formerly East Germany. Because of local labor laws and the impracticality of importing the workforce, a large percentage of the employees would be drawn from the local populace. Their exposure to the latest technology would be limited at this time. What we might take as commonplace technology, such as word-processing or spreadsheet applications, would be foreign to the plant personnel. Think of the impact this would have on implementing more advanced applications such as expert system shells. The implementation at commissioning would consist of simple technologies and manual procedures to provide an initial base to educate and grow the workforce.

An example of taking into account business process understanding would be the commissioning of a textile mill serving mass merchandisers staffed with personnel from the aerospace industry. How can these people make informed decisions about what systems they need if they do not understand how the business should function? They cannot. When teams are forced into this position, it results in a large amount of money being spent two to four years after commissioning to rework the system put in place at commissioning. Although these are extremes, this is quickly becoming reality as more and more companies erect plants in Third World countries.

The design team should also consider the following infrastructure questions, which could impact the information technology plan:

- When will the management group be assembled? If the group is going to be assembled shortly, should the design team wait to finalize their plans to get buy-in from that group? If the group is going to be assembled late in the process, should the design team postpone the implementation of some technologies to give the management team more flexibility on how they want the business to function? What are the skill sets of the management group? Do they understand the capabilities of information technology? Have they worked in this industry before?

- Who will the plant report to? A local division? A corporate group? Will reporting be split? What information technology does the group that the facility reports to use? Can it be leveraged?

- Should we consider centralizing some of the business functions at one of the reporting locations? Can we centralize processing of the information? If the functions and systems are centralized, what interface issues are now created?

During the planning process the team must adopt a definition of what they want to integrate—i.e., technology, activities within a business process, players in the supply chain, organizational structure, culture, etc. Almost every publication stresses the need for integration and puts forth a definition depending on the author's perspective: information technology, supply chain, etc. It is important that the team focus on the business objective of integration, not technical objectives. The objective of integration is to supply more value to your end customers and their customers than the competition. By accepting this focus, integration can be achieved by eliminating activities along the supply chain (the business process perspective) or streamlining the interfaces between systems (the technical perspective).

The business processes to a large extent dictate the information technology solutions. Therefore, any planning process for information technology must either incorporate high-level business process analysis (reengineering) or initiate a separate project to generate the prerequisite input.

Business process analysis should not be confused with information-modeling techniques or activity-based costing (ABC). The intent of business process analysis is to speed up the process by eliminating, combining, out-sourcing, improving, or automating the activities. The techniques employed during this type of analysis focus on the sequence and value of activities which make up the business process and who performs them (see Fig. 14.1). While information technology modeling techniques provide some of the same analysis, their real intent is to promote the effective sharing of information across activities, automated or manual. The models also provide the base for code and database generation. The intent of ABC is to assign accurate costs to a company's activities, services, and products. ABC can then be used as a tool to determine the impact of a new technology on the cost of a company's activities, services, and products.

The focus for each of these analysis techniques is the activities a company performs. Yet, surprisingly, each technique usually becomes its own separate project. This means the same business processes are analyzed multiple times. The design team needs to use all three techniques for designing a new facility. By good project management and constant coordination across the diverse groups (functional, MIS, accounting) performing the analysis, the team can eliminate the redundant work. The result is a single decomposition of the business into its activities. In this way the activities in each of the three models are the same. The benefits of doing this are immense. Imagine the power of reengineering a business process and knowing cost and time saving, the impact to product cost, and the changes required to the system design.

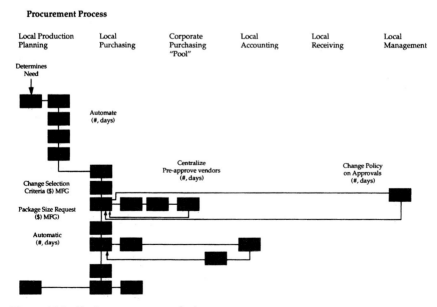

Figure 14.1 Business process analysis.

The team should address the issue of leverage up front (see Fig. 14.2). There are two levels of leverage that should be considered:

Business level: The business level is based on the user's perspective. At this level one should attempt to reuse the procedures, policies, and applications from facility to facility. Trying to leverage solutions at this level provides the greatest potential business benefit. Employees or application analysts can move from plant to plant with little retraining required for them to become productive. The resources expended for modifications can be amortized across multiple sites since the modifications can be physically shared. The current life cycles of application software are significantly longer than that of hardware.

Technical level: This level focuses on the hardware, networks, operating systems, and program languages. Although most companies place the majority of emphasis in this area to achieve leverage, it does not represent as large a potential value to the company as the business level because:

- The information systems community responsible for these areas is relatively small in comparison to the people who have hands-on contact with the business applications.
- The solutions have shorter life cycles which are exponentially shrinking.
- The solutions are not generally portable across locations.

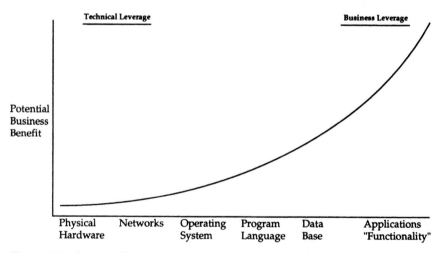

Figure 14.2 Impact of leveraging solutions across organizations.

- "Enabler" software has emerged. Enabler software handles the complexity of a diverse hardware, database, operating system, and network environment for the programmer or user.

The design team should quickly state where it will place its emphasis: at the business or technical levels. It should be careful to define which facilities it will leverage from. Should the team only consider leveraging solutions from similar facilities? Any plant in the supply chain? Or just any plant the company owns?

The team must also put the new plant in perspective with the rest of the existing facilities so as the existing facilities grow or become more complex the migration path is defined (see Fig. 14.3) and the new plant's solution sets leveraged.

The team will need to establish or identify existing, agreed upon, standardized solution sets and technical standards to facilitate leverage and integration. *Standardization* is the agreement to use specific vendors' solutions. *Technical standards* define how application, hardware, physical connections, and so forth should function.

Information standards are not limited to hardware and networks. The following is a list of standards which are typically missed that should also be established:

Computer-aided design (CAD): Drawing standards, electronic media

Modeling: Tools, techniques, and methodology

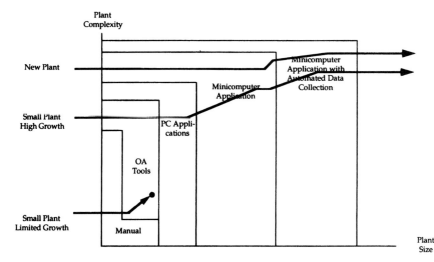

Figure 14.3 Migration and solution set paths.

Documentation: Code structure, user documentation, technical documentation

Code: Error handling, hard-coding information, structure

Ownership: Data files, systems

Process and machine diagnostics: Diagnostic functions, vendor error, and status mapping

It is important that the design team think through the ramifications of each of these standards. For example:

- Should a contractor be required, at the expense of efficiency and cost, to use standard methodologies and computer-aided systems engineering (CASE) tools rather than its own methodologies and tools that produce the same results?

- What if the equipment vendor supplies the best process equipment but cannot supply CAD drawings? When can they be converted? And who pays?

The Planning Process

This section outlines an approach for developing an information technology strategy for a new facility (see Fig. 14.4). This approach is based on several assumptions. The first assumption is that an information technology plan for a new facility should address all technical architectures (see Fig. 14.5), from Level 4 (i.e., business systems) to the interface between Level 1 and Level 0 (i.e., the network the drive motors utilize). The second is that the team that develops the plan is made up of middle managers from each functional area of the business (i.e., accounting, production, sales, R&D, etc.). This is essential for two reasons:

- It promotes buy-in of the vision and solutions which in turn speeds up and helps sustain the implementation.

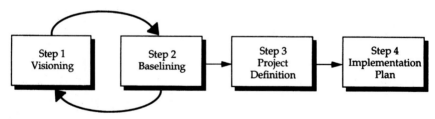

Figure 14.4 Information technology process.

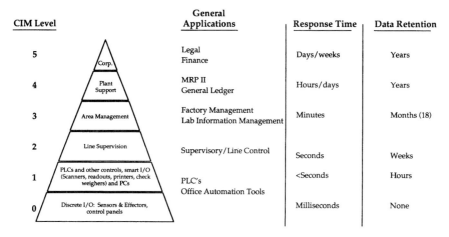

CIM Level		General Applications	Response Time	Data Retention
5	Corp.	Legal Finance	Days/weeks	Years
4	Plant Support	MRP II General Ledger	Hours/days	Years
3	Area Management	Factory Management Lab Information Management	Minutes	Months (18)
2	Line Supervision	Supervisory/Line Control	Seconds	Weeks
1	PLCs and other controls, smart I/O (Scanners, readouts, printers, check weighers) and PCs	PLC's Office Automation Tools	<Seconds	Hours
0	Discrete I/O: Sensors & Effectors, control panels		Milliseconds	None

Figure 14.5 Information technology architectural levels.

- No one person is an expert in all areas of the business.

- It promotes a culture focused on understanding how the various business processes are interrelated, which results in an integrated business.

- It provides an education point for a core group of users on the potential of information technology.

Step 1

The first step in the process of developing an information technology plan is to establish the strategic objective of the facility. This step documents (if not already available) the market and product strategies and their requirements for each of the business processes of the new facility. The key steps in this section are:

- Develop the parameters by which you are going to compete in the marketplace by documenting (if not available) the market and product strategies.

- Determine a high-level vision of the future. The vision should use examples which describe how the business will function internally and externally in the future (see Fig. 14.6). This is usually accomplished by high-level brainstorming of different ways of reengineering the business processes to achieve a new best practice. It should address questions such as: Are there centralized organizations? Where are mixed pallet loads created? Who bills the customer? This should not be confused with benchmarking, which is

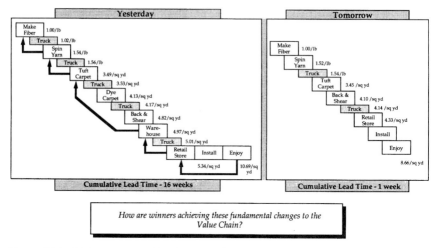

Figure 14.6   Visioning example—Carpet industry value chain.

the determination of how you stack up against your competition or other industries, not how they achieve a higher level of performance (best practices). Benchmarking should not be used, since it does explain how your business processes should change so as to one-up your competition.

- Determine the planning parameters for each business process based on supporting the market and product strategies.

- Prioritize the planning parameters based on their importance in supporting the strategies. Their prioritization should highlight what the facility must do, what it can achieve, and what might be a stretch given the future facility infrastructure of people and process technology (see Fig. 14.7).

- Develop operational measures for each of the key business processes and business objectives, e.g., yield and time to market.

- Establish the potential influence the business processes have on achieving the facility's market and product strategies.

Step 2

During the second step of this approach the team must determine the baseline and the constraints that it needs to plan within. The key steps in this section are:

- Document influencing factors such as:

  - Quality of the plant people

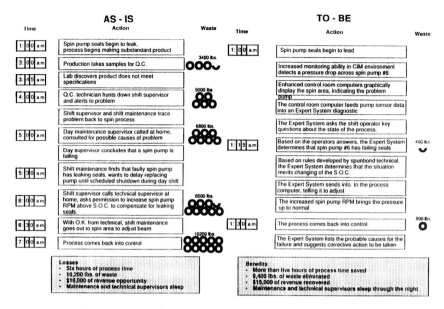

Figure 14.7   Visioning example.

- Location of the new facility
- Timing of the management team
- Use of new technologies

- Develop volume estimates for key design parameters—such as number of product variations, number of new product introductions, number of shipments by method, number of vendors, number of people, and number of ship-to locations—at commissioning, postcommissioning (two years), and maturity.

- Develop planning models for business process and information technology using the IT methodologies. It is essential to manage the level of detail these models are taken to. Since these are planning models, the detail should be limited to just enough to understand the unique features of the business and provide a basis for determining the physical application architecture. This includes developing the following key deliverables:

- Business functions and processes definitions (see Fig. 14.8)
- Context diagram
- Data dictionary
- Entity relationship diagram
- Business process-to-entity matrix (see Fig. 14.9)

Planning Objectives

Planning Objectives	Focus of Importance	Primary Influence
Cost		
Reduce material purchase costs	3	External
Reduce direct processing costs	6	Plant
Reduce overhead costs	6	Plant
Reduce distribution costs	3	External
Reduce working capital employed (inventory and receivables)	3	Plant
Increase Yield	9	Plant
Quality		
Reduce scrap and waste	9	Plant
Increase product consistency	9	Plant
Improve Product Attributes	3	Plant
Service		
Increase service level (shipment reliability and fill rate)	9	Plant
Reduce order lead time and variability	9	Plant
Reduce new product introduction lead time	6	Plant
Increase flexibility	6	Plant
Other		
Comply/Better than Government Environmental Regulations	9	Plant

Figure 14.8   Planning objectives.

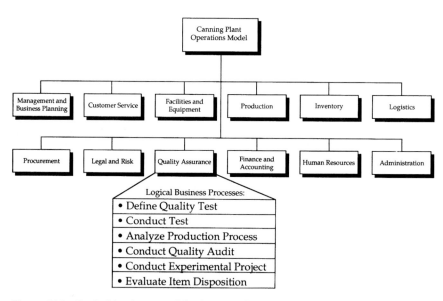

Figure 14.9   Logical business model—functional.

- Business process-to-organization matrix (see Fig. 14.10)
- Logical information systems architecture (see Fig. 14.11)

■ Develop high-level business process flows and functional analysis. This task provides key baseline performance information which

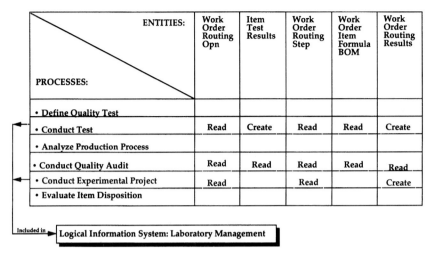

PROCESSES:	ENTITIES:	Work Order Routing Opn	Item Test Results	Work Order Routing Step	Work Order Item Formula BOM	Work Order Routing Results
• Define Quality Test						
• Conduct Test		Read	Create	Read	Read	Create
• Analyze Production Process						
• Conduct Quality Audit		Read	Read	Read	Read	Read
• Conduct Experimental Project		Read		Read		Create
• Evaluate Item Disposition						

Included in ▶ | Logical Information System: Laboratory Management |

Figure 14.10   Logical business process-to-information matrix.

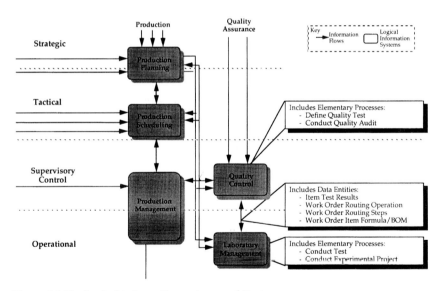

Figure 14.11   Logical information systems architecture.

can be used later to justify the information technology costs and focus people on the gaps.

■ Develop ABC model. This task also provides information for evaluating the impact of reengineering the information technology plan.

| | Technical Standards | | | | Standardized Solutions | | | |
Computing Level	Networks	Operating System	Data Base	Appl.	Networks	Operating System	Data Base	Applications
Level 4								
Level 3								
Level 2								
Level 1								

Figure 14.12   Standards and standardized solution.

- Document technical standards and standardize across the supply chain (see Fig. 14.12).

Step 3

The next step is to determine how the facility will achieve the vision set forth above. The key steps in this section are:

- Define leverage, and scope to which solutions should be leveraged.
- Determine high-level list of projects and initiatives around business process/activities or core competencies (i.e., problem solving). This includes developing the scope, potential benefits, planning parameters, and key requirements.
- Establish the high-level criteria to be used to evaluate the potential alternatives for each project. Although this seems extremely straightforward, it is not. For example, it is simple to say that all solutions should be evaluated on their fit and flexibility with future needs, vendor strength and direction, fit with technical standards, easy access to information, and so on. However, the design team should consider the following technical and management selection considerations:

  - *Evolving technical standards:* Will the key vendors ever agree on a single standard? More than likely no. This implies that you should be somewhat flexible when setting technical standards, since they may not be the ones you want tomorrow.

- *Shrinking technology life cycles:* Since the technology is changing so fast today, the reality of your choice is that it will very quickly be outdated. Therefore you must define how far you want to wed yourself to any one solution long-term. Establish how much you want to risk on emerging technology and plan to constantly enhance and replace what you just implemented.

- *"Legacy" systems:* The characteristics of these types of systems are that they will not meet the needs of the new facility and behind them is a tremendous push to leverage the system in the new facility. Unfortunately, the logic behind the argument to leverage the system is based on the fact that the system has been around forever, someone has a large political stake in the system, or it cost a small fortune to develop. Never let a system dictate how the business is going to function. Learn to manage and reduce the cost of complexity so that the business can utilize the best possible solution.

- *Organizational span of control:* Always know your audience. Typically a new facility has a large, diverse audience which has a number of different agendas. Look for the common theme.

- *Individual versus supply chain focus:* There are trade-offs between solutions which might be great for one business process but hurt three other equally important business processes.

- *Time to commissioning:* The shorter the time to commissioning, the more you have to utilize existing solutions regardless of their fit with the long-term direction.

- Determine the activities that each application should perform by mapping applications to the IT models. This will also help to determine overlaps and missed processes (see Figs. 14.13 and 14.14). Where data are captured, stored for everyday decision making, and archived should also be determined. For example, detailed inventory information such as bin location, quality status, etc., would be maintained by the warehouse application, whereas the MRP application might maintain only summary information.

- Evaluate alternative solutions for each project (see Fig. 14.15). The level of detail applied to the evaluation should be proportional to the importance of the project and sequence in the implementation plan.

- Document and evaluate alternative technical infrastructures which are required to support the application software (see Fig. 14.16).

- Evaluate the collective solution set of both applications and their supporting technical enablers. One method is shown in Fig. 14.17.

Inventory and Work Order Processes

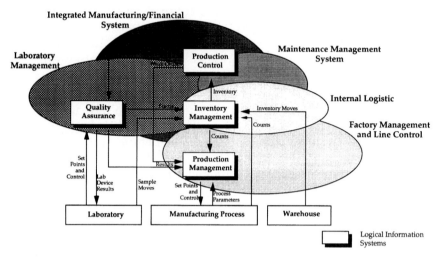

Figure 14.13 Application mapping.

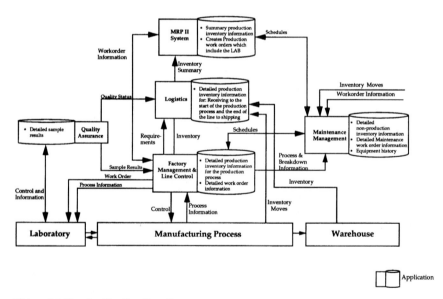

Figure 14.14 Application functions.

ALTERNATIVE APPROACH	O P T	VENDOR/PRODUCT SOLUTIONS	POSITIVE FACTORS	NEGATIVE FACTORS	APPROX COST
1. Install Supervisory Computer System	A	Texas Instrument's TISTAR	• TISTAR is designed to reside over a netowrk of TI PLC systems such as exists in D-8 • TISTAR AR is in final stages of approval • Provides required data collection, analysis, and reporting tools • Supports 802.3 LAN interface	• Integration with other types of process control systems and business systems may require custom hardware and software • TISTAR may not conform to Plant's computer system standards (when established) • Process data resides in a proprietary data base which cannot be accessed directly	
2. Install a different supervisory system	A	Bradley Ward's PMIS	• This alternative may provide a system which will integrate more easily with other process control and business systems • This alternative may permit selection of a system which will conform to Plant's computer system standards (when established)	• This alternative will delay project implementation • Interface to TI PLC's will be inferior to TISTAR's • Will require extensive customization to meet project requirements	

Figure 14.15   Project alternatives.

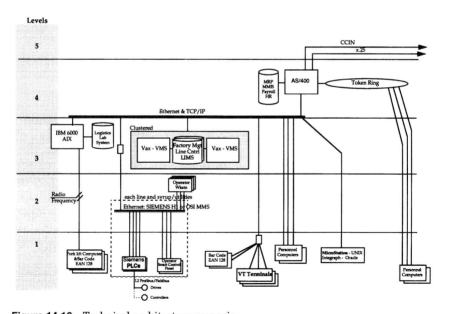

Figure 14.16   Technical architecture scenario.

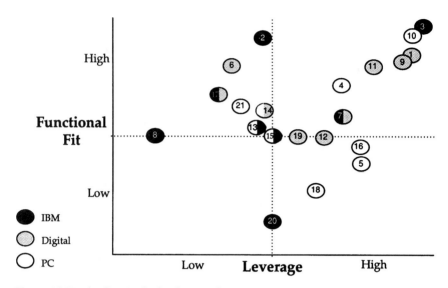

Figure 14.17   Application/technology evaluation.

Step 4

The next step is to develop the implementation schedule. The ke
steps in this section are:

- Determine the business processes which require automation a
  commissioning. This can be determined by the volume and con
  plexity of the process and the ability of the process to meet the pe
  formance objectives established in the first step of the approach.

- Link the projects outlined above to the business process or plan
  ning objectives using the quality technique shown in Fig. 14.18. B
  using this technique the team will be able to reduce the politic
  influence which tends to creep into the setting of priorities.

- Estimate the time and resources required to successfully impl
  ment each project.

- Highlight the risks for individual projects. These would includ
  such things as limited resources with knowledge of this technol
  gy, we have never successfully implemented a system in th
  department, and so forth.

- Develop a high-level program evaluation and review techniqu
  (PERT) chart for the project and the respective applications. Th
  logical information system architecture developed as part of the
  modeling defines all the potential interfaces between the futu
  applications.

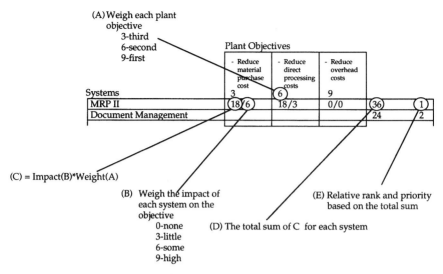

Figure 14.18 Process to set implementation priorities.

- Develop the links between nonsystem initiatives such as TQM and JIT.
- Develop a high-level implementation time line (see Fig. 14.19). This implementation time line should extend three to five years after commissioning. One of the biggest impediments for new facilities in constant pursuit of improvement is gaining approval for funds. If you know you will need it in the future, budget it now, plan it now, and manage expectations accordingly. Designing a facility does not end with the commissioning.

Since the investment analysis of information technology plans varies widely and is largely dependent upon the criteria set forth by some unknown corporate group, there are a number of techniques which can be used to supplement the typical analysis.

ABC provides unique insight into how an investment in the business functions impacts the cost of services and products. Nonfinancial measures should also be incorporated in the analysis because they force people to understand the impact of each project to their specific area and commit to measurable performance. Nonfinancial measures include such things as the following:

- Yield
- Setup time
- Down time, planned or unplanned

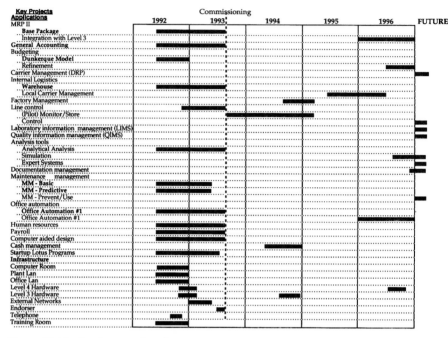

Figure 14.19   Implementation plan.

- Accident rates

- Equipment utilization

There are also a number of problems caused by the typical investment analysis imposed upon an organization. Why should shared enablers like local area networks or computers be cost-justified or split among project requests for capital? The answer is that they should not. By splitting them up you cannot reach the optimum solution. Shared enablers should be treated as unique projects. Their implementation should be linked to the projects which they support.

Most companies require projects to be cost-justified regardless of whether the result of the project can benefit other facilities. Cost-justifying a project based on the benefits achieved by a single facility places an artificially high hurdle and results in missed opportunities for the company as a whole.

The building of new facilities also provides an ideal time to pilot new technologies and ideas. Again, cost justification of these projects will result in missed opportunities since it creates risk-averse behavior. The design team will inherently choose solutions which they are familiar with. Also note that when new technologies are employed they should be bench-tested first.

Conclusion

Because of today's fast-paced competitive environment, we can no longer plan and design new facilities considering only one component at a time. The design team must consider the strategic ramifications of its decisions, the impact of the plant personnel on the implementation plan and solutions, and the activities within each business process, all within an iterative approach that focuses on how the vision will be implemented and the benefits sustained.

Design of Commercial and Industrial Buildings and Structures

David M. Griffith
Vice President, Burns & McDonnell
Engineers-Architects-Consultants
Kansas City, Missouri

Randy D. Pope
Associate, Burns & McDonnell
Engineers-Architects-Consultants
Kansas City, Missouri

Periodically in the management of a facility will come the opportunity for a major expansion of an existing facility or the construction of a new facility. These projects may include one or all of the individual functions described in other chapters of this handbook. Depending upon the product of the plant or facility, the project may encompass receiving, warehousing, process equipment, laboratory, offices, and other support areas, such as packaging and shipping.

The stages of development of the project are planning, site selection, design, and construction. The facility manager's role in each stage is important but is especially critical in the planning and design stages of the project. The success of the new facility hinges on the requirements of the end user being incorporated into the planning and design of the facility.

This chapter deals with the design stage of commercial and industrial buildings and structures. We will describe the process of design of these types of facilities, describe the team of professionals required to implement the planning decisions into the final design, suggest what form that team structure might take, and examine how members of that team can effectively communicate with one another to execute a successful design project.

The different stages of the design will be described. We will discuss whether to use an *architect/engineer consultant* (A/E), the selection of the A/E, the conceptual design, final design, and construction services phase.

Because a major expansion or new construction project is only a portion of the total responsibilities of the facilities manager, and because such projects are infrequent, the process may be to some degree unfamiliar to the manager. For this reason this chapter will acquaint the reader with the portions of the design process for commercial and industrial facilities and the mechanics of interfacing with an A/E.

Motivation to Select an Outside Architect/Engineer

One of the important decisions to be made during a major project is whether or not to hire an outside A/E. Reasons for selecting an A/E for the design of a project are:

- Insufficient numbers of in-house staff.
- A/Es are familiar with codes and regulations affecting design and construction.
- A/Es are familiar with latest design/construction technology.

Economic demands for reduced overhead costs dictate that many companies do not permanently employ adequate staff to design a major construction project. Therefore, when the facility manager is responsible for a major project, it is likely that an outside consultant must be hired to prepare the design documents. (The contract with the consultant and the required documents are described later in this chapter.)

In addition, increasingly rapid changes in regulations governing the process and the construction industries make it difficult for individual organizations to keep abreast of new regulations, codes, and standards which affect a major project. Design professionals, who are involved in the design of projects on a full-time basis, are in a better position to stay abreast of changes in regulations and how they affect the design process.

Not only are regulations changing but also technology is changing at an increasing rate. Processes and methods of construction may be available today to meet the project requirements which were not previously applicable. An outside consultant experienced in the type of process or facility has a better chance of staying current in technology relevant to major projects than the facility manager.

Architect/Engineer Selection

Once the decision has been made to hire an outside consultant, the facility manager must then make the crucial decision as to which consultant to choose. Selection guidelines have been written by several organizations and are useful tools to use in the selection process. Guidelines obviously are not mandatory for the private corporation. Many corporations may have an existing relationship with an outside A/E whom they frequently use for projects at their facilities. In some cases these relationships have been built up over a number of years such that the A/E becomes familiar with corporate expectations and requirements and likewise the corporation has become familiar with the quality of the A/E's service. In this case the A/E may already be well along the learning curve in terms of how to serve the corporation in the typical types of projects it pursues.

On the other hand the corporation or organization may need to select an A/E and may have their own guidelines in selection of A/E services. Government organizations, for example, are required by state and federal laws to follow a predetermined selection process. Practically all of these guidelines require selection of an A/E and state the required qualifications. This process is outlined in Fig. 15.1.

- **Develop scope of project**

- **Develop selection criteria**

- **Issue requests for qualifications**

- **Develop short list**

- **Conduct interviews**

- **Rank firms**

- **Negotiate**

Figure 15.1 Steps in selection of an A/E.

The first step in selecting an A/E is to define the scope of the project. This scope should include a statement of the objectives to be accomplished by the construction of the project, the title given to the project, and any information that has already been determined for the project. This might include the project location, the project schedule, the physical size of the project, and the budget range established for the project.

The next step toward selection of the A/E is to develop the criteria which will be used in the process. Commonly used criteria include:

- Specialized experience of the A/E with respect to the project requirements
- The capacity to perform on schedule
- Performance history in terms of meeting budgets
- Quality of performance
- Geographical location and/or experience with your company
- Familiarity with the project location

The specialized experience and technical competence for the specific project should be demonstrated by the A/E. For example, the facility requirements are different for a steel mill than for a high-rise office building. The A/E selected should have experience uniquely suited to the project.

Architect/engineers who have designed similar projects are farther along on the learning curve for these types of projects. They know common pitfalls to be avoided as well as the features, materials, and methods which work well for the specific project. A quality A/E firm will have learned from past mistakes and experiences and will apply that acquired knowledge to your project.

The consultant should also be able to demonstrate the capacity and capability to perform design work on the project within schedule. He or she should have enough qualified professionals on staff to complete the project as well as any current project already under design. In addition, the consultant should not be dependent on this one project for the livelihood of the design firm. The consultant should be able to demonstrate the financial stability and wherewithal to complete the project. The owner's project manager should evaluate the current project workload of the consultant and the requirements it will impose on members of the consultant's staff who would be assigned to the owner's project.

The performance history of the A/E on similar past projects should be investigated. The consultant should be able to show successful completion of similar projects on schedule and within budget. The specific scope of services provided by the consultant for each referenced project should be described. The A/E should indicate on which projects the proposed personnel have served. References should be solicited from the design professionals and these references should be contacted. References can confirm the design services provided by the A/E and the level of professionalism exhibited by the firm.

Another major criterion in the evaluation and selection of an A/E is the firm's proximity to the location of the owner's firm and familiarity with and proximity to the proposed construction site. Construction technical and regulatory requirements vary substantially with the project location, especially for corporations planning international construction projects. For example, structural design may be based on concrete framing construction in the Middle East, while in the United States for the same facility the structural framing system could be structural steel or perhaps wood. Another example: While seldom seen in the United States, exterior ceramic tile is commonly used in construction in the Far East.

Other selection criteria important to the organization can also be used. All of the selection criteria should be weighted so that more important criteria for the specific project are given greater consideration.

After the scope of the project and the selection criteria have been determined, the owner can issue requests for A/E qualifications. This request can be a public request issued through newspapers or a private request issued to prequalified firms. Firms with which the organization has worked in the past and which have provided quality services would be a starting point for requests for qualifications. Other firms in the same industry can be contacted for recommendations of design professionals who have served on similar projects in the past.

The qualifications submittals are then evaluated in terms of the selection criteria already discussed. Firms would be short-listed to the three or four firms most qualified to perform the design. These firms would be interviewed in order to meet the principals of the firm responsible for the project as well as the project manager and other key personnel who would be assigned. During the interview the firm should demonstrate working knowledge and expertise applicable to the specific project. The A/E should also be able to discuss her or his understanding of the scope of the project. It is suggested that at least two hours be allowed for the presentation interview.

The facility engineer should visit each consultant's office to become acquainted with the work ethic and culture of the organization, to observe the level of activity, and to confirm the capabilities of the organization.

After conducting the interviews and office visits, the firms should be rated again based on the criteria weighted in accordance with importance to the project.

Negotiations should then be conducted with the number 1 ranked firm to conclude the details of the scope of services for the project, fees, and terms and conditions of the contractual agreement. Fee is an important consideration in the selection of an A/E. The fee is based on the understanding of the scope of the project and the scope of services to be provided by the A/E firm. In order to save time of the owner's staff in dealing with the price issue, it is most efficiently handled at this stage in the selection process.

If suitable fee, scope of services, and contractual terms can be negotiated, the most qualified firm will be awarded the contract for design of the project. If not, then negotiations may be discontinued and the number 2 rated firm in terms of qualifications should be brought in for negotiation. This happens infrequently, however, as the owner is normally able to negotiate an acceptable contract with the most highly qualified firm.

Forms of Agreement

The owner-A/E relationship is legally temporary, and just as with any business relationship which requires responsibilities and action of two or more organizations, a formal contract is a necessity.

This contract can take many forms. Sometimes it is a simple, two-page letter agreement to provide some defined services for a defined compensation. Other times it is a longer, more formal legal document with more formal definition of terms, conditions, compensation, and other requirements. The standard contract forms are available from professional societies such as the American Institute of Architects (AIA) or the National Society of Professional Engineers (NSPE). Some owners' organizations have their own standard form of agreement between owner and contractor, and some A/Es have their own standard forms of agreement as well.

As with any legal document the A/E agreement should be carefully reviewed by legal counsel before being signed by either party. If a standard document is used, particular attention should be given to any modifications to the standard.

No matter which form of contract is used the agreement should define the parties entering into the agreement and list the representative of each party who will be responsible for fulfilling the agreement. It should provide an address for each representative, to be used for notices.

The agreement should also clearly and in detail describe the scope of services to be rendered by the A/E. The more concise the definition of the scope of services, the less room there is for later misunderstandings regarding what is to be provided to the owner.

It is also important that the agreement describe, fully, the terms for compensation. This would include the total amount of compensation and schedule for payments. There are several forms of compensation commonly used for A/E contracts. A fixed lump-sum fee is used when the project can be clearly defined in terms of deliverables and services. This is the most basic type of compensation form.

Some owners will prefer an hourly billing rate with a guaranteed maximum fee. In this case the A/E is compensated as effort is expended on the project. However, the maximum risk, in terms of fee to the owner, is limited to a specified amount. This structure also includes definition of scope in terms of services and deliverables so that the A/E can estimate and guarantee his or her maximum fee.

Another less common approach is reimbursing the A/E as a percentage of construction cost. This compensation form is not generally preferred as it tends to reward the A/E for a higher construction cost.

Other portions of the A/E contract should address:

- Responsibilities of the owner
- Provisions for modifying the contract, including termination
- Requirements for insurance coverage
- Responsibilities for errors and omissions
- Alignment of responsibilities
- Assignment of the contract
- Subconsultants
- Time of completion
- Suspension of work
- Settlement of disputes
- Governing law
- Options for additional services
- Penalties for late payments

In addition, contracts dealing with international projects will oftentimes address:

- Force majeure
- Additional tax implications
- Payment of compensation in a specified currency
- Advance payment of fees to the A/E

- Telegraphic transfer of funds
- Language and measure

Communication

One of the most important functions of the owner's project manager during the design of the commercial or industrial facility is to facilitate communications among the various parties involved in the project. These parties include, on the owner's side, the end user of the facility and the owner's corporate management. They also include the design professional and the construction contractor. A contractor may be involved during the initial design phases or not until the construction phase of the project. Other interested parties include outside interests such as the governmental regulatory organizations and the public in general.

Effective communications are needed during the design of the project in order to:

- Incorporate the proper features to the project
- Monitor and control the expected construction cost of the project
- Encourage the public's understanding of a project if required
- Ensure that code requirements are met and that proper permits are obtained
- Monitor and control project schedules

Key lines and elements of communication are shown in Fig. 15.2.

The owner's project manager can have a substantial amount of exposure to his or her own corporate management during a significant construction project. During the project he or she will be responsible for seeing that the owner's bottom line interests are protected. The project manager will also be responsible for seeing that corporate management's goals and objectives for the project are related accurately to the design professionals. There may be other lines of communication from the owner's corporate management to the design professional; however, the owner's project manager must be sure that the design professional understands and appreciates the corporate objectives regardless of whether or not the design professional has had any communications directly with the corporate management.

In addition, it is the owner's project manager's responsibility to see that his or her management is kept apprised of key elements of the design progress. The project manager should periodically report on the project's status, schedule, budget, design features, and critical problem areas. Again, he or she can make use of communication

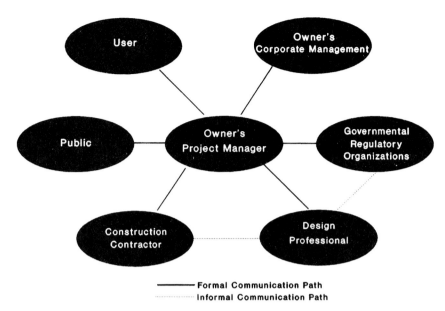

Formal Communication Path
Informal Communication Path

Figure 15.2 Project communications.

directly from the design professional to corporate management, but in any case it is his or her responsibility to be sure that corporate management is fully aware of what's going on with the project.

It is also crucial that the people who will be responsible for using and maintaining the facility be allowed to input their needs and requirements to the project.

For the facility to be successful it must meet the needs of the using group. Their involvement should be solicited early in the design process and continued throughout each stage of review. However, it is important that the user be made aware that the user's primary input will be during the initial design stages of the project. Changes in user requirements during final design can result in costly design changes. Moreover, user changes generated during construction can result in even costlier construction changes.

It is also important that the owner's project manager evaluate user requirements from a value engineering standpoint, that is, determine if a user requirement can be met by some more economical facility feature than that suggested by the user. This goes hand in hand with controlling the cost of the project by keeping user requirements realistic.

Once he or she has received the input from corporate management and the user group, the owner's project manager must communicate these objectives to the design professional. The project manager can motivate the design professional to perform better by communicating the vision of the project. He or she can illustrate the importance of the project to his or her own organization and to the design professional's organization. Most design professionals react positively to the prospect of a long-term professional relationship with the owner.

During the design process the owner's project manager must be kept informed of the design progress. A formal system needs to be in place to monitor the schedule and key milestones of the design. Depending on the size of the project an appropriate reporting procedure should be established and followed. Also the engineer's estimated construction cost for the project should be monitored at various stages of the design. This will help control the project cost and maintain the project within budget.

One key aspect of cost control is requiring all communications from the user to come through the owner's project manager before going to the designer. A designer cannot be allowed to make any design changes at the user's request without the project manager's approval.

Likewise, the owner's project manager motivates the construction contractor to give the best possible performance. It is crucial that very formal lines of communication, recording methods, and compliance procedures be established throughout the construction phase of the project. The reader should refer to Chap. 16, "Managing a Facilities Construction Project," for a detailed discussion of this phase of the project.

Depending on the type of project, various forms of communication with governmental regulatory organizations are required. At a very minimum, typically a construction permit is required. Sometimes environmental permits are required. International projects will require some understanding of export/import regulations as well as specific design and local code requirements. In any case, local officials appreciate being informed about the project early in the design so that they can be prepared to perform the required design review. Their input should also be solicited early as to specific local requirements as they apply to the specific project.

Those projects in the public sector have always been the subject of public scrutiny. More and more private sector projects are also coming into the public eye. Environmental concerns, economic effects, and governmental financing of private projects have all increased the public's awareness of construction projects. The owner's project manager should draw up a strategy for dealing with public communications. Communications should always be accurate and timely. The public may not need to know every detail of a project, but certainly has a right to know and understand certain aspects of the project and how they will affect the public interest.

In order to meet all the communication objectives for the project, several tools can be used. These include:

- The definitized project description as previously discussed
- Design progress review meetings
- Monthly progress reports
- Letters
- Memoranda
- Telephone call documentation
- Facsimile (fax) transmission

The design progress review meetings should be held to inform the client of how the A/E is responding to the client's needs and objectives set forth in the project description. The documents submitted show how the A/E has interpreted the owner's requirements and how he or she has incorporated them into the design. At these meetings the designer can also submit alternative solutions and value engineering options to the owner. These meetings should be documented by design progress review meeting notes issued by the designer shortly after the meeting. The notes should reflect comments from the owner and decisions made during the meeting by the owner-A/E team.

Depending on the size and duration of the design project, monthly progress reports can be the key tool for owner's project manager and upper management to monitor the designer's progressing according to schedule. This progress report should include information concerning:

- The original schedule
- The current status
- Summary of deviations from the original schedule, including recommended corrections if the design process should be behind schedule

Standard forms of business communication should be used to convey information to the designer or from the designer to the owner in a timely manner. This includes letters and memoranda issued by the party responsible for supplying the needed information.

Telephone call communication is certainly an effective way of keeping all parties informed. However, the designer should be tasked with documenting all important points in the telephone conversation in the form of a telephone call memo and issuing this memo to all the concerned parties. A copy of these memos as well as all other documented communications should be kept in the project file by the designer, the owner, and, if appropriate, the contractor.

Transmitting information from letters and memos can be expedited by using the fax machine. However this information should then be transmitted in hard-copy form through the mail to serve as an official record.

The Design Process

The design process logically flows from beginning to end with an opportunity for owner review at certain logical stages. Most often these stages are:

- Schematic design
- Design development
- The final design stage

During the schematic design phase the basic criteria for the project are established. These will include the user requirements, regulatory codes and regulatory compliance criteria, and standard industry practice requirements. A basic outline of the facility is commonly defined along with a number of alternative layouts for the owner's review. A recommended as well as alternative site layout is also commonly provided during this phase. Key design elements of the facility are described, and a preliminary estimate of probable construction costs is determined.

Normally the schematic design is submitted and presented to the owner for review. During this stage it is important that the owner and the using group be involved *in detail* in reviewing the basic criteria and the features of the facility. The schematic design documents should be back-checked against the original program definition and users' functional requirements.

Once the schematic design has been accepted by the owner and choices have been made between any alternative layouts, the A/E is authorized to proceed with design development. During this stage the facility site plan, layout, and design features are further refined and detailed to incorporate all the criteria and owner comments into more graphic form and with greater definition. The estimate of probable construction cost is updated and refined to reflect the improved detail. Once again this package is presented to the owner for review and approval.

After acceptance by the owner of the design development submittal, the A/E is authorized to move into final design. During this phase detailed technical specifications and drawings are prepared to fully define the scope and finishes of the project. This detail is reflected in the construction documents described in another section of this chapter.

After the construction documents have been coordinated by the A/E, and presented and reviewed with the owner, they are bound and issued for bids to interested contractors. More detailed definition of the design process is given in *The Architect's Handbook of Professional Practice.*[1]

Contract Document Organization

The key product of the design phase of the project is the issuance of the construction contract documents. These documents include the specifications and design drawings for construction. The project manual, which is often called the *specifications,* may include the construction contract that is the agreement between the owner and the contractor, general conditions, and special conditions, as well as the technical requirements for the project. The technical specifications are typically divided into numbered divisions. The divisions normally cover requirements for each discipline, such as:

- Civil
- Structural
- Architectural
- Special construction
- Mechanical
- Electrical

The project manual, or specifications, support the contract design drawings. A well-organized set of construction drawings will include:

- Cover sheet
- Drawing index
- Vicinity map showing the project site
- Civil drawings
- Architectural drawings
- Structural drawings
- Mechanical drawings
- Fire protection drawings
- Electrical drawings

The civil drawings will normally include:

- Site layout plan

- Site utility drawing
- Grading plan
- Civil details

The architectural drawings will generally include:

- Architectural floor plans
- Furniture and equipment layouts
- Architectural detail and millwork details

The structural drawings will indicate:

- Foundation design
- Structural steel design
- Structural concrete frame design
- Structural wood frame design

Structural details will be included which show dimensions, materials, and the design intent for structural complements.

The mechanical drawings will define:

- Heating, ventilation, and air-conditioning (HVAC) systems layouts, including ductwork
- Plumbing design, including system riser diagrams
- Routing and sizing of piping systems
- Detailed drawings will be provided which further indicate the location and size of equipment, ductwork materials, and equipment clearances

Depending upon the project, fire protection design will be included in the mechanical section or may be provided as an independent section of drawings and specifications.

Electrical design drawings will include:

- Lighting layouts
- Power wiring layouts
- Fire detection layouts
- Communication systems, including telephone
- Data and paging systems
- One-line diagrams

- System diagrams
- Panelboard schedules
- Electrical details

A separate electrical site plan may or may not be provided.

Designing In Flexibility

Flexibility in design can cost more than planning for current needs only. Accommodations for future uses, changes, or additions generally add 5 to 10 percent to the construction cost of a new facility. These accommodations can include the following:

- Additional structural capacity in footings and building frame for vertical expansion
- Additional structural capacity and additional land for horizontal expansion
- Additional constructed space for future utility equipment
- Main utility distribution elements sized for additional capacity
- Modular or grid-type utility systems, in place or master-planned
- Interstitial utility spaces between floors
- Vertical utility corridors

The degree to which we can plan for future needs depends on the accuracy of projected needs.

Remodels

The proper application of the planning criteria can be better achieved with new construction than by remodeling an existing structure. Remodel work is dependent on the nature of the existing facility. An existing building more than ten years old, or any building not originally designed for a desired function, will present many obstacles to achieving an optimal finished product. The available clear space between floors may pose a physical constraint. Major changes to or complete replacement of HVAC equipment and the distribution system may be necessary.

Existing buildings are difficult to adapt to special system needs. Special systems include fire protection, laboratory waste, and special exhaust. Designing for overall flexibility to react to future modifications in buildings is very difficult.

The design team, with the end users, should survey any existing building considered for use for a new function. Requirements for utility distribution, utility flexibility, modular design, special systems, environmental control, and separation of functions should be considered during the study. This study should be done before making the final decision to remodel a facility.

Construction-Phase Design Services

The designer's role in the project typically continues through the construction phase. The following activities need to be completed for the project. Depending on staffing capabilities of each organization these activities can be done by the owner's staff, the designer's staff, or the contractor's staff. The activities listed below are those which directly relate construction to the design process. Other construction activities are discussed in Chap. 16.

The construction-phase activities that relate to a design can be segregated into two groups. One group are those activities done at the designer's office, and the second group of activities are those performed by members of the designer's staff in the field.

Construction-phase design services typically performed in the designer's office include:

- Answering requests for information from the contractor
- Issuing design changes requested by the owner
- Checking of compliance submittals to be in conformance with the construction documents

These activities directly relate to the original design and are most appropriately done by the original designers of the project.

Another set of activities during construction are those to be completed at the job site. A *resident construction representative* (RCR) should be assigned to be responsible for the completion of these tasks. The RCR may be a member of the owner's staff or a member of the designer's staff. His or her responsibilities will include the following: The RCR provides on-site management contract administration. The RCR establishes and implements coordination and communication procedures.

Construction administration procedures

The RCR establishes and implements procedures for submittals, material samples, change orders, and payment requests, as well as other procedures and maintains daily job reports, logs, files, and other necessary documentation.

As the client's representative at the construction site, the RCR is the party through which change orders, payment requests, requests for information, submittals, and other information will be processed and communicated from the contractor to the client and back to the contractor.

Construction site meetings

Periodically, the RCR conducts construction site meetings with each contractor, and overall coordination meetings with the contractor and the client. The RCR records, transcribes, and distributes minutes to all attendees, the client, and all appropriate parties.

Coordination of other independent consultants

Technical inspection and testing provided by the A/E team or other third parties is coordinated by the RCR. All technical inspection reports are in a format approved by the RCR and are received by the RCR on the day following the inspection or testing.

Progress reports

The RCR reviews the payment requests submitted by the contractor for accuracy and determines whether the amount requested generally reflects the progress of the contractor's work. The RCR makes appropriate adjustments to each payment application and then prepares and forwards to the client a certificate for payment.

Nonconforming work

The RCR, in conjunction with the A/E team, makes recommendations for corrective action on nonconforming work. The RCR makes recommendations to the owner in instances where the RCR observes work that does not conform to the contract documents.

Punch list

In conjunction with the A/E team, the RCR prepares a list of incomplete or defective work, a *punch list,* prior to occupancy or operational completion.

When incomplete work or defective work has been remedied, the RCR advises the owner of project completeness and will issue a certificate of substantial completion. In the event of remaining punch list items, the RCR issues a certificate of substantial completion with exceptions noted.

Claims

The RCR reviews the contents of any claim submitted, assembles information concerning the claim, reviews the alleged cause of the claim, and makes recommendations to the client with respect to the claim. The RCR further negotiates the claim with the contractor on behalf of the owner. The RCR makes a final recommendation to the owner concerning settlement or other appropriate action.

Operation and maintenance materials

The RCR receives all written materials pertaining to the project, including, without limitations, operations and maintenance manuals, warranties, and guarantees for all materials and equipment installed in the project.

Quality review

The RCR establishes and implements a program to monitor the quality of the construction to assist in guarding the owner against defects and deficiency in the work of the contractor.

Safety

The RCR reviews safety programs of the contractor and makes appropriate recommendations.

Monitoring schedule compliance

The RCR monitors and expedites the progress of the work as provided in the general conditions of the construction contract. The RCR reviews the progress of construction of each separate construction contract on a periodic basis and evaluates the percentage complete of each construction activity as indicated in the detailed construction schedule, and reviews these percentages with the contractor.

Reports

The RCR prepares and distributes reports concerning change orders, schedules, project cost summaries, and progress payments.

As-built documents

The RCR coordinates and expedites submittals of information by the contractor for as-built preparation and coordinates and expedites the transmittal of record documents to the client.

Organization and indexing of operation and maintenance material

Prior to the final completion of the project, the RCR compiles material, such as manufacturer's operations and maintenance manuals, warranties, and guarantees, in an organized manner, prepares an appropriate index for the owner's use, and binds such documents.

Occupancy permits

The RCR assists the owner in obtaining necessary occupancy permits. This task includes accompanying government officials during the inspections of the project, preparing and submitting proper documentation to the appropriate approving agencies, coordinating final testing, and other activities.

Final payment

The RCR, at the conclusion of corrective action of all punch list items, makes a final inspection of the project along with the A/E team and owner, and prepares a report on the final inspection for the owner and makes recommendations to the owner as to the final payment to the contractor.

Closeout reports

At the conclusion of the project, the RCR prepares final project accounting and closeout reports for all the above indicated report systems.

Design Project Organization Structure

The organization or team assembled to provide for the design of the project should have a structure which best allows it to complete the objectives of the project. A typical structure is shown in Fig. 15.3 and discussed below.

The *owner's upper management* will be represented by those persons having responsibility and authority to initiate all phases of the project. This would include the authority to enter into contracts with others to provide for design and construction of the project. It would also include the ability to obtain or authorize funding for the project and to authorize notice to proceed with design or construction activities. The owner's upper management may execute their responsibilities directly or through their assigned project manager.

The *owner's project manager* is the person assigned responsibility and authority for the actual execution of the project. The project

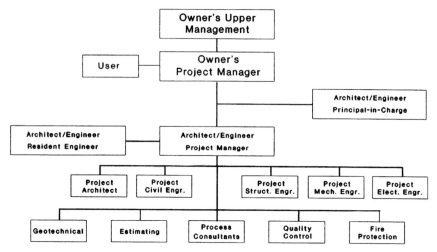

Figure 15.3 Design project organization.

manager is responsible for incorporating goals and objectives as communicated to him or her by the upper management or by the user of the final facility. He or she has responsibility to monitor project activities, schedule, progress, and cost.

The *user* is the person or group within the owner's organization that will utilize the facility upon completion of the construction project. The user is expert on the process or activities which occur within the facility and communicates user requirements to the owner's project manager. The user is not directly responsible for design of the project but is interested in seeing that the design incorporates and facilitates all user objectives.

The *architect/engineer principal in charge* is the senior member of the A/E's management who will be responsible for the firm's performance on the project. She or he is typically the person who signs the contract on behalf of the A/E. She or he has authority and responsibility to resolve any problems arising during the project which are beyond the control of the project manager. She or he is able to involve other groups or organizations within the company outside the project design team to assist in the project.

The *architect/engineer project manager* is responsible for the day-to-day performance of the design activities for the completion of the project. He or she has authority over the senior-level engineers and architects assigned to the project as well as the support groups assisting during the project.

The *project architect, civil engineer, structural engineer, mechanical engineer, and electrical engineer* are senior-level design professionals responsible for the technical quality and constructability of their specific discipline design. The discipline leaders should be able to accept and delegate responsibility for completion of design activities for the project. The A/E firm should provide them with the resources in terms of staff and equipment necessary for them to complete their discipline design.

Geotechnical, estimating, process, quality control, and fire protection expertise may be required to different degrees during the design of the project. These groups shall provide support necessary for the successful completion of the project.

Project Scheduling

Figure 15.4 illustrates graphically the traditional design/bid/build approach to an industrial facility project. The schedule shown is divided into two major phases:

- The design phase
- The construction phase

With the design/bid/build approach the design activities are completed simultaneously. These include:

- Schematic design

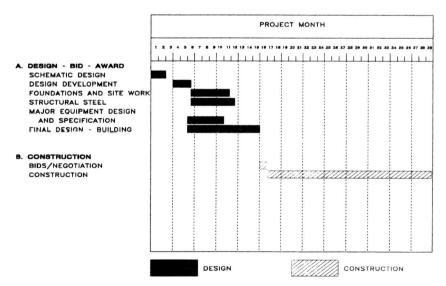

Figure 15.4 Industrial facility preliminary project schedule: Design/bid/build approach.

- Design development
- Foundations and site work
- Structural steel design
- Major equipment design
- Specifications
- Final building design

After design is complete, the construction phase is started. This phase includes bidding negotiation and award of the construction contract and then all of the construction activities required to build the project.

As the schedule shows, with this approach, no construction activities take place until after the completion of all design activities. In this way the project is fully designed and that design is accepted by the owner. For the project illustrated the total project duration is 39 months.

The schedule shown in Fig. 15.5 is for the same industrial facility project using a multiple-prime-contract approach. The schedule is based on the award of multiple prime contracts for:

- Foundations and site work
- Structural steel

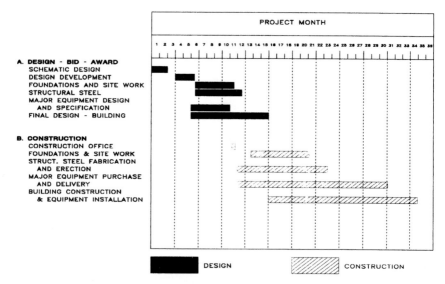

Figure 15.5 Industrial facility preliminary project schedule: Multiple prime contracts.

- Major equipment
- Building construction and equipment insulation

The award of multiple prime contracts will allow construction to begin before the completion of the design for the facility in order to expedite occupation of the facility. By paralleling design and construction activities for a project such as this, it is estimated that the schedule can be shortened by five months for the total design and construction program.

In similar fashion a major project can be divided into phases which correspond to physical areas of the project. This may perhaps allow the owner to occupy some areas of the facility earlier than would be possible if occupancy had to be delayed until the total project was complete. This may be especially attractive for major expansions on existing facilities where other structures are already in place to support the industrial facility.

Conclusion

To assure successful design of commercial/industrial buildings and structures, many finite elements must be carefully attended to by the facility manager. These elements include:

- Definition of the project
- Selection of the A/E if required
- A/E contract
- Communication
- The design process
- Contract document organization
- The construction-phase design services
- The design organization structure
- The project schedule

Each of these factors is uniquely important to the success of the overall major facility expansion or new facility construction. It is understood that these elements are merely a portion of the total responsibilities of a facilities manager. However, by understanding these elements and paying close attention to them during the process, the facility manager can improve the functionality of the new facility and the efficiency with which it meets the user and the owner's needs.

References

1. American Institute of Architects, *The Architect's Handbook of Professional Practice,* AIA, Washington, D.C., 1969.

Further Reading

American Institute of Architects: "Project Checklist," AIA, Washington, D.C., 1973.

American Institute of Architects: Publication B141, "Owner-Architect Agreement," AIA, Washington, D.C. 1987.

Burns & McDonnell: "Construction and Procurement Documents," *Benchmark,* March 1991.

Burns & McDonnell: "Working with an Architecture/Engineering Firm," *Benchmark,* August 1991.

Clark, Ken, P.E., and John Rehak, P.E.: Presentation to Tradeline. "Flexible M/E Design for Multi-Function Buildings," San Francisco, CA, June 13–14, 1990.

Consulting Engineers Council of Missouri: *A Guide to the Procurement of Architectural and Engineering Services,* Consulting Engineers Council of Missouri, Jefferson City, MO, 1979.

Consulting Engineers Council of Missouri: *A Guide to Proper Selection Procedures,* Consulting Engineer Council of Missouri, Jefferson City, MO.

Hilton, Joseph R., Jr.: *Design Engineers Project Management,* Technomic Publishing, Lancaster, PA, 1985.

National Society of Professional Engineers: Publication 1910-1, "Standard Form of Agreement between Owner and Engineer for Professional Services," NSPE, Professional Engineers in Private Practice, Alexandria, VA, 1984.

National Society of Professional Engineers—Professional Engineers in Private Practice Division: *Professional Selection of Professional Engineers,* NSPE, Professional Engineers in Private Practice, Washington, D.C.

Preston, Ralph: "Working with an Architect to Build a Maintenance Facility," *Airline Executive,* July 1987, p. 30.

16

Managing a Facilities Construction Project

Darrell M. Hosler
Executive Vice President
Burns & McDonnell
Engineers-Architects
Kansas City, Missouri

Douglas E. Grimer
Vice President
Burns & McDonnell
Engineers-Architects
Kansas City, Missouri

Construction of facilities is a process that often produces "failures" in the form of cost overruns, missed schedules, or other unrealized expectations. Although many owners have well-qualified facility staffs, they often do not deal with new construction continuously. As a result, each new project can become an experiment—trying something different to avoid problems encountered on the previous project.

Managing a facilities construction project should begin with an understanding that construction projects are not as simple as they seem. They involve many diverse activities, including planning, design, administration, permitting, compliance with regulatory requirements, and, finally, the construction phase.

Planning a project is the first activity a facilities manager faces. Many basic decisions must be made, such as:

- Selecting the construction process from several options
- Establishing a preliminary project schedule for design and construction
- Developing a project budget
- Identifying applicable permits, regulatory agencies, and codes
- Obtaining any requirements of the property insurer
- Obtaining financing or internal budget approvals

For small projects within the owner's in-house design and construction capabilities, a "do-it-yourself" project may be appropriate.

However, many owners elect to retain outside organizations, such as an *architect/engineer* (A/E), construction contractors, or a *construction manager* (CM), to assist in the project.

A large project that is beyond in-house capabilities of the facilities staff requires an early decision on the construction delivery process. The typical options are:

- A traditional design/bid/build approach with either a single general construction contractor or multiple prime contracts
- Design/build
- Construction management

Another decision is the selection of the process to price the construction of the project. Alternatives are:

- Competitive bidding among several construction contractors
- Negotiating with a single contractor
- Using a cost-plus approach, possibly with a guaranteed maximum price

Among the many administrative activities that must be performed by either the owner's facilities staff or the A/E or CM are:

- Shop drawing review
- Construction observation
- Approval of progress payments to the contractor
- Administration of construction change orders
- Preparation of a punch list (list of incomplete work prepared near the end of the project)
- Correction of defective work
- Certificate of substantial completion
- Final payment to the contractor
- Coordinating warranty work/repairs

These topics will be discussed in the following sections.

Responsibilities of the Owner's Facilities Staff

This section deals with the activities and responsibilities of the facilities staff for a major U.S. construction program requiring the services of an outside A/E and the services of a construction contractor. However, most of the activities required for a major project also are required for a smaller project, although the magnitude of effort is reduced.

Planning

In a major construction project, many of the up-front planning activities are accomplished in conjunction with the selected A/E. The A/E's involvement in these decisions depends upon the capability of the facilities staff and the degree of involvement delegated to the A/E. Key elements of the planning phase include:

- *Naming the owner's representative:* A single point of contact should interface with the construction contractor and the A/E. This will facilitate timely decisions and effective communication.

- *Developing the project schedule:* A realistic project design and construction schedule is essential to a project's successful completion. Unrealistic schedules lead to misunderstandings, a contentious relationship among the parties, and, in many cases, increased project costs. Developing a realistic schedule requires the input of persons knowledgeable in the design and construction of projects like the one being planned.

- *Developing the construction budget:* The pitfall to avoid here is developing a budget that is significantly less than that required to build the project. This can cause substantial cost overruns or mandate scope reductions in the middle of the project. The input of the A/E or others experienced in the construction of similar facilities is crucial to developing realistic budget figures.

- *Financing:* The owner's facilities manager may be involved in securing financing or budget approvals for the project. Accurate cost estimates are critical whether the project is financed internally or from external sources.

- *Choosing the construction delivery process:* An early decision should be made regarding the delivery process used for construction activities. The typical options are design/bid/build, design/build, and design/construction management. Before this decision, it is difficult to plan other aspects of the project.

- *Permitting:* The facilities manager must assess the regulations applicable to the project, and the steps to be taken to ensure compliance. These could involve obtaining zoning approvals, adherence to local building codes, and adherence to environmental regulations, including obtaining permits such as air quality permits or discharge permits. Depending on the project scope, expert assistance from qualified specialty consultants may be required.

Construction administration

Administrative activities required for a construction program include:

- *Contract preparation:* A decision must be made on the form of contract to be used. In-house documents may be available. If not, the American Institute of Architects (AIA), the National Society of Professional Engineers (NSPE), and the Associated General Contractors have developed time-tested standard documents which are reprinted at the end of this chapter that are available (see Refs. 1–8). In any case, the administrative portions of such contract forms must be tailored to serve specific requirements of the project.

- *Shop drawing review:* For a major construction project, hundreds, if not thousands, of vendor shop drawings must be reviewed and approved. This can be handled by the owner's facilities staff or by its A/E.

- *Construction observation:* For major construction projects, it is to the owner's advantage to have full-time, on-site construction observation to monitor progress and quality of work completed by the contractor. The size of staff will vary depending upon the complexity of the project. The staff can be provided in-house, or by the A/E or CM. Some owners perform this function, but augment their staffs with help from the A/E or CM.

- *Progress payments:* Usually, the A/E or CM reviews the status of the contractor's work and recommends monthly progress payments for payment by the owner. For lump-sum work, expertise is needed to avoid paying the contractor at a rate that exceeds the value of completed construction activities.

- *Change order preparation:* Change orders to add or deduct cost from the construction contract should be prepared on forms prescribed by the construction contract. Change orders are signed by the owner, but can be prepared by the A/E or a CM.

- *Contract closeout:* As the construction program is nearing completion, these activities are performed:

  - Start-up, testing, and commissioning of equipment
  - Preparation of a final punch list
  - Correction of defective work by the contractor
  - Preparation of the certificate of substantial completion (typically required by the building code authorities having jurisdiction)
  - Final payment

- *Warranty work:* Generally, projects are warranted for one year following the date of substantial completion (established by the certificate of substantial completion discussed above). Defective materials or workmanship discovered in this one-year period is to be corrected by the contractor at no additional expense to the owner. As defective work is discovered, a formal written notice should be given to the contractor and the contractor's commitment to make the necessary repairs obtained.

Construction Delivery Process

There are several construction delivery processes available, including:

- Design/bid/build
- Design/build (turnkey)
- Design/construction management

Each has perceived advantages and disadvantages. Following is an overview of each of these alternatives. Figure 16.1 summarizes the comparison of alternatives.

Design/bid/build

In this traditional approach, the owner's A/E prepares construction drawings and specifications. The A/E then issues drawings and specifications to construction contractors for bidding. The A/E reviews the bids and recommends the owner award the contract to the contractor with the lowest and best bid.

| | Design/Bid/Build (Traditional) | | Design/Build (Turnkey) | Design/Construction Mgmt. | |
	Single General Construction Contract	Multiple Construction Contracts		C.M. contracts with subs	Owner contracts with subs
Prime contracts with owner	2 (A/E + 1 const. contract)	A/E + several const. contracts	1	2 (A/E + CM)	A/E, CM + several const. contracts
Administrative responsibilities of owner	High	Highest	Lowest	Low	Low
Opportunities for owner input into design	High	Highest	Lowest	High	High
Schedule duration	Longest	Medium	Shortest	Medium	Medium

Figure 16.1 Comparison of construction delivery options.

The A/E's responsibilities for design and preparation of bid documents are covered under the terms of the A/E's contract with the owner. (The A/E's responsibilities in this process are discussed in Chap. 15.)

The advantages of the design/bid/build process compared to a design/build approach generally are perceived to be:

- Owners have greater control and input for design, and, therefore, are better able to incorporate their needs into the project.

- Owners have more control over the selection of materials and equipment and in approving construction results; therefore, the project's quality can be controlled.

- Since the detailed design is completed before bidding of construction, the scope of construction work is well defined, enhancing competition.

Construction of a design/bid/build project can be performed by a single general (prime) contractor or by multiple prime contracts. These options are discussed below.

Single general (prime) contractor.   The owner signs a single contract with a general construction contractor. Awarding the construction contract can be accomplished through lump-sum competitive bidding, or the owner may choose to negotiate a lump-sum contract with a preferred contractor.

A less common approach is to negotiate a guaranteed maximum price with the general contractor; the contractor is paid its actual cost plus an agreed-upon fee for profit. Profit can be established either as a percentage of construction costs or a fixed amount.

In general, a single prime contract is used on smaller projects, of up to $20 million. Larger projects are successfully bid as single-general, lump-sum contracts; however, the impact upon project schedule must be considered (refer to Fig. 16.2).

The owner's or A/E's administrative effort for a single prime contract is less than for multiple-contract approaches. However, the project duration is somewhat longer, so the length of time for which administration is required is also longer.

Multiple prime contracts.   In this process, the A/E prepares phased specialty construction and equipment procurement contracts for bidding. This process is often used on large, complex projects costing more than $20 million. Projects costing in excess of $1 billion have been successfully constructed using this approach. In general, these contracts are awarded through lump-sum competitive bidding and are rarely negotiated.

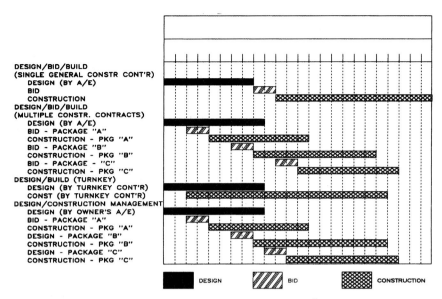

Figure 16.2 Typical phases for construction delivery approaches.

Using this process shortens the time required for construction, because it allows early phases of construction to begin prior to completion of design activities for later phases (refer to Fig. 16.2 for this comparison).

The multiple-contract approach facilitates the use of well-qualified local or regional contractors that are most familiar with the labor and working conditions applicable to the area. Also, such smaller specialty contractors often are quite efficient. These factors can achieve economies for the owner.

The number of contracts to be administered can vary from fewer than 10 to 200 construction contracts, and as many as 100 equipment contracts on very large projects. Administrative effort required for the multiple-prime-contract approach is greater than that required for a single general contract. However, many owners feel that eliminating the overhead markups charged by a single general contractor on subcontracts, and the time saved during construction, compensate for the additional staff that the owner or its A/E must employ.

Design/build (turnkey)

Using this approach, the owner contracts with a single design/build contractor. Design may be performed by the design/build contractor's in-house organization, or the design/build contractor may subcontract with an A/E firm to perform the design. The term *turnkey* implies that by procuring both design and construction from a single contractor, the contractor performs a complete package and at the end of the job, simply turns over the keys to the owner. In some regions, *turnkey* also implies that the firm can provide or arrange project financing.

To form the basis of the contract, the owner prepares a description of the project scope and requirements. A relatively complete and detailed statement of project scope will result in more competitive bids from prospective design/build contractors and a higher likelihood that the owner will receive the features desired. Often, owners employ an A/E to prepare design/build specifications and to administer the design/build contract.

This approach is applicable to virtually all sizes of projects. However, the more complex the project, the more important it becomes to have a very clear definition of the requirements prior to contracting with the design/build contractor.

The process most frequently used to select a design/build contractor is to contact experienced and qualified firms to determine their interest. A request for proposal should require the prospective design/build contractors to prepare a conceptual design in sufficient detail to allow pricing the project. Because of the variations in designs submitted by the bidders, a careful analysis is required to select the lowest and best bid.

Because of the high cost of preparing such turnkey proposals, some well-qualified contractors may decline to submit a bid if more than three to five design/build contractors are invited to bid. Three to five bidders are generally sufficient to obtain a reasonable degree of competition. Alternatively, the contract may be negotiated directly with a preferred contractor.

The contractor's pricing may be based upon:

- Lump-sum amount
- Cost with a guaranteed maximum price
- Cost with a target price and with any savings or overruns shared between the contractor and owner

An advantage to the design/build approach is the speed with which a project can be completed. Normally, design/build construction schedules have shorter durations than other delivery processes used for construction. Another advantage is that the owner signs a single contract for both design and construction of the project. This reduces the administration required by the owner.

A disadvantage to this process is less opportunity for owner input into design. The owner contracts with the design/build contractor based on a conceptual design and outline specifications. Once the contract is executed, additional input from the owner as to the design details usually results in cost increases due to changes in scope.

Another disadvantage is that the quality of design/build projects depends greatly upon the ability and commitment of the contractor. Therefore, it is important that the qualifications and experience of the prospective design/build contractors be thoroughly investigated before they are invited to submit proposals. Depending upon the geographic location of the project, the number of well-qualified design/build contractors may be limited.

Construction management (CM)

In the *construction management* (CM) approach, a construction management firm is retained by the owner to administer construction contracts. The CM may actually subcontract directly with construction contractors (in which case the process resembles the process using a single general prime contractor), or the CM may administer and coordinate contracts that are fashioned directly with the owner (in which case the process resembles a multiple-prime-contract approach).

With CM, detailed design is performed by the owner's A/E. The A/E can be selected and retained by the owner, or the A/E may be selected jointly by the owner and the CM and then retained by the owner. Rarely is the A/E employed directly by the construction manager. It is generally to the owner's advantage to have separate contracts with the A/E and with the CM to avoid subordinating design to construction activities.

There is a wide variation of arrangements that are called "construction management." At one end of the spectrum, the CM, for a fee, represents the owner and effectively serves as the owner's on-site staff in administering construction activities contracted by the owner. This arrangement is essentially the same as the design/bid/build process described above, except that instead of construction observation being performed by the owner's staff or A/E, it is performed by the CM.

At the other end of the spectrum, the CM assumes a part of the risk in completing the construction within the prescribed price. Here, the CM may actually subcontract construction activities or have authority to direct construction contractors' activities. Ultimately, the CM's role approaches that of a single general (prime) contractor.

The CM approach is most often used on very large, complex projects requiring a sophisticated and experienced construction manager, perhaps beyond the capability of the preferred A/E. The process requires that the A/E and CM work together as a team, and it is important that the contract with each reflect this teamwork approach.

A basic difference between the CM approach and the multiple-prime-contract approach is that the CM administers the entire construction program on behalf of the owner, including the construction contracts. In the design/bid/build approach with multiple prime contracts, the A/E or the owner administers the construction contracts. The approach best suited for a particular project depends upon the ability of the owner's facilities staff or A/E to manage the program, and the level of control the owner wishes to exercise.

The typical fee arrangement for the CM depends upon who signs the construction contracts. One approach is to pay the CM a fixed, lump-sum fee. If the owner desires a guaranteed maximum cost of construction from the CM, then the A/E would prepare a conceptual design, on which the CM would base its guaranteed maximum cost. In some cases, the guaranteed maximum cost is bid among several CMs. In others it is negotiated with a preferred CM. These agreements may also include a shared savings if the project is completed for less than a targeted amount. In general, the CM arrangement has been most successful when the CM is paid a fee for its services and then administers, for the owner, the competitive bidding of multiple construction contracts.

There is much diversity among CM firms. Some are pure CM firms that specialize in serving owners in a CM role. Others are primarily general contractors who also offer CM services. Also, many large A/E firms provide CM services, at least for arrangements where the CM coordinates construction packages contracted by the owner.

For major projects, the competition for the CM assignment may be limited. However, in many cases, the owners prefer to select a construction manager with whom they have worked before.

When the CM contracts for construction, the process can complete a project nearly as fast as the design/build process while providing owner control over design. When construction contracts are bid on a lump-sum basis to maximize competition and cost control, the schedule advantages are somewhat less.

Construction quality using CM is generally favorable and the owner's administrative efforts are minimized. By properly structuring the approach, owners can use the CM's and the A/E's experience to perform value engineering so that quality, function, and cost are optimized.

Selecting Construction Contractors

Selecting qualified contractors is critical. Contractors must be experienced, and capable of securing a performance bond (if required).

They also should have a record of successful projects of a similar size and complexity to the one contemplated, and they must possess adequate financial resources. Bidders should be prequalified and selectively invited to submit proposals. Prospective contractors may be prequalified by requiring a statement of qualifications. AIA Document A305[7] is an example of a contractor's qualification statement form.

International Projects

For international projects, it is important to use the services of local architects, engineers, and contractors to assure compliance with local customs, codes, and regulatory requirements. The same construction delivery processes described above can be used for completion of international projects, depending on the local resources available. It is important to understand the procedures normally followed in the host country.

Owners wishing to develop projects overseas should retain the services of firms experienced in the local area. Design may by performed by a U.S.-based A/E that uses a local A/E to review or produce a portion of the design. Alternatively, the owner may retain the services of the local A/E directly. Many major U.S. A/E firms maintain affiliate relationships with A/E firms in other countries.

A similar situation applies to contracting for construction of the project. Many U.S.-based owners prefer to retain the services of a U.S. CM firm and use the CM approach for construction. Others have successfully used the design/build approach.

Conclusion

The construction of major facilities may appear straightforward, perhaps even simple. This is not the case. The processes involved seem low-tech compared to the sophisticated manufacturing, computerization, and other advanced technologies that owners have mastered. Owners and their facilities staffs may be tempted to underestimate the risks and complexities of major construction projects.

Construction differs from manufacturing and other technologies in that it is not a repetitive process. You only contract one time for construction of a particular facility. It is not a continuous, evolving process that can be fine-tuned and debugged over time. Rather, each project is a high-stakes, roll-the-dice-once proposition. And the results, in the form of construction contracting, do not favor the inexperienced, even though well intentioned and intelligent.

Figure 16.3 Bidder's qualification statement.

DOCUMENT 00040 - BIDDER'S QUALIFICATIONS STATEMENT

Project name _____

Contract name _____

Contract no. _____

Submitted to:

_____

_____

_____

Submitted by

    Company _____

    Name _____

    Address _____

    Principal office _____

    Corporation, partnership, individual, joint venture,
       other _____

    Contractor license number, and state

    _____

Experience statement

1. Bidder has been engaged as a General Contractor in construction for _____ years and has performed work of the nature and magnitude of this Contract for _____ years. Bidder has been in business under its present name for _____ years.

2. Bidder now has the following bonded projects under contract: (On a separate sheet, list project name, owner, architect, amount of contract, surety, and estimated completion date.)

3. Bidder has completed the following (three) (_____) contracts consisting of work similar to that proposed by this Contract: (On a separate sheet, list project name, owner, architect, amount of contract, surety, and date of completion) and percentage of the cost of the Work performed with own forces.

BD00040.CM
111490

00040-1

Figure 16.3 (*Continued*) Bidder's qualification statement.

4. Has Bidder ever failed to complete any project? If so, state when, where, and why.

5. Bidder normally performs the following work with his own forces:

_____

_____

_____

_____

6. Construction experience of key individuals in the organization is as follows:

_____

_____

_____

_____

7. In the event the contract is awarded to Bidder, the required surety bonds will be furnished by the following surety company and name and address of agent:

_____

_____

_____

_____

Financial statement

Bidder possesses adequate financial resources as indicated by the following:

1. Assets and liabilities: Attach a financial statement, audited if available, including bidders latest balance sheet and income statements showing the following items:

   a. Current assets (cash, joint venture accounts, accounts receivable, notes receivable, accrued income, deposits, materials inventory, and prepaid expenses).

   b. Net fixed assets.

   c. Other assets.

   d. Current liabilities (accounts payable, notes payable, accrued expenses, provision for income taxes, advances, accrued salaries, and accrued payroll taxes).

00040-1

Figure 16.3   *(Continued)*   Bidder's qualification statement.

e. Other liabilities (capital, capital stock, authorized and outstanding shares per values, earned surplus, and retained earnings).

f. Name of firm preparing financial statement and date thereof:

If financial statement is not for identical organization named herein, explain relationship and financial responsibility of the organization furnished.

1. Current judgements: The following judgements are outstanding against bidder:

Judgment creditors	Where docketed and date	Amount
a. _____	_____	_____
b. _____	_____	_____

Bidder hereby represents and warrants that all statements set forth herein are true and correct.

Date _____ 19_____

(OFFICIAL SEAL)

Name of Organization:

_____

By _____

Title _____

(If Bidder is a partnership, the partnership must be signed, followed by the signature of at least one of the partners. If Bidder is a corporation, the corporate name must be signed, followed by the signature of a duly-authorized office and the corporate seal affixed).

END OF DOCUMENT 00040

BD00040.CM   00040-1
111490

CHANGE
ORDER

AIA DOCUMENT G701

OWNER ☐
ARCHITECT ☐
CONTRACTOR ☐
FIELD ☐
OTHER ☐

PROJECT: Third Bank of Westcreek
(name, address) 202 N. Main Street
                Westcreek MO 95209
TO CONTRACTOR: Buildum Construction Co.
(name, address) 1151 N. Blight Avenue
                 Eastcreek MO 95207

CHANGE ORDER NUMBER: 5

DATE: October 10, 1990

ARCHITECT'S PROJECT NO: 89-2805

CONTRACT DATE: January 5, 1990

CONTRACT FOR: Banking Facility

The Contract is changed as follows:

Change wood paneling from oak to walnut: Add $ 99.00

Change door hardware from chrome to brass: Add $1,201.00

                                                 $1,300.00

Not valid until signed by the Owner, Architect and Contractor.

The original (Contract Sum) (~~Guaranteed Maximum Price~~) was . . . .$ 1,234,567.89
Net change by previously authorized Change Orders . . . .$+   2,000.00
The (Contract Sum) (~~Guaranteed Maximum Price~~) prior to this Change Order was . . . .$ 1,236,567.89
The (Contract Sum) (Guaranteed Maximum Price) will be (increased) (~~decreased~~)
   (~~unchanged~~) by this Change Order in the amount of . . . .$+   1,300.00
The new (Contract Sum) (~~Guaranteed Maximum Price~~) including this Change Order will be . .$ 1,237,867.89

The Contract Time will be (~~increased~~)(~~decreased~~) (unchanged) by   (0) days.
The date of Substantial Completion as of the date of this Change Order therefore is December 1, 1990

NOTE: This summary does not reflect changes in the Contract Sum, Contract Time or Guaranteed Maximum Price which have been authorized by
         Construction Change Directive.

Joe R. Doe and Associates   Buildum Construction Co.   Separated Banks of Missouri
ARCHITECT   CONTRACTOR   OWNER
3207 S. Fox Road   1151 N. Blight Avenue   201 S. Elm
Address   Address   Address
Woodville MO 95208   Eastcreek MO 95209   St. Robert MO 68420

BY Jack Faron, AIA   BY Wilbur Woodpost   BY Joseph B. Day, President

DATE October 10, 1990   DATE October 12, 1990   DATE October 14, 1990

AIA DOCUMENT G701 • CHANGE ORDER • 1987 EDITION • AIA® • ©1987 • THE
AMERICAN INSTITUTE OF ARCHITECTS, 1735 NEW YORK AVE., N.W., WASHINGTON, D.C. 20006   **G701—1987**

Figure 16.4 Change order.

INSTRUCTION SHEET

FOR AIA DOCUMENT G701, CHANGE ORDER

A. GENERAL INFORMATION

1. Purpose

This document is intended for use in implementing changes in the Work agreed to by the Owner, Architect and Contractor. Execution of a completed G701 form indicates agreement upon all the terms of the change, including any changes in the Contract Sum (or Guaranteed Maximum Price) and Contract Time.

2. Related Documents

This document was prepared for use under the terms of AIA general conditions first published in 1987, including AIA Document A201, General Conditions of the Contract for Construction, and the general conditions contained in AIA Documents A107 and A117.

3. Use of Current Documents

Prior to using any AIA document, the user should consult the AIA, an AIA component chapter or a current AIA Documents Price List to determine the current edition of each document.

4. Limited License for Reproduction

AIA Document G701 is a copyrighted work and may not be reproduced or excerpted from in substantial part without the express written permission of the AIA. The G701 document is intended to be used as a consumable—that is, the original document purchased by the user is intended to be consumed in the course of being used. There is no implied permission to reproduce this document, nor does membership in The American Institute of Architects confer any further rights to reproduce them.

A limited license is hereby granted to retail purchasers to reproduce a maximum of ten copies of a completed or executed G701, but only for use in connection with a particular Project.

B. CHANGES FROM THE PREVIOUS EDITION

Unlike the previous edition, the 1987 edition of AIA Document G701 requires the signatures of the Owner, Architect *and Contractor* for validity. Changes to be made over the Contractor's objection (with disputed terms to be settled afterwards) should be effected through the use of AIA Document G714, Construction Change Directive.

C. COMPLETING THE G701 FORM

1. Description of Change in the Contract

Insert a detailed description of the change to be made in the Contract by this Change Order, including any Drawings, Specifications, documents or other supporting data to clarify the scope of the change.

2. Determination of Costs

Insert the following information in the blanks provided, and strike out the terms in parentheses that do not apply:

a) the original Contract Sum or Guaranteed Maximum Price;

b) the net change by previously authorized Change Orders (note that this does not include changes authorized by Construction Change Directive unless such a change was subsequently agreed to by the Contractor and recorded as a Change Order);

c) the Contract Sum or Guaranteed Maximum Price prior to this Change Order;

d) the amount of increase or decrease, if any, in the Contract Sum or Guaranteed Maximum Price, and

e) the new Contract Sum or Guaranteed Maximum Price as adjusted by this Change Order.

3. Change in Contract Time

Insert the following information in the blanks provided, and strike out the terms in parentheses that do not apply:

a) the amount in days of the increase or decrease, if any, in the Contract Time; and

b) the date of Substantial Completion, including any adjustment effected by this Change Order.

D. EXECUTION OF THE DOCUMENT

When the Owner, Architect and Contractor have reached agreement on the change to be made in the Contract, including any adjustments in the Contract Sum (or Guaranteed Maximum Price) and Contract Time, the G701 document should be executed in triplicate by the two parties and the Architect, each of whom retains an original.

1/88

Figure 16.4   *(Continued)*   Change order.

CERTIFICATE OF SUBSTANTIAL COMPLETION

AIA DOCUMENT G704

Distribution to:
OWNER ☒
ARCHITECT ☒
CONTRACTOR ☒
FIELD ☒
OTHER ☐

PROJECT: Third Bank of Westcreek
(name, address) 202 N. Main Street
Westcreek MO 95209

TO (Owner):

┌ Separated Banks of Missouri
  201 South Elm
  St. Robert MO 68420

└

DATE OF ISSUANCE:

PROJECT OR DESIGNATED PORTION SHALL INCLUDE: Entire facility of Third Bank of Westcreek

ARCHITECT: Joe R. Doe and Associates

ARCHITECT'S PROJECT NUMBER: 89-2805

CONTRACTOR: Buildum Construction Co.

CONTRACT FOR: $1,237,867.89

CONTRACT DATE: April 1, 1990

The Work performed under this Contract has been reviewed and found to be substantially complete. The Date of Substantial Completion of the Project or portion thereof designated above is hereby established as **November 29, 1990**

which is also the date of commencement of applicable warranties required by the Contract Documents, except as stated below.

DEFINITION OF DATE OF SUBSTANTIAL COMPLETION

The Date of Substantial Completion of the Work or designated portion thereof is the Date certified by the Architect when construction is sufficiently complete, in accordance with the Contract Documents, so the Owner can occupy or utilize the Work or designated portion thereof for the use for which it is intended, as expressed in the Contract Documents.

A list of items to be completed or corrected, prepared by the Contractor and verified and amended by the Architect, is attached hereto. The failure to include any items on such list does not alter the responsibility of the Contractor to complete all Work in accordance with the Contract Documents. The date of commencement of warranties for items on the attached list will be the date of final payment unless otherwise agreed to in writing.

John R. Doe and Associates   Jack Faron, AIA   Nov. 30, 1990
ARCHITECT   BY   DATE

The Contractor will complete or correct the Work on the list of items attached hereto within   days
from the above Date of Substantial Completion.

Buildum Construction Co.   Wilbur Woodpost   Dec. 1, 1990
CONTRACTOR   BY   DATE

The Owner accepts the Work or designated portion thereof as substantially complete and will assume full possession thereof at 8 a.m.   (time) on December 5, 1990   (date).

Separated Banks of Missouri   Joseph B. Day   Dec 3, 1990
OWNER   BY   DATE

The responsibilities of the Owner and the Contractor for security, maintenance, heat, utilities, damage to the Work and insurance shall be as follows:
(Note—Owner's and Contractor's legal and insurance counsel should determine and review insurance requirements and coverage; Contractor shall secure consent of surety company, if any.)
Owner will maintain and pay for utilities and insurance regarding facilities of first floor. Contractor shall be responsible for all other areas (second floor and basement) until final completion.

Figure 16.5 Certificate of substantial completion.

PROJECT APPLICATION AND PROJECT CERTIFICATE FOR PAYMENT
AIA Document G722 (Instructions on reverse side)

TO (Owner): Separated Banks of Missouri

PROJECT: Third Bank of Westcreek
202 N. Main Street
Westcreek MO 95209

APPLICATION NO: 3

PERIOD FROM: October 1
TO: October 31

ARCHITECT'S
PROJECT NO: 89-2805

PAGE ONE OF   PAGES

Distribution to:
☐ OWNER
☐ ARCHITECT
☐ CONSTRUCTION
   MANAGER
☐

ATTENTION: Joseph B. Day

CONSTRUCTION MANAGER: John Doe, Sr.

PROJECT APPLICATION FOR PAYMENT

The undersigned Construction Manager certifies that to the best of the Construction Manager's knowledge, information and belief Work covered by this Project Application for Payment has been completed in accordance with the Contract Documents, that all amounts have been paid by the Contractors for Work for which previous Project Certificates for Payment were issued and payments received from the Owner, and that current payment shown herein is now due.

Application is made for Payment, as shown below, in connection with the Project. AIA Document G723, Project Application Summary, is attached.

The present status of the account for all Contractors for this Project is as follows:

TOTAL CONTRACT SUMS (Item A Totals) .... $ 1,234,567.89

Total Net changes by Change Orders (Item B Totals) ... $ 3,300.00

TOTAL CONTRACT SUM TO DATE (Item C Totals) .... $ 1,237,867.89

TOTAL COMPLETED & STORED TO DATE .... $ 600,000.00
   (Item F Totals)

RETAINAGE (Item H Totals) .... $ 3,000.00

LESS PREVIOUS TOTAL PAYMENTS (Item I Totals) ... $ 450,000.00

CURRENT PAYMENT DUE (Item J Totals) .... $ 150,000.00

CONSTRUCTION MANAGER

By: _____   Date: _____

State of: _____   County of: _____
Subscribed and sworn to before me this _____ day of _____, 19__
Notary Public:
My Commission expires:

ARCHITECT'S PROJECT CERTIFICATE FOR PAYMENT

In accordance with the Contract Documents, based on on-site observations and the data comprising the above Application, the Architect certifies to the Owner that Work has progressed as indicated; that to the best of the Architect's knowledge, information and belief the quality of the Work is in accordance with the Contract Documents; and that the Contractors are entitled to payment of the AMOUNTS CERTIFIED.

TOTAL OF AMOUNTS CERTIFIED .... $ 150,000.00
(Attach explanation if amount certified differs from the amount applied for.)
ARCHITECT:

By: Jack Faron   Date: November 3, 1990

This Certificate is not negotiable. The AMOUNTS CERTIFIED are payable only to the Contractors named in AIA Document G723, attached. Issuance, payment and acceptance of payment are without prejudice to any rights of the Owner or the Contractor under this Contract.

G722—1980

AIA DOCUMENT G722 • PROJECT APPLICATION AND PROJECT CERTIFICATE FOR PAYMENT • JUNE 1980 EDITION • AIA®
© 1980 • THE AMERICAN INSTITUTE OF ARCHITECTS, 1735 NEW YORK AVENUE, N.W., WASHINGTON, D.C. 20006

Figure 16.6 Project application and project certificate for payment.

INSTRUCTION SHEET
AIA DOCUMENT G722a

A. GENERAL INFORMATION:

AIA Document G722 is a new document to be used in conjunction with AIA Document G723, Project Application Summary. These documents are designed to be used on Projects where a Construction Manager is employed and where multiple Contractors have separate direct Agreements with the Owner. Procedures for their use are covered in AIA Document A201/CM, General Conditions of the Contract, Construction Management Edition, 1980 Edition.

B. COMPLETING THE G722 FORM:

After the Construction Manager has completed AIA Document G723, Project Application Summary, summary information should be transferred to the G722 form.

The Construction Manager should sign the form, have it notarized and submit it, together with G723 and a separate G702, Application, from each Contractor, to the Architect.

The Architect should review it and, if acceptable, complete the lower Project Certificate for Payment on this form. DO NOT SIGN A CERTIFICATION ON EACH G702 SUBMITTED BY THE CONTRACTORS.

The completed form should be forwarded to the Owner. The Owner will make payment directly to each separate Contractor based on the amount due each as noted in Line J of G723.

C. COMPLETING THE G723 FORM:

Each separate Contractor on the Project should complete and sign AIA Document G702, Application and Certificate for Payment, and forward it to the Construction Manager. The Construction Manager will review each separate Contractor's Application for Payment and, if it is acceptable, complete one vertical column for each separate Contractor.

If the Construction Manager does not agree with the amounts requested by any Contractor, the Construction Manager should note the corrected amount in the appropriate location on G723.

One vertical column should be completed for each application period for all Contractors involved in the Project whether or not any amount is due the particular Contractor for the period in question. Each page should be summarized horizontally and all pages summarized once to provide Project totals.

Project totals should be transferred to AIA Document G722, Project Application and Project Certificate for Payment.

AIA Document G702 from each of the separate Contractors should be attached to the G723 and submitted with G722 to the Architect for review and appropriate action.

Figure 16.6   *(Continued)*   Project application and project certificate for payment.

AIA Document A305

Contractor's Qualification Statement
1986 EDITION

This form is approved and recommended by The American Institute of Architects (AIA) and The Associated General Contractors of America (AGC) for use in evaluating the qualifications of contractors. No endorsement of the submitting party or verification of the information is made by the AIA or AGC.

The Undersigned certifies under oath that the information provided herein is true and sufficiently complete so as not to be misleading.

SUBMITTED TO:

ADDRESS:

SUBMITTED BY:	Corporation ☐
NAME:	Partnership ☐
ADDRESS:	Individual ☐
PRINCIPAL OFFICE:	Joint Venture ☐
	Other ☐

NAME OF PROJECT (if applicable):

TYPE OF WORK (file separate form for each Classification of Work):

_____ General Construction   _____ HVAC

_____ Plumbing   _____ Electrical

_____ Other_____
                           (please specify)

A305-1986   1

Figure 16.7   Contractor's qualification statement.

1.   ORGANIZATION

1.1   How many years has your organization been in business as a Contractor?

1.2   How many years has your organization been in business under its present business name?

    1.2.1   Under what other or former names has your organization operated?

1.3   If your organization is a corporation, answer the following:
    1.3.1   Date of incorporation:
    1.3.2   State of incorporation:
    1.3.3   President's name:
    1.3.4   Vice-president's name(s):

    1.3.5   Secretary's name:
    1.3.6   Treasurer's name:

1.4   If your organization is a partnership, answer the following:
    1.4.1   Date of organization:
    1.4.2   Type of partnership (if applicable):
    1.4.3   Name(s) of general partner(s):

1.5   If your organization is individually owned, answer the following:
    1.5.1   Date of organization:
    1.5.2   Name of owner:

Figure 16.7   *(Continued)*   Contractor's qualification statement.

1.6   If the form of your organization is other than those listed above, describe it and name the principals:

2.   LICENSING

2.1   List jurisdictions and trade categories in which your organization is legally qualified to do business, and indicate registration or license numbers, if applicable.

2.2   List jurisdictions in which your organization's partnership or trade name is filed.

3.   EXPERIENCE

3.1   List the categories of work that your organization normally performs with its own forces.

3.2   Claims and Suits. (If the answer to any of the questions below is yes, please attach details.)

    3.2.1   Has your organization ever failed to complete any work awarded to it?

    3.2.2   Are there any judgments, claims, arbitration proceedings or suits pending or outstanding against your organization or its officers?

    3.2.3   Has your organization filed any law suits or requested arbitration with regard to construction contracts within the last five years?

3.3   Within the last five years, has any officer or principal of your organization ever been an officer or principal of another organization when it failed to complete a construction contract? (If the answer is yes, please attach details.)

AIA DOCUMENT A305 • CONTRACTOR'S QUALIFICATION STATEMENT • 1986 EDITION • AIA® • ©1986
THE AMERICAN INSTITUTE OF ARCHITECTS, 1735 NEW YORK AVENUE, N.W., WASHINGTON, D.C. 20006   **A305-1986   3**

WARNING: Unlicensed photocopying violates U.S. copyright laws and is subject to legal prosecution.

Figure 16.7   (*Continued*)   Contractor's qualification statement.

3.4   On a separate sheet, list major construction projects your organization has in progress, giving the name of project, owner, architect, contract amount, percent complete and scheduled completion date.

    3.4.1   State total worth of work in progress and under contract:

3.5   On a separate sheet, list the major projects your organization has completed in the past five years, giving the name of project, owner, architect, contract amount, date of completion and percentage of the cost of the work performed with your own forces.

    3.5.1   State average annual amount of construction work performed during the past five years:

3.6   On a separate sheet, list the construction experience and present commitments of the key individuals of your organization.

4.   REFERENCES

4.1   Trade References:

4.2   Bank References:

4.3   Surety:

    4.3.1   Name of bonding company:

    4.3.2   Name and address of agent:

AIA DOCUMENT A305 • CONTRACTOR'S QUALIFICATION STATEMENT • 1986 EDITION • AIA® • ©1986
THE AMERICAN INSTITUTE OF ARCHITECTS, 1735 NEW YORK AVENUE, N.W., WASHINGTON, D.C. 20006   **A305-1986   4**

WARNING: Unlicensed photocopying violates U.S. copyright laws and is subject to legal prosecution.

Figure 16.7 *(Continued)* Contractor's qualification statement.

5. **FINANCING**

5.1   Financial Statement.

        5.1.1   Attach a financial statement, preferably audited, including your organization's latest balance sheet and income statement showing the following items:

            Current Assets (e.g., cash, joint venture accounts, accounts receivable, notes receivable, accrued income, deposits, materials inventory and prepaid expenses);

            Net Fixed Assets;

            Other Assets;

            Current Liabilities (e.g., accounts payable, notes payable, accrued expenses, provision for income taxes, advances, accrued salaries and accrued payroll taxes);

            Other Liabilities (e.g., capital, capital stock, authorized and outstanding shares par values, earned surplus and retained earnings).

        5.1.2   Name and address of firm preparing attached financial statement, and date thereof:

        5.1.3   Is the attached financial statement for the identical organization named on page one?

        5.1.4   If not, explain the relationship and financial responsibility of the organization whose financial statement is provided (e.g., parent-subsidiary).

5.2   Will the organization whose financial statement is attached act as guarantor of the contract for construction?

Figure 16.7   (*Continued*)   Contractor's qualification statement.

6.   SIGNATURE

6.1   Dated at   this   day of
                                                            19

        Name of Organization:

        By:

        Title:

6.2

M   being
duly sworn deposes and says that the information provided herein is true and sufficiently complete so as not to be
misleading.

Subscribed and sworn before me this   day of
                                                    19

        Notary Public:

        My Commission Expires:

AIA DOCUMENT A305 • CONTRACTOR'S QUALIFICATION STATEMENT • 1986 EDITION • AIA® • ©1986
THE AMERICAN INSTITUTE OF ARCHITECTS, 1735 NEW YORK AVENUE, N.W., WASHINGTON, D.C. 20006   **A305-1986   6**

Figure 16.7   (*Continued*)   Contractor's qualification statement.

References

1. National Society of Professional Engineers, Publication 1910-1, "Standard Form of Agreement Between Owner and Engineer for Professional Services" (1984 ed.), NSPE, Washington, D.C.
2. National Society of Professional Engineers, Publication 1910-1, "Standard Form of Agreement between Owner and Contractor on the Basis of Cost-Plus" (1983 ed.), NSPE, Washington, D.C.
3. National Society of Professional Engineers, Publication 1910-1, "Certificate of Substantial Completion" (1983 ed.), NSPE, Washington, D.C.
4. National Society of Professional Engineers, Publication 1910-1, Robert J. Smith, Esq., "Recommended Competitive Bidding Procedures for Construction Projects" (1987 ed.), NSPE, Washington, D.C.
5. National Society of Professional Engineers, Publication 1910-1, "Owner-Contractor Agreement—Stipulated Sum," NSPE, Washington, D.C., 1987.
6. American Institute of Architects, Publication A101, "Owner-Contractor Agreement—Stipulated Sum" (1987 ed.), AIA, Washington, D.C.
7. American Institute of Architects, Publication A305, "Contractor's Qualification Statement" (1986 ed.), AIA, Washington, D.C.
8. American Institute of Architects, Publication B141, "Owner-Architect Agreement" (1987 ed.), AIA, Washington, D.C.

Facility Management Systems

Jeffrey M. Hamer
President, The Computer-Aided Design Group™
Marina del Rey, California

Perhaps the most important force impacting computers in the real estate industry is *facility management*. The International Facility Management Association (IFMA), only ten years old, has experienced explosive growth, doubling its membership almost every year, to 9000 members in 1992.

Let's explore some of the trends and economics molding these events.

Changing Economics

Fundamental forces are at work changing our economy. Competition that was national now is global. There is every indication that this is only the beginning of a very broad and important trend to more open markets, and especially to greatly increased competition.

The equity markets have experienced unprecedented volatility. Capital has begun to flow across national borders. The effects of increased competition can and will be felt much more quickly.

Major changes are occurring within corporations, institutions, and government. They struggle to be more productive in their newly expanded competitive arenas. They want the markets and investors (both within and outside of their national boundaries) to place maximum value on their assets.

Restructuring is the current buzzword.

A large U.S. investment bank and brokerage firm recently observed in an annual report:

> "[T]he assets of many corporations are undervalued in the marketplace, despite strong equity markets."

"[C]ompanies that do not adequately utilize and restructure their balance sheets, thus weakening equity prices, are now more likely than ever to face unsolicited bids for control."

"[Corporations are] looking for ways to increase profits and productivity and fully utilize hidden assets...."

"[F]inancial entrepreneurs...may seek to acquire undervalued assets through a change in corporate control."

"Corporate leaders are taking a closer look at their existing assets."

"[I]n a period of tight competition and narrowing profit margins, corporations cannot afford to hold onto unproductive assets...."

"Other companies are acquiring assets to increase output and productivity...."

"The competitive vitality of the U.S. depends on rearranging the balance sheet...."

One of the major changes concurrent with these events has been *"asset revaluation."* Corporate raiders and day-to-day managers alike have sought to raise the value and productivity of the asset side of their balance sheets. Robert Anderson, the former Chairman of Atlantic Richfield Company, has put it somewhat more bluntly: "If you can't manage assets to reflect their true value, you're inviting someone to do it for you."

Facilities Assets

Facilities are *the largest asset on the balance sheet.* If you open up the annual report or financial statements of almost any organization, the facilities assets represent the largest single line item. Frequently they are undervalued at that.

Advertisements have been seen in the most prominent U.S. business magazines, such as *Forbes, Fortune,* and *Business Week,* placed by another major U.S. investment bank and brokerage, headlined "Real Estate: Realizing corporate America's most undervalued asset" and including the following quotation: "The most significantly undervalued asset on corporate America's books is being discovered." Managers are beginning to have a very new awareness of the strategic opportunity facility management presents.

Information Systems

A very significant part of this opportunity is facility management *systems:* the *information tools* which enable managers to make informed facilities decisions.

Facility management is information management. Even the entry-level real estate manager (or architect, for that matter) does not physically move walls: he or she moves *information,* which instructs others to move walls. Facility managers manipulate information instructing others in the "how," "when," and "where" to move and build. More senior facility managers also manipulate information that *communicates* the "why" the walls are to be moved.

America's top corporations, organizations such as General Motors, IBM, Shell, Boeing, McDonnell-Douglas, Eastman Kodak, Westinghouse, and many others, recently have implemented *facilities information systems.* Significantly, the world's major computer hardware and software vendors, firms such as IBM, Digital Equipment, and Autodesk, also now offer facility management information products.

These are not computer-aided design (CAD) systems, but *decision support* tools. *Computer-assisted facility management* (CAFM) systems help managers stay current with a detailed functional inventory of buildings and their contents: rooms and spaces, furniture and equipment, cables and leases. These systems are designed to help executives make the right choices about moves, leases, occupancies, and capital equipment usage. The return on this investment has been most impressive.

A Business Case

Virtually all top management teams now are more than aware of the fundamental changes in the economic environment discussed at the beginning of this chapter. Many are even becoming aware of the very large role facilities play in their asset mix. This trend of course will continue, and accelerate. Few managers, however, are aware of the opportunities information systems provide for the creative and proactive management of their facilities assets.

In order to persuade top management, we must first begin to think like top management. That is, we must understand and empathize with their position, and especially with their decision-making needs. Everyone gives lip service to the trite expression "It's lonely at the top," without understanding how much truth there is in it. Top managers daily must make trade-offs and choices amongst many projects, and many potential recipients of funds, management time, and other limited resources.

Almost all of these projects considered by top management are very deserving and would benefit the organization. But resources are finite. Choices must be made. The only way to make the important and convincing case for facility management systems (as for any other business investment) is via a dispassionate and cold-blooded *cost-benefits analysis.* The benefits, costs, and resultant *return on investment* of facilities information systems generally are very impressive: sufficient to justify implementation today in a majority of organizations. Yet it is surprising how few competent business cases are communicated.

To begin, *get management's attention.* A copy of your organization's annual report is the place to start. Present the industry trends and *your organization's specifics* in the terms management understands and cares about: in terms of money. Some generalities follow that can and should be customized for each individual organization.

Background

Most organizations do not think themselves to be in the real estate business, but the reality is that they are.

According to the IFMA, between 10 and 18 percent of annual expenses (total expenses on the organizationwide income or profit/loss statement) are facilities related.

More significantly, if we turn our attention from the income statement to the balance sheet, according to Harvard Real Estate Inc., *between 25 and 50 percent of total assets are facilities related.* This means that between one-quarter and one-half of the organization's balance sheet is facilities. If you look at the annual report of any large organization, these line items are land, buildings, machinery, furniture, and equipment.

How large is 25 percent of the balance sheet? In a 1981 study by Harvard Real Estate Inc., 25 percent of America's corporate balance sheet was 7 billion square feet of owned space plus another 7 billion square feet of leased space, worth almost $1.4 trillion. That was in 1981. In 1992 independent market research estimated this number to be $2.6 trillion.

The size of this number has not been lost on the management and consulting services firms which help large organizations to manage better. Price Waterhouse, the international accounting and consulting firm, says that in 1986 there was over $1.5 trillion in "undermanaged" corporate real estate assets in America, and that there are three reasons for this:

- *Historical cost basis.* Traditionally, real estate and fixed assets are carried on the books on a *cost basis.* That is, they are carried on the balance sheet at cost—at the price the organizations paid to acquire them. Generally this understates their value by a large degree. Real estate generally appreciates, while generally accepted accounting principles call for depreciation. Over time the disparity between the "book" value (purchase price less depreciation) and the "market" value (what these facilities assets would bring if sold) only becomes greater.

- *"G & A" mindset.* Historically real estate and facility-related expenses have been part of general and administrative (G&A) expenses. These often have been intransigent and/or impervious to cost-cutting measures. There is a mindset on the part of many executives that facility-related expenses are part of administrative overhead and thus difficult to analyze, much less to control.

- *Failure to focus.* Management teams historically have not identified facilities as a significant part of their operation. (Organizations do not think they're "in the real estate business.") As noted above, this is changing. Price Waterhouse suggests that executives think of facilities assets as strategic resources, right along with people, capital, and technology.

There's ample evidence of this historical undermanagement and failure to focus:

- According to Harvard Real Estate Inc., *one organization in five has no real property inventory.* That means none whatever: not even on pieces of paper. Assembling a comprehensive or centralized inventory of all real property for a large organization is more difficult than it seems, especially if the property is spread out in many cities or possibly countries, and under many different operating entities or subsidiaries. And *keeping it current,* as groups move, leases and options expire, and ownerships and occupancies change, makes even the difficulties of the initial inventory look easy.

- Two organizations out of three have *no real property information system.* But this too is rapidly changing.

Also there is a phenomenon which IFMA calls *churn.* The term refers to the change rate due to moves, construction projects, etc. The average churn rate of IFMA organizations is greater than 30 percent per year. This does not necessarily mean that one third of the organization moves or changes each year; sometimes it can mean that some unhappy 5 percent of the organization moved six times. It is not uncommon for project-driven organizations (such as aerospace, defense, and engineering firms) or organizations in industries experiencing rapid change due to phenomena such as deregulation (such as transportation or financial services) routinely to experience churn rates well in excess of 100 percent per year!

Facility management executives also by definition are in the business of trying to anticipate the (facility needs and requirements) future. This of course is very difficult for any business activity.

Trends

There are a few very significant trends happening in response to all of the above. One is that people are being hired. According to Real Estate Research Corporation and Equitable Real Estate Inc., in the four-year period 1982–1986, more than three-quarters (76 percent) of real estate staffs increased. This by itself is not very interesting. If however we look at the 76 percent that did increase, almost half (49 percent) grew by more than 400 percent. Clearly something is happening here. Many "bodies are being thrown at the problem." Perhaps this is because top management is beginning to realize the large value of the assets involved.

Another relatively recent and very powerful trend is the wave of restructurings and takeovers, both friendly and hostile, in the public equity markets, and business in general. As discussed at the beginning of this chapter, *organizations are being restructured in order to get at (and revalue) major assets.* Keep in mind that the largest single asset on the balance sheet is facilities. The two major trends top management is recognizing that have fueled the big changes happening in facility management information systems are:

- *Facilities costs are increasing over time* (and increasing as a percentage of total operating costs).

- The *cost of automation is decreasing over time.*

As a result there now is greater justification for providing facilities executives with information via the use of automation tools. We can argue over "when" and "how much" is cost effective, but no longer over "whether."

Costs and Benefits

Now let's talk about the (economic) benefits of facilities information systems. The various benefits can be neatly and qualitatively categorized into two major types:

- Benefits in *reduction of* (total facility-related) *cost.*

- Benefits in *increased facilities performance and responsiveness* to users' requirements (the best fit between building and occupancy).

Quantitatively, the benefits come from four places. They are listed below in ascending order of dollar magnitude.

- *Facility management group labor.* This is a "people productivity" benefit, as with a CAD system. A very easily measurable and provable benefit, but on the labor of a relatively small number of people.

- *Implementation costs.* This is a reduction in the number of moves, number of construction projects, and similar expenses. These are relatively small percentage savings, but on very large moving, construction, and implementation budgets.

- *Resource costs.* This is a reduction in the amount of space itself, as well as in the amount of furniture and number of pieces of equipment used. Even a very small percentage saving here is very large in absolute dollar amount.

- *Operations costs.* This is potentially the largest benefit, although difficult to measure and prove. It results from the organization functioning and operating better due to departments, activities, and processes that need to be near each other being located near to each other; from functional, technology, and service needs being better met; and so forth.

It is important to keep in mind that facility management is a *decision support function.* Large economic benefits come pfrom beautiful drawings or color high-resolution three-dimensional perspectives. Large economic benefits come from a few bright ideas once in a while, a few smart decisions made from time-to-time.

Answering Questions

We achieve economic benefits for management by answering questions. Questions such as:

How many people do they have?

What equipment do they have in their department?

Where is that process located now?

Where have we made exceptions to our guidelines?

What space do they need?

What services and environmental support does that activity need?

What organizations should be adjacent to each other?

Does this group fit in the building?

What is the total cost to move to this area?

How do we look there five years from now?

What happens if...?

Databases Are Key

This is done by creating *a single, definitive information source: the database.* The database should be central, current, accessible, accurate, and without redundancy. All departments in the organization need to "play with the same deck" of information.

An exciting database concept for facility managers is what we call *"one information base from many."* It's exciting because virtually *all of the information* that facility managers need in order to achieve these very large benefits *already exists* in a majority of organizations. A facility management database is built from information already existing in databases such as:

Strategic planning

Architecture/design

CAD

Payroll

Phone

Real estate

Manufacturing resource planning (MRP)

Purchasing

Fixed assets

Accounting

Personnel

Production planning and control

Budget

Another exciting concept is what we call *bridging. Well over 90 percent of the information already exists in machine-readable form* in the above databases. This means that it can be transferred electronically from existing databases. This is a powerful idea, as the biggest single cost of any computer application usually is neither hardware or software, but *data:* gathering, verifying, and entering it into the system.

Goldman Sachs notes:

"Today, people use real estate to enhance shareholder values...[and as] takeover defense."

"Companies are looking at real estate...as potential value to be created through long-range corporate strategy."

The bottom line is that facilities assets are perhaps the largest untapped profit potential on the world's balance sheet. Facilities information systems are the key.

18

Facilities Maintenance

Kenneth Knott, Ph.D., P.E.
Department of Industrial and Management Systems Engineering
The Pennsylvania State University
University Park, Pennsylvania

Timothy P. Woodworth, M.S.
Griffin Data Systems, Inc.
Cochranton, Pennsylvania

Introduction

A major portion of the total assets of any organization is likely to be represented by land, buildings, and equipment. The condition of these assets has a direct impact on:

1. The net worth of the company
2. The ability of the company to generate profit

Maintenance and net worth

The book value of the buildings and equipment is subject to depreciation. If, due to ineffective maintenance, these facilities are allowed to deteriorate at a rate greater than that of depreciation, their actual value will be less than the book value. When this occurs, the real value of the company is reduced.

It is a fact that the rate of deterioration of buildings and equipment can be reduced by effective maintenance. Therefore, the financial strength and profitability of the company can be directly affected by the maintenance activity.

Maintenance and profitability

The effect of maintenance on the profitability of a company can be classified into two groups:

1. System life costs
2. System availability costs

System life cost. The *system life cost* of equipment is reflected in the ratio of the total cost to maintain the facility over its life to the acquisition cost of the facility.

$$\text{Life cost ratio} = \frac{\text{total maintenance cost}}{\text{acquisition costs}} \tag{18.1}$$

This ratio is usually much greater than 1. A value of 1500 for this ratio has been quoted[1] for a piece of military equipment. The same reference indicates this value is extraordinarily high. The value is usually between 10 and 100.

A broad view should be taken of what is to be included in the system life cost. To provide maintenance support to a single item in a facility could require a major investment in specialized test equipment. Beyond this, there are costs incurred in maintaining the specialized equipment. Both of these should be included in the system life cost.

The costs of personnel involved in maintenance activities can represent as much as 75 percent of the total system life cost. Clearly, control of cost and performance of maintenance labor is important.

Management of replacement and repaired parts used in maintenance of facilities provides a major challenge in controlling the system life costs. Excessive stock levels will increase the cost of investment, storage, and maintenance of inventory.

System availability costs. When a system is inoperative, either for maintenance or repairs to be carried out, a cost is incurred. The *system availability cost* of equipment is a function of its operational availability. A measure of this operational availability, commonly used, is

$$A = \frac{\text{MTBF}}{\text{MTBF} + \text{MDT}} \tag{18.2}$$

where A = System availability
     MTBF = Mean time between failures
     MDT = Mean downtime

System availability costs are incurred when the system is nonoperational due to maintenance or repair demands. Overmaintenance will increase the system availability costs and the system life costs. On the other hand, undermaintenance will increase the rate of deterioration of the plant and, possibly, decrease the value of the company.

Optimizing the maintenance effectiveness

A number of measures have been used to describe the effectiveness of a maintenance activity. All of these eventually relate to the system life cost and the system availability cost. Therefore, to optimize maintenance, these two costs must be balanced, one against the other.

Maintenance issues have been exhaustively investigated. Some of the general areas of interest which have been considered in the open literature include design and operating conditions. These effects are combined under a general measure, *reliability*.

Preventive maintenance

The purpose of *preventive maintenance* (PM) is to keep equipment in good operational condition. Successful PM presents many possible advantages, which will be discussed below (see under "Planned Maintenance"). The dependency of the frequency of failure upon the reliability of the equipment and components has already been noted.

Some of the advantages to be gained from using PM include:

1. Reduction in maintenance costs
2. Reduction in the number of large-scale repairs required
3. Reduction in the cost of repairs
4. Reduction in unit manufacturing costs
5. Reduction in operational downtime
6. Reduction in the overtime worked by maintenance personnel
7. Improved control of maintenance spare parts
8. Improved safety

Types of Maintenance

One dichotomy of forms of maintenance is shown in Fig. 18.1. This figure has been adapted from one developed by the British Standards Institution.[2] With one exception,[3] all of the definitions of the different forms of maintenance are from this same publication.[2]

The first division of types of maintenance is whether the work is performed on a *planned* or on an *emergency* basis. In many organiza-

Maintenance All actions necessary for retaining an item in or restoring it to a specified condition.[a]		
Planned Maintenance Work organized and carried out with forethought, control and records.[a]		Emergency Maintenance (This means unplanned emergency maintenance). Work necessitated by unforseen breakdowns or damage. [a]
Preventive Maintenance (This means planned preventive maintenance). Work which is directed to prevention of failure of facility. [a]	Corrective Maintenance (This means planned corrective maintenance). Work undertaken to restore a facility to an acceptable standard. [a]	
Running Maintenance (This means planned preventive maintenance). Work which can be carried out while the facility is in service.[a]	Shut-Down Maintenance (This means planned preventive running maintenance). Work which can be carried out when the facility is out of service. [a]	Breakdown Maintenance (This means planned corrective breakdown maintenance). Work which is carried out after a failure, but for which advance provision has been made, in the forms of materials, labor and equipment.[a]

Figure 18.1   A dichotomy of the common forms of maintenance.

tions, equipment and facility maintenance is ignored until there is a catastrophic failure. Work is then carried out on an emergency basis. No provision has been made for planning the labor, materials, or equipment needed to bring the equipment back to an operational level in as short a time as possible.

Planned Maintenance

From Fig. 18.1, it will be seen that *planned maintenance* can be divided into two broad groups.

1. Corrective maintenance

2. Preventive maintenance

Corrective maintenance

The purpose of a *corrective maintenance program* is to correct an item or piece of equipment which is not performing satisfactorily. The need for corrective maintenance can usually be attributed to three basic causes, as follows:[4]

1. *Errors:* Human, test equipment, and procedures

2. *Failure:* Wear-out, primary and secondary

3. *Miscellaneous:* Performance deterioration, handling damage, environmental effects, and nonoperational defects

Corrective maintenance tasks are usually performed on an "as needed" basis, but with forethought and planning. Here the availability of the labor, materials, and equipment is planned and the work is carried out before catastrophic failure occurs.

Corrective maintenance can be further subdivided into *shutdown maintenance* or *breakdown maintenance*.

Preventive maintenance

Preventive maintenance (PM) uses information or early detection and correction of failures to maintain equipment in an operating condition. Seven categories of PM have been identified, as follows:[5]

1. Replacement

2. Servicing

3. Alignment

4. Calibration

5. Visual checks and inspection

6. Modification

7. Confidence testing

Reliability and Preventive Maintenance

The success of a PM program depends directly on the frequency with which inspections and repairs are carried out. The frequency of failure is related to the reliability of the equipment. In this chapter the relationship between reliability and maintenance is explored.

Accelerated Test: A test in which the applied stress level is chosen to exceed that stated in the reference conditions in order to shorten the time required to observe the stress response of the item, or magnify the response in a given time. To be valid, an accelerated test must not alter the basic modes and/or mechanisms of failure.

Availability: The probability that the time is operational at the time of need.

Burn-In: The operation of items prior to use in their intended application in order to stabilize their characteristics and to identify early failures.

Constant Failure Period: The period during which failures occur at an approximately uniform rate.

Failure: The termination of the ability of an item to perform its required function.

Failure, Catastrophic: Failures that are both sudden and complete.

Failure, Degradation: Failures that are both gradual and partial.

Failure, Misuse: Failures attributable to the application of stress beyond the stated capabilities of the item.

Failure, Random: Any failure whose cause and/or mechanism make its time of occurrence unpredictable.

Failure, Secondary: Failure of an item caused either directly or indirectly by the failure of another item.

Failure, Wear-Out: A failure that occurs as a result of deterioration processes or mechanical wear and whose probability of occurrence increases with time.

Failure Mechanism: The process that results in failure.

Failure Mode: The physical process or processes that occur or combine their effects to generate a failure.

Failure Rate: The number of failures of an item per unit of measurable life. Life can be measured in units such as miles, cycles, time, and events.

Instantaneous Failure Rate (Hazard Rate): The rate of change of the number of items that have failed divided by the number of items surviving at a particular time.

Mean Time Between Failures (MTBF): The mean time between failures of a repairable complex item computed from its design considerations and from the failure rates of its components for the intended conditions of use.

Mean Time To Failure (MTTF): The total operating time of a number of items divided by the total number of failures.

Redundancy: The existence of more than one means of performing an item's function.

Reliability: The probability that an item (a system, a subsystem, or a component) will carry out its assigned mission under specified conditions for a stated time period.

Steady State Condition (statistical): The condition where the probability of being in a specific state is independent of time.

Stress: Any factor (i.e. temperature, environment, electric current) that tend to produce a failure.

Useful Life: The length of time an item operates with an acceptable failure rate.

Figure 18.2   Standard reliability terminology.

Terminology of reliability and maintenance

Several standard terminologies have been published on maintenance and reliability.[2,3,6] A summary based on the terminology published by the Institute of Electrical and Electronics Engineers (IEEE)[6] is given in Fig. 18.2. This terminology will be used in this section.

Reliability measures for maintaining equipment

The emphasis of reliability studies has been directed toward electrical and electronic equipment. Information on mechanical equipment is not so readily available, but it will be referred to here.

A successful maintenance scheme gives increased equipment reliability and availability. This availability is dependent on the probability of the component failing. If a maintenance scheme decreases the probability a unit or component will go out of service, by definition, its availability will increase. In many cases, mechanical equipment cannot be described by a constant failure rate. This is due to its nonconstant degradation. Loading, wear effects, interaction with other components, environmental effects, vibration, and many other factors cause the failure rate to vary over time.[6,7]

Failure rate definition

The *failure rate* of a piece of equipment is an important measure of reliability when studying maintenance. Figure 18.3 shows the "bathtub" curve, frequently used to describe a failure rate.[8] This plot of failure rate versus time in use shows:

1. Break-in period
2. Steady-state period
3. Wear-out period

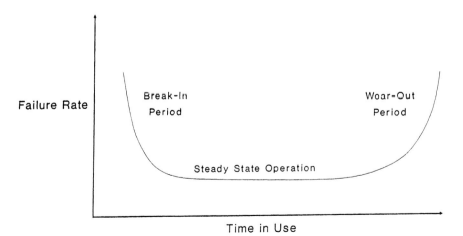

Figure 18.3 Traditional bathtub curve.

Manufacturing defects and improper installation are the principal causes of failure in the break-in period. The steady-state life period shows the expected operating life, where the probability of failure is independent of time. The failure rate is essentially constant since the item is performing as designed under allowable stresses. Failures in this period may be caused by human error and/or purely random effects.[9]

Figure 18.3 is a conceptual view of equipment failure rates. A failure rate will vary in detail with the type of unit. Not all types of equipment will display the idealized periods in Fig. 18.3. Electrical and mechanical hardware have very different failure rate plots.

Typical failure characteristics of mechanical equipment are shown in Fig. 18.4. The initial break-in period is followed by a gradual increase in the failure rate. The most noteworthy cause of the increasing failure rate is mechanical wear.

The wear-out period shows an exponential increase in the failure rate. Component interactions, wear, fatigue, vibration, and other failure modes cause the unit to fail at an increasing rate. Figure 18.5 shows the typical failure characteristics of electrical equipment. The break-in period reflects the start-up effects such as manufacturing defects and installation errors. The steady-state life period is common for electrical components.

Published reliability databases

Data handbooks, such as MIL-HDBK-217B,[10] describe the behavior of electrical components in varying operating conditions. These handbooks typically contain specifications of equipment testing procedures and the resulting component reliability.

There are handbooks for mechanical equipment, such as *Non-electronic Parts Reliability Data.*[11] However, they must be used very

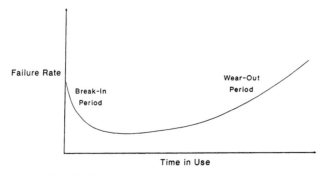

Figure 18.4 Failure rate of a typical mechanical item.

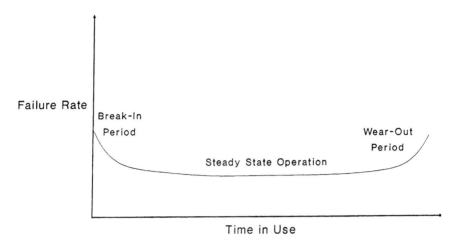

Figure 18.5 Failure rate of a typical electrical item.

carefully. The largest source of uncertainty with these data sources is the lack of detail about the operating conditions. The handbooks describe the probability of a part failing in a particular operational environment. Common phrases used to describe environmental conditions are "ground fixed," such as a generator, or "ground mobile," such as an armored fighting vehicle.[7,10,11]

These handbooks do not attempt to quantify component reliability in an assembly, or estimate the effects of variation in materials or workmanship. They do not define what a *failure* is, nor define the operating conditions surrounding the failure. Since the handbooks cannot cover all of the possible conditions which affect equipment reliability, failure definition is difficult. The definition of a valve *failing* in a nuclear facility, for example, is far stricter than that for a valve in a washing machine. Certainly the conditions and results surrounding the two failures are vastly different.

Published failure rates often have extreme ranges in magnitude. These handbooks have limited use in making reliable predictions for mechanical equipment, but they provide a starting point. The U.S. Department of Defense is the largest source of handbook data, methodologies in life testing, and reliability experiment design.[12–14]

Failure mode assessment

Good estimates of failure rates are critical for a successful PM program. This estimation is one of the most difficult aspects of PM. Often it is not possible to obtain a failure rate for an entire system, such as a computer or a locomotive. Instead, reliability estimates of subsystems or individual components are made, then used to build a system-level reliability estimate. Methods for tracing component interaction are used to build a schematic of an entire system. These interaction networks enumerate types of failure, or *failure modes.*

Failure modes and effects analysis (FMEA), event trees, and fault trees can be used to analyze a system from component level.[10] These methods are based on a similar methodology, determining system reliability by identifying all possible failure modes and the causes of each. *Failure modes and criticality analysis* (FMECA) enhances this idea by including the effects of each failure.[15]

These methods organize information and build causal relationships, usually in the form of a parent-child tree structure, which lead to a system reliability level. An example of a fault tree is shown in Fig. 18.6.

Each block label in Fig. 18.6 corresponds to the reliability of that component. Combining these groups in series and in parallel will provide a system reliability. The key to developing system reliabilities is to develop sound component reliabilities.

Derivation of Reliability Equations

Concepts of mathematical reliability

Reliability is defined as "the probability that a component, device, equipment, or system will perform its intended function for a specific period of time under a given set of conditions." Some of these condi-

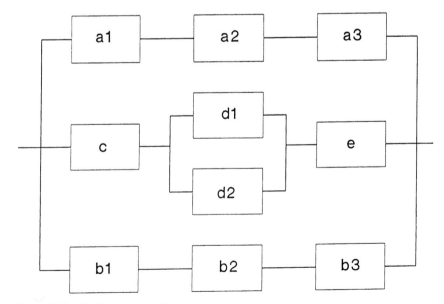

Figure 18.6   Fault tree example.

tions are environment, materials, operating cycle, quality and frequency of maintenance, time in service, and manufacturing defects.

Mathematically, reliability can be expressed by using the *probability distribution function* (PDF) of the random variable t, the time until system failure. The PDF, $f(t)$, can be stated as:

$$f(t)\,\Delta t = p\{t < r \le t + \Delta t\} = \text{probability the system}$$
$$\text{fails between time } t$$
$$\text{and } t + \Delta t \qquad (18.3)$$

The cumulative density function (CDF), $F(t)$, is defined as the probability that failure occurs at a time less than t, namely:

$$F(t) = p\{r \le t\} \qquad (18.4)$$

Basic probability theory states that the CDF can be derived from the PDF as:

$$F(t) = \int_0^t f(r)\,dr \qquad 0 \le t \le + Inf \qquad (18.5)$$

Finally, reliability is expressed as the probability that failure occurs at a time greater than time t. That is

$$R(t) - p\{r > t\} \qquad (18.6)$$

This is the exact complement of Eq. (18.4). Therefore,

$$R(t) = 1\ F(t) \qquad (18.7)$$

or by substituting Eq. (18.5) in Eq. (18.7) for $F(t)$,0

$$R(t) = 1 - \int_0^t f(r)\,dr \qquad (18.8)$$

Changing the limits of integration to remove the constant yields,

$$R(t) = \int_t^{-Inf} f(r)\,dr \qquad (18.9)$$

Solving for $f(t)$,

$$f(t) = -\frac{d}{dt} R(t) \qquad (18.10)$$

These probability relationships lead to a measure which is widely used in reliability analysis, the failure rate $\lambda(t)$. The *failure rate* is described as the probability that the system will fail at some time

$r < t + \Delta t$, given that it has survived until time t. This is a conditional probability, expressed as

$$\lambda(t)\Delta t = p\{r < t + \Delta t \mid r > t\} \tag{18.11}$$

By using the definition of a conditional probability:

$$p\{r < t + \Delta t \mid r > t\} = \frac{P\{< t + \Delta t\}}{p\{r > t\}} \tag{18.12}$$

The numerator of Eq. (18.12) can be seen in Eq. (18.3) and the denominator appears in Eq. (18.6). Making these substitutions yields

$$\lambda(t)\Delta t = \frac{f(t)\Delta t}{R(t)} \tag{18.13}$$

Finally, canceling Δt from both sides gives

$$\lambda(t) = \frac{f(t)}{R(t)} \tag{18.14}$$

Reliability can be expressed in terms of the failure rate. This removes the need to solve the potentially complex Eq. (18.14). The derivation of this expression starts with an alternate form of (t), found by substituting Eq. (18.10) into Eq. (18.14), and resulting in

$$\lambda(t) = \frac{-1}{R(t)} \frac{d}{dt} R(t) \tag{18.15}$$

Multiplying both sides of Eq. (18.15) by dt gives

$$\lambda(t)\,dt = \frac{-dR(t)}{R(t)} \tag{18.16}$$

Integrating both sides from 0 to t,

$$\int_0^t \lambda(r)\,dr = \ln[R(t)] \tag{18.17}$$

Solving for $R(t)$ yields

$$R(t) = \exp\left[-\int_0^t \lambda(r)\,dr \right] \tag{18.18}$$

Another useful relationship, derived from the reliability, is the *mean time to failure* (MTTF). Since $R(t)$ is the complementary CDF of the random variable t, the time until failure, the MTTF, can be expressed as a function of t. First, MTTF is found using the PDF of t and by the definition of the mean of a continuous random variable, or

$$\text{MTTF} = \int_0^{-Inf} t f(t)\, dt \tag{18.19}$$

Substituting Eq. (18.10) in Eq. (18.19) and integrating by parts produces

$$\text{MTTF} = \int_0^t \frac{-dR(r)}{dr}\, dr = tR(t)\big|_0^{+Inf} + \int_0^{-Inf} R(t)\, dt \tag{18.20}$$

Since

$$tR(t)\big|_0^{+Inf} = 0$$

the MTTF reduces to

$$\text{MTTF} = \int_0^{-Inf} R(t)\, dt \tag{18.21}$$

Predictive maintenance and reliability measures

Assuming perfect maintenance and a time-dependent failure rate, a system operating under a predictive maintenance policy will display an increased reliability. Perfect maintenance implies that any maintenance action performed is complete and correct. A time-dependent failure rate is synonymous with the life of a piece of equipment changing over time. If the conditional probability of a part failing does not change over time, there is no point in performing predictive maintenance. That it has no effect on reliability is illustrated in Fig. 18.7.

The reliability of a maintained system shows the property that the reliability is unchanged before the first maintenance action, or

$$R_M(t) = R(t) \qquad 0 < t < T$$

and that the reliability is greater than the reliability of a nonmaintained system after the first maintenance action and thereafter. This is given by

$$R_M(t) > R(t) \qquad T < t < NT \qquad \text{for } N = 2, 3, 4,\ldots$$

The expected benefit of predictive maintenance is that the decay in the probability of survival is slowed as maintenance actions are per-

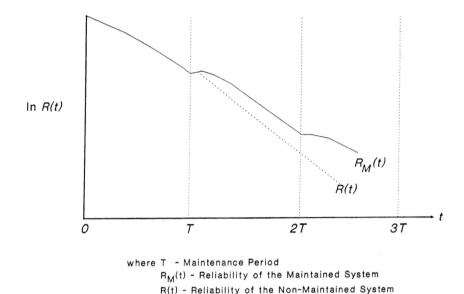

where T - Maintenance Period
R$_M$(t) - Reliability of the Maintained System
R(t) - Reliability of the Non-Maintained System

Figure 18.7   Reliability improvement under predictive maintenance.

formed. Mathematically, this is shown by the following derivation. It is assumed that maintenance restores the system to a good-as-new condition at time T. This implies the system has no "memory" of the accumulated wear during the period $0 \leq t \leq T$. The new reliability $R_M(t)$, at some point in the second period, is the product of the probability that the system survives the current time period $T < t \leq 2T$ or $R(t - T)$ and the reliability of the previous period $R(T)$. The first-period reliability is calculated at the last time interval of the period T. The second-period reliability is calculated from the same reference point T, so the expression $t - T$ is some portion of the second time period.

Let $R_M(t)$ be the probability of the equipment or component surviving the two periods. This can be expressed as the probability of surviving the first period and the second period. It is expressed as

$$R_M(t) = R(T)\, R(t - T) \qquad T < t \leq 2T \tag{18.22}$$

The second maintenance period, between time $2T$ and $3T$, is dependent on the system surviving the first and second periods and the current period, or

$$R_M(t) = R(T)^2\, R(t - 2T) \qquad 2T < t \leq 3T \tag{18.23}$$

In a general form, $R_M(t)$ can be expressed as

$$R_M(t) = R(T)^N R(t - NT) \qquad NT < t \le (N + 1)T$$

$$N = 0, 1, 2... \tag{18.24}$$

Mean time to failure under predictive maintenance, MTTF_M, is found by substituting the results of Eqs. (18.22) through (18.24) into Eq. (18.21) and simplifying. The derivation of the MTTF_M begins with a restated MTTF from Eq. (18.21). This expression is

$$\text{MTTF}_M = \int_{N=0}^{-Inf} R_M(r)\, dr \tag{18.25}$$

The interval $[0 + Inf]$ can be broken down into equal time intervals of length T as

$$\text{MTTF}_M = \sum_{N=0}^{-Inf} \int_{t=NT}^{(N-1)T} R_M(r)\, dr \tag{18.26}$$

Substituting Eq. (18.24) for $R_M(t)$ results in

$$\text{MTTF}_M = \sum_{N=0}^{-Inf} \int_{t=NT}^{(N-1)T} R(T)^N R(r - NT)\, dr \tag{18.27}$$

Simplifying the integral and setting

$$r^l = r - NT$$

yields

$$\text{MTTF}_M = \sum_{N=0}^{Inf} R(T)^N \int_{t=0}^{T} R(r^l)\, dr^l \tag{18.28}$$

The infinite summation reduces to a closed form as

$$\sum_{N=0}^{Inf} R(T)N = \frac{1}{1 - R(T)^l} \tag{18.29}$$

thereby leaving the final form of the mean time to failure under predictive maintenance:

$$\text{MTTF}_m = \frac{\displaystyle\int_{t=0}^{T} R(r^l)\, dr^l}{1 - R(T)} \tag{18.30}$$

Failure rate analysis model

The most important reliability relationships derived above were:

- Various forms of reliability and failure rate calculations
- The benefits of predictive maintenance
- MTTF with and without predictive maintenance

One measure can be seen as the basis of all of the reliability derivations above, namely, the failure rate.

Maintainability

Many definitions of *maintainability* have found acceptance. In this chapter, the one proposed by the U.S. Department of Defense[16] will be used. This definition is: "Maintainability is a characteristic of design and installation which is expressed as the probability that an item will conform to specified conditions, within a given period of time, when maintenance actions are performed in accordance with prescribed procedures and resources."

Maintainability is designed into a piece of equipment and is one of the two major contributors to equipment availability. The second characteristic is the reliability of the equipment.

Design for manufacturability

Maintainability is a characteristic of the design process; however, a study of the design process[17] showed it may have to be performed "out-of-phase" with detail design. This is due to the need for system knowledge before any predictions can be made.

The following objectives were given with respect to the maintainability of aircraft.[17]

1. Identification of areas of poor design for maintainability in the early stages
2. Recommendations for elimination of these problems, concept changes, and modifications, compatible with current maintainability disciplines
3. Provision for adequate accessibility to all aircraft zones, subsystems, and components, for both visual and manipulative tasks
4. Significant reduction of support equipment
5. Simplicity of design
6. Compliance of the various tasks with human engineering factors

7. Use of standard equipment, where improvement on design was not possible, compatible with high maintainability

8. Continual analysis and assessment of the achieved maintenance levels for the individual systems

9. Applications of general maintainability prediction methods in aircraft design

10. Identification of inadequacies which require further investigation

These objectives can easily be extended to any area of design or engineering activity.

Indices of maintainability

Indices of maintainability are used to determine whether the specified maintainability for a system has been achieved. Measures of maintainability which have proved to be the most useful and reliable are those related to time. In specifying these indices, the reference is the period the system is operable. The indices compare the "downtime," when the system is inoperable, against this reference.

Downtime can be categorized as (1) *active downtime* or (2) *delay downtime*. When the downtime is spent performing inspection, testing, repair, replacement, and other support activities, it is classified as *active*.[18-20] On the other hand, if the inoperability of the system is the result of the unavailability of some type of support system, it is classified as *delay*.

Two important, basic elements found in maintainability indices are the repair time R_t and the corrective action time $\bar{M}_{ct}$. The *repair time* is the active downtime required to return the system to operational status, after a failure has occurred. The *corrective action downtime* is the time required to locate, isolate, and correct the fault. It includes the time needed to calibrate and test the system to ensure the fault has been corrected.

Four of the most widely used maintenance indices are considered briefly below.

Mean time to repair (MTTR)

Mean time to repair (MTTR) is the statistical mean of sample repair times. It is defined as

$$\text{MTTR} = \frac{\sum_{i=1}^{n} R_{t1}}{n} \qquad (18.31)$$

where $n = $ number of sample failures

$R_{t1} = $ time required to repair failure i

This measure is synonymous with the mean corrective time $\overline{M}_{ct}$.

Mean preventive action time ($\overline{M}_{pt}$)

Regular preventive maintenance is performed on most systems. This will include cleaning, lubrication, adjustment, etc. The time required to perform this maintenance is an identifiable portion of the total downtime of the system. It can be calculated as follows:

$$\overline{M}_{pt} = \frac{\sum\limits_{i=1}^{n} M_{pt_1}}{n} \qquad (18.32)$$

where $n = $ number of preventive maintenance actions

$M_{pt_{i1}} = $ time required for task i

Mean active corrective and preventive action time ($\overline{M}$)

The downtime of the system due to corrective and preventive tasks is measured by this index. It is calculated as follows:

$$\overline{M} = \frac{\overline{M}_{ct}f_d + M_{pt}f_p}{f_c + f_p} \qquad (18.33)$$

where $f_c = $ number of corrective actions

$f_p = $ number of preventive actions

Any downtime due to administrative actions is excluded from this measure.

Mean downtime (MDT)

This measure indicates the operational availability of the system. It is the sum of the mean active corrective and preventive action time and the mean delay time. The delay time is a function of administrative and supply factors. Since these factors are totally beyond the control of the designer of the system, this measure is of little consequence in designing unavailability.

System availability

The overall operating cost of a system is a direct function of the length of time which it operates. To maximize the overall availability of the system is, therefore, a prime responsibility of the maintenance department. *Indices of availability* are used to measure the level of effectiveness in terms of this characteristic. Three of these measures are considered below.

Inherent availability (A_i)

The *inherent availability* of a system is the probability that when used under ideal conditions, without any consideration for preventive action, the system will operate satisfactorily at any time.

$$A_i = \frac{\text{MTBF}}{\text{MTBF} + \text{MTBR}} \qquad (18.34)$$

where MTBF = mean time between failures
MTBR = mean time between repairs

The ideal environment exists when all tools, parts, equipment, skills, and other support are available. Therefore, A_i excludes any consideration of ready time, preventive maintenance, or downtime a system may require.

Achieved availability (A_a)

The achieved availability of a system is calculated as follows:

$$A_a = \frac{\text{MTBM}}{\text{MTBM} + \bar{M}} \qquad (18.35)$$

where MTBM = mean time between required actions, which results from MTBF and mean time between preventive actions
$\bar{M}$ = mean active downtime resulting from both preventive and corrective actions

The *achieved availability* is the probability a system will operate satisfactorily, at any time, under stated conditions, in an ideal environment. Its only difference from the inherent availability is the inclusion of the effect of the preventive actions. There is no consideration of supply or administrative downtime in this measure of availability.

Operational availability (A_o)

The probability that a system will operate satisfactorily, at any time, under stated conditions, in an actual support environment is calculated as follows:

$$A_o = \frac{\text{MTBM}}{\text{MTBM} + \text{MDT}} \qquad (18.36)$$

where MDT = mean downtime

Comparison of the availability indices

These three system availability indices serve the system designer and user. A_i and A_a provide the designer with a measure of the extent to which reliability and maintainability have been achieved in the system. The indices A_i and A_a are also useful to the potential user of the system, providing quantitative targets by which systems can be compared.

The index A_o is particularly useful to the user of the system. Since A_o includes delay times, it can be of value in estimating the total system cost.

Maintenance Support Analysis and Control

For a maintenance program to be effective, support in the correct quantity and of the correct type must be available. To plan the required support system, the following must be established.

1. Documentation of the requirements to be met by the maintenance system

2. Testing, corrective, and predictive procedures which are to be performed

3. Controls to evaluate the behavior of the system

Mean time to repair (MTTR)

The MTTR is the principal feature controlling the maintainability of a system. This is influenced by the downtime of the various subsystems. Continuous examination and adjustment of these downtimes is needed to maintain effective maintenance.

Allocating downtime

A procedure for allocating downtime consists of the following steps. (1) Determine the contribution of the downtime of each subsystem to the total. (2) Evaluate these against the MTTR for the system. The following example illustrates the procedure:

Assume an operational availability A_o of 95 percent and an MTBF of 150 hours. The inherent availability A_i is unknown, but must be

Sub-System	No.	Failure Rate	Contribution to Total Failures	Proportion Contribution to Total Failures	Average Corrective Action Time	Contribution to Total Corrective Action Time
	Q	F_r	$C_i = (Q)(F_r)$	$R = C_i/\Sigma C_i$	$\overline{M}_{ct}$	$C_m = C(C_i)(\overline{M}_{ct})$
A	1	0.0030	0.0030	0.1395	2.8	0.0084
B	2	0.0050	0.0100	0.4651	5.7	0.0570
C	1	0.0010	0.0010	0.0465	4.4	0.0044
D	1	0.0020	0.0020	0.0930	2.2	0.0044
E	1	0.0040	0.0040	0.1860	1.5	0.0060
F	1	0.0015	0.0015	0.0698	3.6	0.0054
TOTAL			0.0215	1.0000		0.0856

$$MTTR = \frac{\Sigma (C_i)(\overline{M}_{ct})}{\Sigma C_i} = \frac{\Sigma C_m}{\Sigma C_i} = \frac{0.0856}{0.0215} = 3.98 \text{ hours}$$

Figure 18.8   Determination of downtime allocation.

greater than A_o. At this stage of the analysis it is assumed to be 98 percent. The MTTR is calculated by

$$MTTR = \frac{MTBF\,(1 - A_i)}{A_i}$$

$$= \frac{150\,(1 - 0.98)}{0.98}$$

$$= 3.06 \text{ hours} \qquad (18.37)$$

Each subsystem can be considered, using either estimates or historical data. This data is tabulated as shown in Fig. 18.8. Assume that these data are based on historical data in the example.

There are five subsystems in the total system. The frequency with which each occurs is entered in the column Q. The failure rate per 1000 hours of each subsystem is determined and entered as F_r. The contribution of each subsystem to the total failure rate C_f is the product of Q and F_r. The proportion which each subsystem makes to the total failure R is then determined. From the historical records it is possible to estimate the average time taken to correct a failure ($\overline{M}_{ct}$) in a subsystem. The contribution of each subsystem to the total corrective action time is the product of C_f and $\overline{M}_{ct}$. The MTTR can be calculated as

$$\text{MTTR} = \frac{\sum (C_f)(\overline{M}_{ct})}{\sum C_f} = \frac{\sum C_m}{\sum C_f} \qquad (18.38)$$

In the example, the MTTR is 3.98 hours. This is significantly more than was anticipated in the previous calculation, Eq. (18.37). If this target of 3.06 hours MTTR is to be achieved, there are three alternative courses of action, as follows:

1. Decrease failure rates.
2. Decrease average corrective action times.
3. Decrease either or both on a trade-off basis.

It is left to the reader to investigate the behavior of this type of analysis. The ways in which this approach can be extended should also be explored. Instead of considering a subsystem, a type of failure could be used in the analysis.

Equipment Maintenance/Replacement Models

In this section selected models which can be used in making decisions on maintenance are considered. They are a selection from hundreds which are available. For identification, the models are numbered in sequence. There is no significance to these numbers.

In this section, *maintenance* and *repairs* are being used interchangeably. Therefore, the problem being considered is the *maintain/replace decision*.

Model I—Maintain/replace decision based on cost and MTBF

A simple basis for the maintain/replace decision relates to the MTBF of new and repaired equipment. The MTBF for these two scenarios can be estimated, or obtained from historical data. The decision model is

$$\text{If} \left(\frac{\text{MTBF}_2}{\text{MTBF}_1} \right) * F < \frac{(C_L + C_M)}{C_P} \quad \text{then discard} \qquad (18.39)$$

where $\text{MTBF}_1 = $ MTBF of new equipment

$$\begin{aligned}
\text{MTBF}_2 &= \text{MTBF of repaired equipment} \\
F &= \text{an arbitrary constant} \\
C_L &= \text{cost of labor to repair equipment} \\
C_M &= \text{cost of materials to repair equipment} \\
C_P &= \text{cost to purchase new equipment}
\end{aligned}$$

The factor F might be influenced by such things as the confidence in the values of MTBF_1 and MTBF_2, the cash flow position of the company, and the lead time on replacement equipment. The following example illustrates this model.

Assume, in a particular environment, the following applies:

$$\begin{aligned}
\text{MTBF}_1 &= 740 \text{ hours} \\
\text{MTBF}_2 &= 1000 \text{ hours} \\
C_L &= \$80.00 \\
C_M &= \$120.00 \\
C_p &= \$450.00
\end{aligned}$$

The ratio of cost in this case will be calculated by:

$$\frac{(C_L + C_M)}{C_p} = \frac{80 + 120}{450} = 0.44 \qquad (18.40)$$

The maintain/replace decision is governed by the value of F, the effect of which is shown in Fig. 18.9. The cost ratio is shown as the

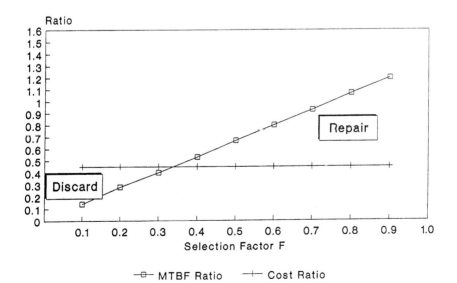

Figure 18.9 Effect of selection factor on repair/discard decision.

dotted horizontal line on the graph. The ratio of MTBF is the solid sloping line.

Model II—Determination of number of standby units

In the process industries, such as petrochemicals, it is essential to have standby units. Without sufficient standby equipment, failures have catastrophic effects. The objective of this model is to predict the number of standby units required to prevent stock-out over a specified run period. It is calculated to meet this goal with a specified probability, based on the normal distribution.

$$S_n = \frac{(F_r)}{1000} t_m = Z\left[\frac{(F_r)}{1000} t_m\right]^{1/2} \tag{18.41}$$

where S_n = number of spares required
F_r = failure rate per 1000 hours of operation
t_m = time horizon of prediction
Z = a function of the normal distribution

The following examples will illustrate this model. Assume a pump has a failure rate of 0.5 failures per 1000 hours of running time. If the predicted time is 4000 hours and a confidence level of 0.975 is required, determine the number of pumps for no "stock-out."

$F_r = 0.7$
$t_m = 4000$
$Z = 1.96$ (from standard tables of the normal distribution function)[21]

$$S_n = \left(\frac{0.7 * 4000}{1000}\right) + 1.96\left[\left(\frac{0.7}{1000}\right) * 4000\right]^{1/2} = 6.079 \tag{18.42}$$

The safe practice would be to have six pieces of similar equipment.

Model III—Optimum equipment life based on minimum annual total cost

The optimum life of the equipment is determined for minimizing the annual total cost by this model. The model is based on the mean total cost T_k, which has three components. They are (1) mean maintenance cost, (2) mean operating cost, and (3) mean investment costs. T_k is computed as follows:

$$T_k = \left(\sum_{i=1}^{2}\right) + \frac{C_1}{t} + \frac{(t-1)}{2\,(i_{oc} + i_{mc})} \tag{18.43}$$

where T_k = mean total cost
K_1 = operational cost for year 1
K_2 = maintenance cost for year 1
C_1 = cost of the investment
t = life of the equipment in years
i_{oc} = increase in operational costs per year
i_{mc} = increase in maintenance costs per year

Differentiating Eq. (18.43) with respect to t,

$$\frac{dT_K}{dt} = \frac{1}{2}(i_{oc} + i_{mc}) - \frac{C_1}{t^2} \tag{18.44}$$

If Eq. (18.44) is set to zero and rearranged,

$$t^* = \left(\frac{2_{c_1}}{i_{oc} + i_{mc}}\right)^{1/2} \tag{18.45}$$

where t^* is the optimum replacement interval.

Substituting Eq. (18.45) in Eq. (18.44) results in

$$T_K^* = \sum_{i=1}^{2} K_i \tag{18.46}$$

when $K_i = K_1 + K_2$

As an example of this model, consider the case where the investment cost is \$15,600.00, operational costs increase by \$1000.00 per year, and maintenance costs increase by \$700.00 per year. Find the optimum replacement interval:

$$t^* = \left(\frac{2 * 15000}{1000 + 700}\right)^{1/2} = 4.2 \text{ years} \tag{18.47}$$

Model IV—Number of parallel units

To ensure safety or profitability, some processes need parallel systems. A parallel system consists of N identical machines connected in parallel and the output is fed into the next stage of the process. For the system to operate, at least one of the machines must be operating. This model is concerned with determining the optimal number of parallel units, to minimize the total cost per unit time in operating and downtime losses.[21]

The model considers the average proportion of unit time U_s that the parallel system is unavailable for service in terms of queueing theory.

$$U_s = \left(\frac{\lambda}{\lambda + \mu}\right)^N \tag{18.48}$$

where $\lambda =$ constant failure rate of a single machine
$\mu =$ constant repair rate of a machine

The total cost K_{TS} is expressed as

$$K_{TS} = U_s K_{DT} + NK_O \tag{18.49}$$

where $K_{DT} =$ downtime cost per unit time
$K_O =$ operational cost per unit time of a single machine

Substituting Eq. (18.48) into Eq. (18.49),

$$K_{TS} = \left(\frac{\lambda}{\lambda + \mu}\right)^N K_{DT} + NK_O \tag{18.50}$$

Differentiating Eq. (18.50) with respect to N results in

$$\frac{\partial K_{TS}}{\partial N} = \left(\frac{\lambda}{\lambda + \mu}\right)^N K_{DT} \ln\left(\frac{\lambda}{\lambda + \mu}\right) + K_O \tag{18.51}$$

Setting Eq. (18.51) to zero and solving for N gives

$$N^* = \ln\left[\frac{-K_O}{K_{DT} \log U}\right] \cdot \frac{1}{\log U} \tag{18.52}$$

where $N^* =$ optimum number of machines in parallel
$U = \lambda/(\lambda + \mu)$

As an example of this model consider the case where the following information was obtained from historical data:

$K_{DT} =$ \$6,000.00
$K_O =$ \$250.00
$\lambda =$ 15 failures/1000 hours
$\mu =$ 45 failures/1000 hours

Substituting,

$$U = \frac{\lambda}{(\lambda + \mu)} = \frac{15}{15 + 45} = 0.25 \tag{18.53}$$

Using Eq. (18.52)

$$N^* = \ln\left(\frac{-250}{6000 \ln 0.25} \right) \frac{1}{\ln 0.25}$$

$$= 2.562 \qquad\qquad (18.54)$$

Three machines will be needed.

Model V—Maintain/repair based on investment and increased cost to repair

A fundamental problem in maintenance is estimating the life of a piece of equipment. Beyond what time limit should the equipment be disposed of rather than repaired? A model used to determine this is [22,23]

$$L_c = \left[\frac{2(C - S)}{B} \right]^{1/2} \qquad\qquad (18.55)$$

where L_c = economic life of the equipment in years
C = initial cost of the equipment, installed
S = scrap value of the machine
B = annual increase in the cost of repairs

To illustrate this model, consider the case where the initial cost of a machine, including installation, is $45,000.00. The estimated scrap value of this equipment is $6000.00. The estimated increase in yearly maintenance cost is $1400.00. Beyond what time limit is the cost of repairing the machine not justified?

$$L_c = \left[\frac{2\,(45,000 - 6,000)}{1,400} \right]^{1/2}$$

$$= 7.46 \text{ years} \qquad\qquad (18.56)$$

Labor standards for maintenance tasks

The control of costs must be based on reliable time standards. Maintenance tasks are no exception to this rule. Unfortunately, there is a widely held opinion that time standards cannot be set for maintenance-type tasks. The reasons put forward to support this opinion are that the work content of these types of tasks is highly variable and difficult to predict. In fact, there have been several excellent methods developed and used successfully for setting time standards for maintenance tasks.

The most successful systems have been based on the concept of comparative estimating. Two of these approaches are described below.

Comparative Estimating

Comparative estimating has been most widely used in developing time standards for maintenance tasks. It is likely that this is the basis of systems which use other proprietary names. The technique first achieved prominence when used by the U.S. Navy in developing time standards for dockyard operations.[24–26]

> *Definition of comparative estimating:* A work measurement technique in which the time for a job is evaluated by comparing its work content with the work content of a series of benchmark jobs, the work content of which have been measured.[27,28]

> *Description of comparative estimating:* This approach can best be studied after its overall operation is appreciated. A maintenance department in a hypothetical plant will be used for the example, given below.

A wide variety of nonrepetitive jobs are carried out in this department. However, there are eight jobs with which all maintenance personnel and management are familiar. These eight jobs are candidates to be used as *benchmark jobs*. Some work measurement technique is used to set the standard time of these jobs. They can then be arranged in tabular form to create a *spreadsheet*. A sample spreadsheet is shown in Fig. 18.10.

Each job is contained in a small section of the spreadsheet. The job is described briefly, but in sufficient detail to allow plant personnel to identify the job. The standard time for job is shown below the description. The identification number or code for the job is also shown. In the case of the job described in the block at the top left-hand corner of the spreadsheet, the standard time is 2.25 hours and the job identification number is VTA 764.

Time boundary values are shown along the top edge of the spreadsheet. It will be noticed that time values for the jobs in each column lie between the upper and lower boundary values for the column. These columns are referred to as *time slots*. A job is located in the time slot whose upper and lower boundaries embrace the standard time for the job.

The *slot time* for each slot is given along the bottom edge of the spreadsheet. The *slot time* is the time allocated to any job which falls into the slot. For example, the jobs falling into the first slot would all be planned to be completed in 2.85 hours. This is in spite of the fact that the actual times are known. The method used to establish the boundary times and the slot time will be considered below. When the time for a new, unmeasured job is to be estimated, the procedure illustrated in Fig. 18.11 is used.

Slot	A	B	C	D
Task Description	Clear 250'(3" dia.) flow tubes with mechanical punch. Blow clear with compressed air. (2.25 hr.) No. VTA 764	Wash convertor tank with pressurized water. Remove residue. Spray plaster coat over inside of tank. (6.22 hr.) No. VTA 764	Chip solid residue from seal area and reface seal with hand tool. Remove and replace gasket. Test sealing pressure. (8.34 hr.) No. VMA 663	
				Neutralize excess fluid. Test reactivity. Filter tank output. Clean and replace all filters. Heat seal joints. (11.91 hr.) No. VTA 446
	Clear 100' (1.5" dia.) flow tubes with hydraulic punch. Clean with pressurized water. Seal joints. (4.6 hr.) No. VMB 442			
		Scrape convertor tank. Remove residue. Wash and neutralize inside area of tank. (7.63 hr.) No. VTA 264	Remove and replace (1) safety valve. Test valve performance. (9.62 hr.) No. VTA 224	Remove filters from (6) filter housings. Reline each housing. Make (6) gaskets. Replace (6) filters and (6) gaskets. (13.60 hr.) No. VMA 044
Slot Time	2.85	6.97	7.90	11.60
Time Boundary	>0 To 6	>6 To 8	>8 To 10	>10 To 14

Figure 18.10   Sample spreadsheet.

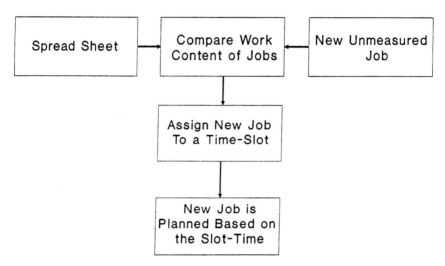

Figure 18.11   Concept of comparative estimating.

The comparative estimating procedure should be carried out by a person skilled in the work area in which the estimate is being made. Usually, the work to be done on the unmeasured job can only be described in words. This description is in the same form as the descriptions on the spreadsheet. By comparing the descriptions with the benchmark jobs, these new jobs can be allocated into the appropriate time slot. If, from a description, the estimator judged the new job to have a work content less than job No. VMB 442, it would be placed in slot A. On the other hand, if the work content was judged to be less than job No. VTA 264 but greater than job No. VTA 016, the new job would be allocated to time slot B.

The most likely place for a problem to arise is where the new job has a work content between the minimum benchmark job time in one slot and the maximum benchmark job time in the next lowest time slot. The experience of the estimator usually overcomes this problem. In any case, this is not so great a problem as might at first be imagined.

Planning and control for each job are on the basis of the slot time. The variance in the slots can give rise to some errors; however, this does not prove to be a great problem. The time-slotting system should be designed so that errors due to using the slot time will be averaged out over a specified time period. This specified time is known as the *balance time*. Variance analysis over the balance time can be used to eliminate these spurious values.

The theory of time slotting

In the original presentation of comparative estimating,[24-26] a series of time-slot boundaries and slot times were suggested. No real justification was given for this particular choice. Using computer simulation, the behavior of a specific work area was investigated by several groups.[29-31] The results of these studies did not answer the fundamental problem of designing the system to meet a priori requirements.

One approach to designing a system used *cluster analysis*.[32] Here, given a distribution of job times, the time boundaries and slot times can be determined to give an a priori criterion. This might be, for example, the absolute percent error calculated over a time period of one week.

$$\text{Absolute percent error} = \frac{\text{actual standard time} - \text{slot time}}{\text{actual standard time}} \times 100$$

$$(18.57)$$

To calculate slot times and boundary times, one method proposed using a geometric progression.[33] The attraction of this approach is its mathematical simplicity. The approach has two important disadvantages: first, the resulting real times and boundary times; second, the errors introduced by the comparative estimating procedure cannot be delimited.

The success of comparative estimating relies on the stochastic properties of the system. An analytical approach using this characteristic has been proposed[34] and explored.[34-36] This discussion of the theory of time slotting is based upon an analytical approach.[33-36]

Elements of time-slotting systems.   Investigations have shown that the distributions of job times in practice are not normal, but have a "positive skew." A typical distribution of job times is shown in Fig. 18.12. A simple slotting scale has been constructed onto this distribution.

The disposition of a time slot is defined by the values of its time boundaries. The disposition of time slot i is defined by t_{i-1} and t_i. The slot time for slot i is designated as T_i.

The minimum time considered by the time-slotting system is t_0, where $t_0 \geq 0$. The maximum time considered by the time-slotting system is t_n, the upper boundary time of time slot n. This does not imply that t_0 and t_n are the minimum and maximum job times which might be encountered in the work area. Jobs outside of the range $t_n - t_0$ are excluded by the system and treated as special cases. This can give rise to the effect known as *truncation*. This effect is demonstrated for times greater than t_n in Fig. 18.12.

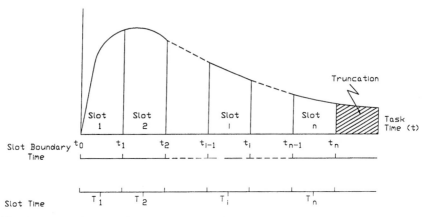

Figure 18.12   Elements of a time-slotting scale.

The theory of time slotting assumes the probability distribution of the work task times can be expressed as a function of the work task time t.

Identifying the errors in time slotting.   Three errors can be identified in time slotting; they are:

1. *Systematic errors* which arise as the result of using improper assumptions and/or design procedures

2. *Random errors* which are the result of using a single time value for all of the jobs falling into a time slot

3. *Allocation errors* which result from the estimator allocating the unmeasured job to the wrong time slot

The systematic errors are referred to as the *accuracy of the system*. The random errors are referred to as the *precision of the system*. The allocation error is called the *estimating error*.

Systematic error of a single time slot.   The time value selected for T_i should be selected so that no long-term bias is generated in the prediction of jobs allocated to time slot i.

A job should be allocated to slot i if the job time t is greater or equal to t_{i-1} or less than t_i. That is,

$$t_{i-1} \leq t \leq t_i \qquad (18.58)$$

Intuitively, the mean of t_{i-1} and t_i might be chosen. The only condition under which this would be acceptable is when the distribution of job times is uniform. Elsewhere, the distribution of the times in the time slot should be considered. Take, as an example, the distribution of the job times in the time slot shown in Fig. 18.13.

Let the area of the plane figure $ABCD$ in Fig. 18.13 be a_i. Then

$$P(t_{i-1} \leq t < t_i) - a_i - \int_{t_{i-1}}^{t_i} f(t)\,dt \tag{18.59}$$

The position of the centroid and, consequently, the value of T_i, can then be determined by

$$T_i - \frac{\displaystyle\int_{t_{i-1}}^{t_i} t \cdot f(t)\,dt}{\displaystyle\int_{t_{i-1}}^{t_i} f(t)\,dt} \tag{18.60}$$

When some value other than T_i is used, a bias, or systematic error, is introduced into the time slot i. If this incorrect slot time is designated $\widehat{T}_i$ and the resulting systematic error is E_i, then

$$E_i - (T_i - \widehat{T}_i) \int_{t_{t-1}}^{t_i} f(t) \cdot dt \tag{18.61}$$

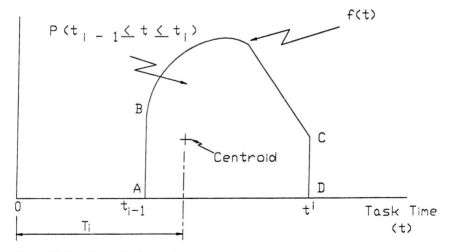

Figure 18.13   Centroid of time slot i.

System accuracy. If a time-slotting system is designed to have a single time slot, then the single slot time would be T'. The value of T' would be calculated as follows:

$$T' = \frac{\int_{t_a}^{t_m} t\,[f(t)\,dt]}{\int_{t_a}^{t_m} f(t)\,dt} \qquad (18.62)$$

The system accuracy E' is determined by combining Eqs. (18.60) and (18.61), resulting in

$$E' = \frac{\sum_{i=1}^{n} E_i}{T'} * 100 \qquad (18.63)$$

Random errors and system precision. The coefficient of variation is used to express the effect of random errors and system precision. The suitability of this measure has been demonstrated in other applications by several investigators.[26,36,37]

The variance in a particular time slot will be a function of the width of the time slot and the slot time. Thus, for a particular time slot i whose width is $t_i - t_{i-1}$, the variance will be different if $\hat{T}_i$ is used rather than T_i. Nevertheless, whichever value is used, the form of the resulting equation will be the same. Therefore, the value T_i will be used in the following equations.

The variance σ_i^2 for slot i is determined by the following:

$$\sigma_i^2 = \int_{t_{i-1}}^{t_i} (T_i - t)^2 f(t)\,dt \qquad (18.64)$$

The variance for the whole system is a weighted sum of the variances of separate time slots. From this, the standard deviation for the time system is σ'. This is calculated as follows:

$$\sigma' = \left(\sum_{i=1}^{n} \int_{t_{i-1}}^{t_i} (T_i - T)^2 f(t)\,dt \left[\int_{t_{i-1}}^{t_i} f(t)\cdot dt \right] \right)^{1/2} \qquad (18.65)$$

The system precision p' is the coefficient of variation of the system expressed as a percentage of the single slot time T'. Thus, combining Eqs. (18.62) and (18.64),

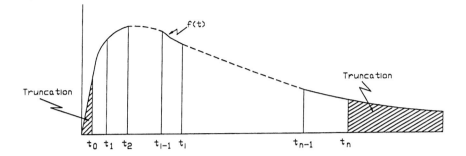

Truncation of Slotting Scales

Figure 18.14   Truncation of slotting scales.

$$p' = \frac{100\,\sigma'}{T'} \tag{18.66}$$

The effect of truncation.   The range of job times which might occur in a work area is often truncated as a matter of convenience. That is, for job time $t < 0$ or $t > t_n$, special methods are used to plan and control the job. A time-slotting system using truncation is illustrated in Fig. 18.14.

The proportion of the total job in the time range t_0 to t_n is $P(t_0 \leq t < t_n)$. Thus,

$$P(t_0 \leq t < t_n) = \int_{t_0}^{t_n} f(t)\,dt \tag{18.67}$$

This value appeared as the denominator in Eqs. (18.60) and (18.62). When there is truncation of a system, $P(t_0 \leq t < t_n) < 1$. This new value should be used in any calculations.

Defining the estimating error.   The investigation of time slotting using simulation[29–32] assumed the estimator would always be able to allocate a job to the correct time slot. One paper[36] dealt specifically with this source of error. In this paper the estimating error p was defined as

$$p = \left| \frac{t - t_e}{t} \right| \tag{18.68}$$

where p = estimating error

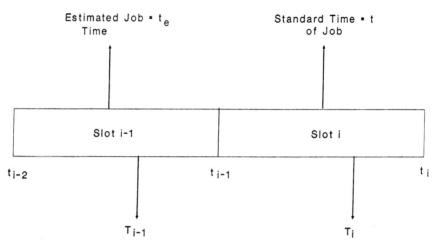

Figure 18.15 Concept of misallocation.

$t =$ correct standard time for a particular job
$t_e =$ estimated time for a particular job

In the treatment which follows, the distribution of errors of the estimated time for a particular job is assumed to be normal. This assumption is purely for convenience. It is not known whether this assumption could be justified. Using an estimating error of p, the range is $\pm pt$. The 3σ limits in this case are bounded by $t(1 - p)$ and $t(1 + p)$.

Misallocation occurs when the estimator decides the time for a particular job is $t_{i-2} \leq t \leq t_{i-1}$, when it is $t_{i-1} \leq t < t_i$. In which case, the slot time T_{i-1} is used instead of T_i. The misallocation error which results is $T_i - T_{i-1}$. The concept of misallocation is shown in Fig. 18.15.

Statistical nature of misallocation. The statistical nature of misallocation in time slotting is illustrated in Fig. 18.16. From this diagram, it can be seen that

$$V_1 \leq t_e \leq t_{i-1} \qquad \text{where allowed time} = T_{i-1} \qquad (18.69)$$

and

$$t_{i-1} < t_e \leq t_i \qquad \text{where allowed time} = T_i \qquad (18.70)$$

To estimate the proportion of misallocations in the system, the following equation is used:

$$(T_{i=1} - T_i) * P(V_1 \leq t_e < t_{i-1}) \qquad (18.71)$$

As t approaches t_1, there will be a misallocation. In this case, the error will be

$$(T_i - T_{i+1}) * P(t_i \le t_e \le V_2) \tag{18.72}$$

The effect of these errors will depend on the job times estimated by the slotting system and the distribution of standard times of the jobs. For this reason, it is necessary to make a reasonable assumption or determine the distribution of these job times.

Effect of estimating error over the total slotting system. The effect of misallocation in a single time slot was considered above. Over the total system being analyzed, the possibility of misallocation into more than one time slot must be accounted for. The effect of this extended misallocation is illustrated in Fig. 18.17.

The total effect on the system of the separate misallocation errors is the *total estimating error* E_A. A nine-step procedure[36] has been developed to calculate this value.

Step 1: The minimum value of $t_e = V_1$. Set t equal to V_1.

Step 2: Determine the limits $V_1 = (1 - p)t$ and $V_2 = (1 + p)t$, where

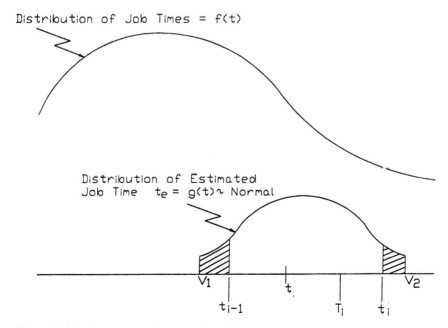

Figure 18.16 The statistical nature of misallocation.

p is the estimating error calculated in Eq. (18.70).

Step 3: Examine the time range $V_1 \le t \le V_2$, which will identify the time slots $(i - k)$ and $(i + k)$.

Step 4: Using the equations given in Fig. 18.17, calculate the values $A_{i-k} \cdots A_{i+k}$.

Step 5: Calculate the expected value of the square of the misallocation error using Eq. (18.73).

$$W_t = \sum_{j=i-k}^{i-k} (T_i - T_j)^2 \cdot A_j \qquad (18.73)$$

Step 6: If the accuracy of task times is in the order σ, calculate value Q_t.

$$Q_t = \int_{t-\frac{2}{\sigma}}^{t+\frac{2}{\sigma}} f(t)\,dt \qquad (18.74)$$

where $f(t)$ is the density function of the distribution of work task times.

Step 7: Set $t = t + \sigma$. If $t < t_n$ go to Step 2. Otherwise, go to Step 9.

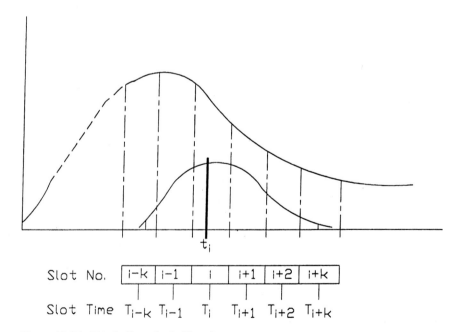

Figure 18.17   Total effect of misallocation.

Step 8: For all values of *t,* calculate the expected value of (correct slot time − incorrect slot time)² using the following equation:

$$W - \sum_{allt} W_t Q_t \tag{18.75}$$

Step 9: Calculate the total estimating error E_A using Eq. (18.76).

$$E_A = \frac{\sqrt{W}}{T'} \tag{18.76}$$

where T' is calculated using Eq. (18.62).

Distribution of job times.   The distribution of job times will affect the results of the calculation of these errors. Investigators[29-31, 34-36] have explored several alternatives. For details of these results, the original papers should be referred to. To use the equations given above, it is necessary to determine, or assume, a probability distribution.

Disposition and width of time slots.   Two assumptions can be used for developing the time slots. They consider the time slots can have either equal widths or equal probability. The preferred method will depend upon the distribution, the number of slots required in the system, and the truncation used. The results of one analysis[35] are illustrated in Fig. 18.18. The analysis was performed on data with an assumed log-normal distribution, with mean job time of 2.08 hours and standard deviation of 2.91 hours.

Several points should be noted in this figure. First, the precision of the system improves as the number of time slots increase. Second, increasing the degree of truncation at the higher job times improves the system precision. Third, as the degree of truncation increases, the preferred method of slot design, based on system precision, appears to have changed. Fourth, the crossover from equal probabilities to equal widths as the preferred system might also be a function of the number of time slots in the system.

The probability of misallocation increases as the widths of the time slot decrease. This is of interest when the number of slots are decreased to simplify the slotting system.

Size of database.   To design the time-slotting system, it is necessary to develop a database of standard times for jobs carried out in the work area being applied. This database is used to estimate the type

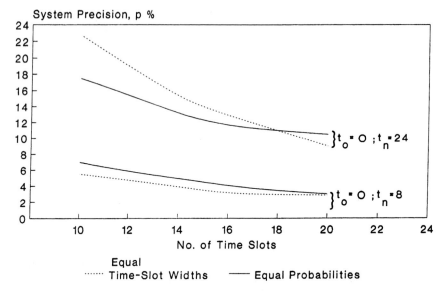

Figure 18.18   Effect of slot design on system precision.

of distribution and its parameters. In addition, the benchmarks used to construct the spreadsheets are chosen from these jobs.

The question which must be asked is: What should be the size of this database? It could be expected that a statistical approach would have been developed to answer this question. In fact, no such analysis seems to have been published. Two applications[35,36] gave empirical estimates to answer the question. These estimates are summarized in Fig. 18.19.

The cost involved in developing this database can be a deterrent to smaller companies adapting comparative estimating for measuring indirect activities.

Work Area	No. of Jobs
Fitting	300
Joinery	150
Civil	170
Pipefitting	150
Electrical	250
Welding	50
Instrument Repair	2199

Figure 18.19   Empirical number of jobs for designing a slotting system.

Time gaps in comparative estimating

Apart from the problem of database size, noted above, there are two other significant disadvantages which can be identified with respect to time slots.

First, two jobs which clearly have different work contents are allocated the same time estimate. This can lead to operators viewing the results with suspicion. The planning and control process is also less sensitive than perhaps it should be.

Second, the allocation of a job into a particular time slot at the upper end of one slot and the lower end of the other slot can produce inconsistencies. Time slotting, effectively, uses two gauge points to identify the separation.

To overcome these two problems and reduce the financial investment in developing the database, the *time-gap method* was developed.[34] The design specifications set for the approach were:

1. The number of sample tasks times which need to be developed must be kept to a minimum.

2. The time gaps should be defined automatically, as part of the setting-up procedure.

3. The accuracy of the resulting time-gap system should be better, or at least as good, as a similar time-slotting system.

Description of the time-gap approach. A different terminology is used in the time-gap approach to that used in time slotting. The reader is advised to compare the terminologies of the two systems.

An example of a time-gap system is shown in Fig. 18.20. The same information used in Fig. 18.10 is used in this figure. What were called *benchmark jobs* in time slotting would be *time marks* in time gapping. They are referred to as *time marks* because they mark the bounds of the time gaps.

The method used to estimate job times using time gaps is somewhat similar to that used with time slotting. Using a written description of the unmeasured job, the estimator compares its work content to those of the time-mark jobs. From this comparison, the unmeasured job is allocated into a time gap. Estimating of times, based on the beta distribution, has been performed successfully in applying PERT[35] to project control. By using the same approach to the time gaps, it was argued that improved time estimates would result. Therefore, when a job has been allocated to a time gap, the optimum time T_o for performing the job is assumed to be the time mark at the lower end of the gap. Similarly, the pessimistic time T_p is taken to be the time mark at the higher end of the time gap. The estimator is

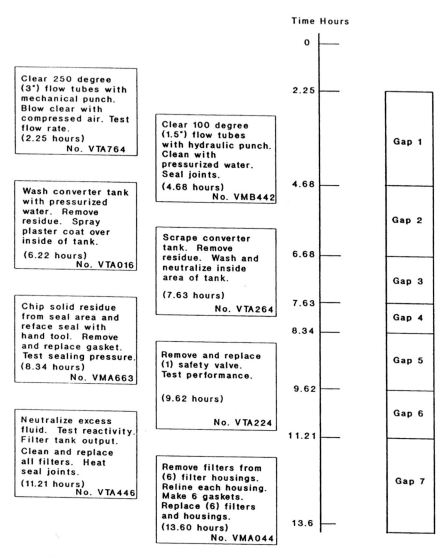

Figure 18.20 Sample time-gap system.

required to estimate the most likely time T_m to perform the job. The expected job time T_e can be calculated as follows

$$T_e = \frac{(T_o + 4T_m + T_p)}{6} \qquad (18.77)$$

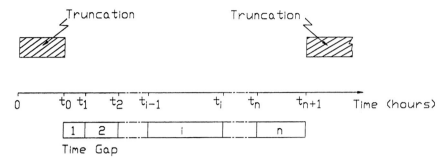

Figure 18.21 General format of time-gap scale.

General form of time gaps. A general form of a time-gap scale is shown in Fig. 18.21. Several variations can be made which will extend its use and give different interpretations.

For n time gaps, $n + 1$ time marks are needed. A truncation occurs at the upper end of the time-gap scale. At the lower end of the scale, t_0 can be taken as zero hours, which means that there is no truncation.

If the value of time mark t_0 is set to a value greater than zero, but less than the time of the first measured time mark, certain advantages can be achieved. First, the degree of truncation at the lower end of the scale can be set. Second, the lower end of the true scale can be set rather than established by chance. This will mean that in time gap number 1, the estimator will have only one time-mark description to work to.

In a similar way, t_n can be set arbitrarily. Similar advantages and restrictions will then apply at the upper end of the scale.

Factors considered in setting the time-gap scale. Three factors influence the design of the time-gap scale. They are:

1. Total range of time to be covered by the system
2. The number of time gaps to be used in the system
3. The minimum width of time gap

Range of time-gap scale. The control period used is normally one week, or multiples thereof. This will be governed by the method used to calculate wage payment in a particular company. If the time scale used equals or exceeds the control period, day-to-day control of operations becomes unwieldy and ineffective. If, for example, the normal working day was eight hours, the maximum time mark would probably be set to that value. This is arbitrary, but the designed level of control is the guide which should be used to set the time.

Number of time gaps. There is no published, systematic approach to determining the number of time gaps which should be used in a system. A simulation study[36] of the time-gapping approach indicated that the best number of gaps should be somewhere between 15 and 20. It must be noted that this was based upon a limited sample. Nevertheless, there are strong indications that this number can be reduced and still produce superior results to a time-slot-based system.

Minimum width of time gap. Time gaps are generated as sample jobs and measured in the work area. The jobs will occur at random; therefore, two restrictions are placed on their selection as time-mark jobs.

If the standard time of a sample job is the same as a job already forming an existing time mark, one of them is excluded. The time-mark job retained is the one with which the estimators are more familiar.

The second restriction is the width of the time gap. As the width of the time gap is reduced, resolution between the work content of two jobs becomes increasingly difficult. Consequently, the probability of misallocation increases. The designer must, therefore, specify a minimum width for the time gap. If this is (say) 15 minutes, then

$$t_i - t_{i-1} \geq 15 \qquad (18.78)$$

The choice of the minimum time-gap width is arbitrary. No study has been made on the effect of time-gap width on the performance of the system.

Comparison of time gaps and time slots in measuring indirect jobs. The time-gap approach was shown to be superior to time slotting on two accounts.[36]

Using comparable designs and data the time gap resulted in a system performance equal to or better than time slotting. This was in spite of estimator errors being included in the results of the time-gap approach.

The most impressive of the advantages was the economic gain by using time gapping. The database needed to establish the time slots was shown in Fig. 18.19. A comparison of the two approaches, using computer simulation, indicated the number of sample jobs in the database for time gaps was approximately 1.5 times the number of time gaps being used. Thus, if 20 time gaps are required, approximately 30 task times would be needed. The immense gain over time slotting is obvious.

Activities in indirect activities

Service and maintenance are typical of what are referred to as *indirect activities*. It is characteristic that the total work content of these activities is made up of more than the operations. In maintenance operations, for example, the total activity is made up of:

1. Prepare for operation
2. Travel to work site
3. Set up at work site
4. Perform task
5. Clean up

Times for each of these separate activities must be accounted for. The method used will vary from case to case. It is likely, however, that some small modification in the methods described above will prove to be satisfactory.

References

1. Gerce, John, and Walker W. Holler, *Manufacturing Engineering,* U.S. Department of Commerce, National Bureau of Standards, Washington, D.C., 1965.
2. British Standards Institute, *Glossary of General Terms Used in Maintenance Organization—British Standard 3811:1964,* British Standards Institution, London, 1964.
3. *Industrial Engineering Terminology,* ANSI Z94.0, 1982, Institute of Industrial Engineers, Atlanta, GA, 1982.
4. Dhillon, B. S., *Mechanical Reliability Theory, Models and Applications,* AIAA Education Series, American Institute of Aeronautics and Astronautics, Inc., Washington D.C., 1988.
5. Gabrielson, D. N., "Frequency of Maintenance," *Proceedings of Fifth Reliability and Maintenance Conference,* AIAA, New York, July 1966, pp. 428–433.
6. Naresky, J. J., "Reliability Definition," *IIE Transactions on Reliability,* vol. 19, pp. 198–200, 1970.
7. Collins, J. A., "Failure Experience Feedback in Design Improvement," *Failure Prevention and Reliability,* ed. S. B. Bennet, A. L. Ross, and P. Z. Zemaamick, ASME, New York, 1978, pp. 245–249.
8. Patton, J. D., Jr., *Preventive Maintenance,* Instrument Society of America, Publisher's Creative Services, 1983.
9. Lewis, E. E., *Introduction to Reliability Engineering,* Wiley, New York, 1987.
10. MIL-HDBK-217B, *Reliability Prediction of Electronic Equipment,* U.S. Department of Defense, Washington D.C., 1974.
11. Rossi, M. J., *Non-electronic Parts Reliability Data,* Reliability Analysis Center, Rome Air Development Center (RADC), Griffins Air Force Base, NY, Report NPRD-3, 1985.
12. MIL-STD-765B, *Reliability Modelling and Prediction,* U.S. Department of Defense, Washington D.C., 1981.
13. MIL-STD-2077A (Navy), *General Requirement Test Program Sets,* U.S. Department of Defense, Washington D.C., 1987.
14. MIL-HBX-472, *Maintainability Prediction,* U.S. Department of Defense, Washington D.C., 1966.
15. MIL-STD-1629, "Failure Modes, Effects and Criticality Analysis," U.S. Department of Defense, Washington D.C., 1976.

16. Department of the Navy, Navidocks, *Engineering Performance Standards,* Washington, D.C., Bureau of Yards and Docks.
17. Maynard, H. B., and A. J. Stegemerten, "Universal Maintenance Time Standards," *Factory Maintenance and Management,* November 1955.
18. Lewis, B. T., *Developing Maintenance Time Standards,* Industrial Education Institute, Boston, 1978.
19. *Industrial Engineering Terminology,* American Standard ANSI Z94.0, 1982, Institute of Industrial Engineers, Atlanta, GA, 1983.
20. British Standards Institution, *Glossary of Terms Used in Work Study—British Standard 3138:1969,* British Standards Institution, London, 1969.
21. Ferri, F., and B. W. Niebel, "Universal Maintenance Standards," *Industrial Engineering,* February 1976.
22. Enscore, E. E., and B. W. Niebel, "Reliable Indirect Labor Standards Achieved through Computerized Computation of Mean Slot Values of the Gamma Distribution," *Third National Conference on Computers in Industrial Engineering,* October 22–24, 1987, Atlanta, GA.
23. Enscore, E. E., K. Knott, and B. W. Niebel, "An Alternative Time Slotting Method for Indirect Time Standards," *Proceedings Institute of Industrial Engineering Conference,* New Orleans, May 23-27, 1982.
24. Enscore, E. E., K. Knott, and B. W. Niebel, "Cluster Analysis—Alternative to Statistical Distributions for Determining Slotting Schemes Used in Determining Indirect Work Standards," *Computers in Indirect Engineering,* vol. 12, no. 1, pp. 1–7, 1987.
25. McKenzie, H. A., "The Selection and Design of Slotting Scales," *Management Services,* December 1977, pp. 10–15.
26. Knott, K., "An Examination of the Theory of Time-Slotting," *International Journal of Production Research,* vol. 25, no. 3, pp. 253–262, 1987.
27. Knott, K., Jeya Chandra, and E. Emory Enscore, Jr., "Time-Slotting Based Upon an Assumed Log-Normal Distribution of Sample Work Task Times," *International Journal of Production Research,* vol. 25, no. 4, pp. 487–512, 1987.
28. Richard L. Shell (ed.), Principles and Practice of Work Measurement, Institute of Industrial Engineers, Atlanta, GA, 1986, pp. 263–273.
29. Chandra, Jeya, K. Knott, and E. Emory Enscore, Jr., "Precision and Accuracy of Time Slotting Based on an Assumed Log Normal Distribution of Work Task Times," Conference on Computers and Industrial Engineers, Orlando, FL, March 19-21, 1986.
30. Brinkloe, W. D., and M. T. Coughlin, *Precision Analysis of MTM-1 and MOST,* University Research Institute, Pittsburgh, PA, 1975.
31. Hancock, Walter W., "The System Precision of MTM-1," *Journal of Methods Time Measurement,* vol. XV, no. 4, 1980.
32. Gates, R. E., "Comparative Estimating for Indirect Work Evaluation," *Work Study and Management Services,* February 1972, pp. 85–90.
33. McKenzie, H. J., "A Good Hard Look at Comparative Estimating," *Management Services,* February 1968, pp. 16–19.
34. Knott, K., "A Simplified Approach to Comparative Estimating," IMSE Working Paper 82-124, Department of Industrial and Management Systems Engineering, The Pennsylvania State University, University Park, 1982.
35. PERT, Summary Report, Phase I, Navy Special Project Office, U.S. Government Printing Office, Catalogue No. D217.2: P94/958, Washington, D.C., 1968.
36. Pena, Manuel F., *Using the Properties of the Beta Distribution to Estimate the Standard Time of Indirect Jobs,* Unpublished Master of Science thesis, Department of Industrial and Management Systems Engineering, The Pennsylvania State University, University Park, 1984.

Index

ABOUT THE EDITORS

WILLIAM WRENNALL is President of the Leawood Group, Ltd., an internationally known industrial engineering and management consulting company. He received his bachelors degree at Durham University in England and his Masters at Macquarie University in Australia. His biography appears in directories such as *Who's Who in Finance and Industry*. Mr. Wrennall has held positions as production manager, plant manager, and general manager. As manager, consultant, and academic on several continents he has received numerous accolades for his services to productivity science. He is Past President of the World Confederation of Productivity Science, Founding Fellow of the World Academy of Productivity Science, and Honorary Fellow of the Indian Productivity Academy. Mr. Wrennall was awarded the Queen's Silver Jubilee Medal and the Faraday Medal for his contribution to productivity. Wrennall has authored many articles, presented conference papers, conducted seminars, and consulted in Facilities Design for over 30 years.

QUARTERMAN LEE is Vice President of the Leawood Group, Ltd. He is management consultant, educator, and author in industrial management. He has assisted clients worldwide to improve their profitability and global competitiveness through the application of innovative management methods. Mr. Lee has conducted seminars for The Institute of Industrial Engineers, The Society of Manufacturing Engineers, corporations, and universities. His magazine and handbook contributions have made him internationally known in the fields of industrial management and engineering.